Springer Series in Information Sciences 26

Editor: Thomas S. Huang

Springer Series in Information Sciences

Editors: Thomas S. Huang Teuvo Kohonen Manfred R. Schroeder
Managing Editor: H. K. V. Lotsch

1 **Content-Addressable Memories**
By T. Kohonen 2nd Edition

2 **Fast Fourier Transform and Convolution Algorithms**
By H. J. Nussbaumer 2nd Edition

3 **Pitch Determination of Speech Signals**
Algorithms and Devices By W. Hess

4 **Pattern Analysis and Understanding**
By H. Niemann 2nd Edition

5 **Image Sequence Analysis**
Editor: T. S. Huang

6 **Picture Engineering**
Editors: King-sun Fu and T. L. Kunii

7 **Number Theory in Science and Communication** With Applications in Cryptography, Physics, Digital Information, Computing, and Self-Similarity By M. R. Schroeder
2nd Edition

8 **Self-Organization and Associative Memory** By T. Kohonen
3rd Edition

9 **Digital Picture Processing**
An Introduction By L. P. Yaroslavsky

10 **Probability, Statistical Optics, and Data Testing** A Problem Solving Approach By B. R. Frieden
2nd Edition

11 **Physical and Biological Processing of Images** Editors: O. J. Braddick and A. C. Sleigh

12 **Multiresolution Image Processing and Analysis** Editor: A. Rosenfeld

13 **VLSI for Pattern Recognition and Image Processing** Editor: King-sun Fu

14 **Mathematics of Kalman-Bucy Filtering**
By P. A. Ruymgaart and T. T. Soong
2nd Edition

15 **Fundamentals of Electronic Imaging Systems** Some Aspects of Image Processing By W. F. Schreiber
2nd Edition

16 **Radon and Projection Transform-Based Computer Vision**
Algorithms, A Pipeline Architecture, and Industrial Applications By J. L. C. Sanz, E. B. Hinkle, and A. K. Jain

17 **Kalman Filtering** with Real-Time Applications By C. K. Chui and G. Chen
2nd Edition

18 **Linear Systems and Optimal Control**
By C. K. Chui and G. Chen

19 **Harmony: A Psychoacoustical Approach** By R. Parncutt

20 **Group-Theoretical Methods in Image Understanding** By K. Kanatani

21 **Linear Prediction Theory**
A Mathematical Basis for Adaptive Systems By P. Strobach

22 **Psychoacoustics** Facts and Models
By E. Zwicker and H. Fastl

23 **Digital Image Restoration**
Editor: A. K. Katsaggelos

24 **Parallel Algorithms in Computational Science**
By D. W. Heermann and A. N. Burkitt

25 **Radar Array Processing**
Editors: S. Haykin, J. Litva, and T. J. Shepherd

26 **Signal Processing and Systems Theory**
Selected Topics
By C. K. Chui and G. Chen

27 **3D Dynamic Scene Analysis**
A Stereo Based Approach
By Z. Zhang and O. Faugeras

Charles K. Chui Guanrong Chen

Signal Processing and Systems Theory

Selected Topics

With 38 Figures

Springer-Verlag
Berlin Heidelberg New York
London Paris Tokyo
Hong Kong Barcelona
Budapest

Professor Charles K. Chui
Department of Mathematics, and Department of Electrical Engineering, Texas A&M University, College Station, TX 77843-3368, USA

Dr. Guanrong Chen
Department of Electrical Engineering, University of Houston, Houston, TX 77204-4793, USA

Series Editors:
Professor Thomas S. Huang
Department of Electrical Engineering and Coordinated Science Laboratory, University of Illinois, Urbana, IL 61801, USA

Professor Teuvo Kohonen
Laboratory of Computer and Information Sciences, Helsinki University of Technology, SF-02150 Espoo 15, Finland

Professor Dr. Manfred R. Schroeder
Drittes Physikalisches Institut, Universität Göttingen, Bürgerstrasse 42–44, W-3400 Göttingen, Fed. Rep. of Germany

Managing Editor: Helmut K. V. Lotsch
Springer-Verlag, Tiergartenstrasse 17, W-6900 Heidelberg, Fed. Rep. of Germany

ISBN 3-540-55442-4 Springer-Verlag Berlin Heidelberg New York
ISBN 0-387-55442-4 Springer-Verlag New York Berlin Heidelberg

Printed in the United States of America

54/3140-5 4 3 2 1 0 – Printed on acid-free paper

Dedicated to the memory
of our friend
Professor Xie-Chang Shen
(1934 – 1991)

Preface

It is well known that mathematical concepts and techniques always play an important role in various areas of signal processing and systems theory. In fact, the work of Norbert Wiener on time series and the pioneering research of A. N. Kolmogorov form the basis of the field of communication theory. In particular, the famous sampling theorem, usually attributed to Wiener's student Claude Shannon, is based on the Paley-Wiener theorem for entire functions. In systems engineering, the important work of Wiener, Kalman, and many other pioneers is now widely applied to real-time filtering, prediction, and smoothing problems, while optimal control theory is built on the classical variational calculus, Pontryagin's maximum principle, and Bellman's dynamic programming.

There is at least one common issue in the study of signal processing and systems engineering. A filter, stable or not, has a state-space description in the form of a linear time-invariant system. Hence, there are common problems of approximation, identification, stability, and rank reduction of the transfer functions. Recently, the fundamental work of Adamjan, Arov, and Krein (or AAK) has been recognized as an important tool for at least some of these problems. This work directly relates the approximation of a transfer function by stable rational functions in the supremum norm on the unit circle to that of the corresponding Hankel operator by finite-rank bounded operators in the Hilbert space operator norm, so that various mathematical concepts and methods from approximation theory, function theory, and the theory of linear operators are now applicable to this study. In addition, since uniformly bounded rational approximants on the unit circle are crucial for sensitivity considerations, approximation in the Hardy space H^{∞} also plays an important role in this exciting area of research.

This monograph is devoted to the study of several selected topics in the mathematical theories and methods that apply to both signal processing and systems theory. In order to give a unified presentation, we have chosen to concentrate our attention on discrete-time methods, not only for digital filters, but also for linear systems. Hence, our discussions of the theories and techniques in Hardy spaces and from the AAK approach are restricted to the unit disk. In particular, the reader will find that the detailed treatment of multi-input/multi-output systems in Chap. 6 distinguishes itself from the bulk of the published literature in that the balanced realization approach of discrete-time linear systems is followed to study the matrix-valued AAK theory and H^{∞}-optimization.

The selection of topics in this monograph is guided by our objective to present a unified frequency-domain approach to discrete-time signal processing and systems theory. However, since there has been considerable progress in these areas during recent years, this book should be viewed as only an introductory treatise on these topics. The interested reader is referred to more advanced and specialized texts and original research papers for further study.

We give a fairly rigorous and yet elementary introduction to signals and digital filters in the first chapter, and discrete-time linear systems theory in the second chapter. Hardy space techniques, including minimum-norm and Nevanlinna-Pick interpolations will be discussed in Chap. 3. Chap. 4 will be devoted to optimal Hankel-norm approximation. A thorough treatment of the theory of AAK will be discussed in Chap. 5. Multi-input/multi-output discrete-time linear systems and multivariate theory in digital signal processing will be studied in the final chapter via balanced realization.

The first author would like to acknowledge the continuous support from the National Science Foundation and the U.S. Army Research Office in his research in this and other related areas. To Stephanie Sellers, Secretary of the Center for Approximation Theory at Texas A&M University, he is grateful for unfailing cheerful assistance including making corrections of the TEX files. To his wife, Margaret, he would like to express his appreciation for her understanding, support, and assistance. The second author would like to express his gratitude to his wife Qiyun Xian for her patience and understanding.

During the preparation of the manuscript, we received assistance from several individuals. Our special thanks are due to the Series Editor, Thomas Huang, for his encouragement and the time he took to read over the manuscript, Xin Li for his assistance in improving the presentation of Chap. 5, and H. Berens, I. Gohberg, and J. Partington for their valuable comments and pointing out several typos. In addition, we would like to thank J. A. Ball, B. A. Francis, and A. Tannenbaum for their interest in this project. Finally, the friendly cooperation and kind assistance from Dr. H. Lotsch and his editorial staff at Springer-Verlag are greatly appreciated.

College Station
Houston
August 1991

Charles K. Chui
Guanrong Chen

Contents

1. Digital Signals and Digital Filters

Digital filtering techniques are widely used in many scientific and industrial endeavors such as digital telephony and communications, television and facsimile image processing, electrical and speech signal processing, radar, sonar, and space control systems. The major advantage of digital techniques over the classical analog version is the availability of modern digital equipment for high-speed processing with high signal/noise ratio and fast computation with high accuracy, and at the same time with a decrease in the implementation and computation costs.

This chapter is essentially a review of some of the basic properties of digital signals in the time and frequency domains, z-transforms, digital filters, and design methods of both optimal and suboptimal digital filters. Our presentation of these concepts and results is in some sense not traditional, and several approaches which are not well known in the literature are included to enhance the mathematical flavor of the subject.

1.1 Analog and Digital Signals

To understand signal processing and systems theory, it is essential to have some knowledge of the basic properties of analog and digital signals or controllers. Since one of the main objectives of this text is to provide a unified approach to signal processing and systems theory, terminologies from both areas will be introduced.

1.1.1 Band-Limited Analog Signals

A continuous-time signal, sometimes called an *analog signal*, $u(t)$, is a piecewise continuous function of the time variable t, where t ranges from $-\infty$ to ∞. In systems theory, when a control system is being considered with $u(t)$ as its input, $u(t)$ may also be called a *controller*. In any case, the function $u(t)$ is said to be *band-limited* if it has the following integral representation:

$$u(t) = \int_{-\omega_0}^{\omega_0} \sigma(\omega) e^{\mathrm{j}\omega t} d\omega \tag{1.1}$$

where $\mathrm{j} = \sqrt{-1}$, $\sigma(\omega)$ is some function in $L^1(-\omega_0, \omega_0)$, and ω_0 is some positive number. If the function $u(t)$ happens to be in $L^1(-\infty, \infty)$, then its *Fourier transform* defined by

$$\hat{u}(\omega) = \int_{-\infty}^{\infty} u(t)e^{-j\omega t}dt \tag{1.2}$$

is given by

$$\hat{u}(\omega) = \begin{cases} 2\pi\sigma(\omega) & \text{for } -\omega_0 < \omega < \omega_0 \\ 0 & \text{otherwise}. \end{cases}$$

We will say that the Fourier transformation takes $u(t)$, defined on the *time domain*, to $\hat{u}(\omega)$, defined on the *frequency domain*. Here and throughout the text, ω will be reserved for the frequency variable. Note that the Fourier transform $\hat{u}(\omega)$ of $u(t)$ satisfying (1.1) has a compact support on the frequency domain. The length of the smallest subinterval of $(-\omega_0, \omega_0)$ outside of which $\sigma(\omega)$ vanishes identically, is called the *bandwidth* of $u(t)$. Hence, the bandwidth of $u(t)$, as defined in (1.1), does not exceed $2\omega_0$.

It should be remarked that a band-limited signal $u(t)$ as in (1.1) can be extended from the time domain to the complex z-plane $\mathcal{C}$, where $z = t + js$, and the extension $u(z)$ is analytic everywhere and satisfies

$$|u(z)| \leq Ae^{\omega_0|z|}$$

for all z in $\mathcal{C}$ with

$$A = \int_{-\omega_0}^{\omega_0} |\sigma(\omega)|d\omega < \infty .$$

In other words, a band-limited function $u(t)$ given by (1.1) can be extended, by replacing the real variable t with the complex variable z, to an *entire function of exponential type* ω_0.

The most useful and elegant result for band-limited analog functions (or signals) is the following theorem, see, for example, Rudin [1966]:

Theorem 1.1. (Paley-Wiener's Theorem)

A complex-valued function $u(z)$ is an entire function of exponential type ω_0 and its restriction to the real axis is in $L^2(-\infty, \infty)$ if and only if $u(t)$ satisfies (1.1) with $\sigma(\omega)$ in $L^2(-\omega_0, \omega_0)$.

Example 1.1. The analog signal

$$u(t) = \frac{\sin \omega_0 t}{\omega_0 t}, \quad -\infty < t < \infty,$$

is band-limited with bandwidth $2\omega_0 > 0$.

This statement can be verified by letting

$$\sigma(\omega) = \begin{cases} \dfrac{1}{2\omega_0} & \text{for } -\omega_0 < \omega < \omega_0\,, \\ 0 & \text{otherwise}. \end{cases}$$

In doing so, on one hand, we have

$$u(t) = \int_{-\infty}^{\infty} \sigma(\omega) e^{j\omega t} d\omega\,,$$

and, on the other hand, we have

$$\begin{aligned} \hat{u}(\omega) &= \int_{-\infty}^{\infty} u(t) e^{-j\omega t} dt \\ &= \int_{-\infty}^{\infty} \frac{\sin \omega_0 t}{\omega_0 t} e^{-j\omega t} dt \\ &= \begin{cases} 2\pi\sigma(\omega) & \text{for } -\omega_0 < \omega < \omega_0 \\ 0 & \text{otherwise}. \end{cases} \end{aligned}$$

Example 1.2. The analog signal

$$u(t) = e^{-t^2/2}$$

is not band-limited.

The reason is that the Fourier transform of $u(t)$ is given by

$$\hat{u}(\omega) = \sqrt{2\pi} e^{-\omega^2/2}\,,$$

which is positive for all real values of ω.

1.1.2 Digital Signals and the Sampling Theorem

Any analog signal or controller $u(t)$ can be converted to a *discrete-time* signal or controller u_n $(n = \cdots, -1, 0, 1, \cdots)$ by first sampling $u(t)$ periodically with sampling time $t_0 > 0$ and then quantizing it by rounding off the values of $u(nt_0)$, as shown in Fig. 1.1.

It is important to note, however, that if the sampling time t_0 is not chosen small enough, then the analog signal will not be well represented.

Example 1.3. Consider the two band-limited analog signals

$$u_1(t) = \frac{\sin \pi t}{\pi t} \qquad \text{and} \qquad u_2(t) = \frac{\sin 2\pi t}{2\pi t}\,,$$

and discuss their sampling time.

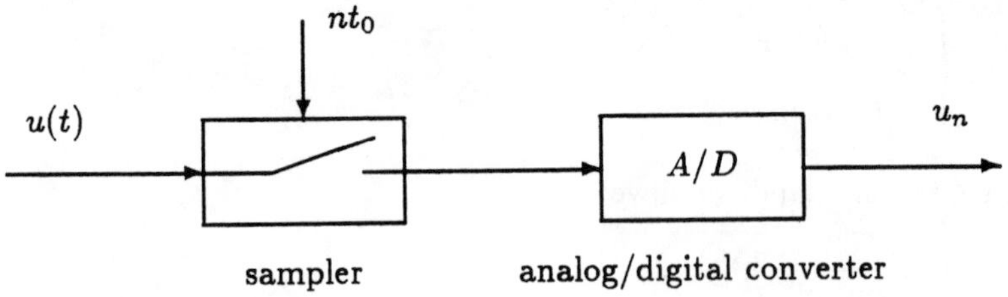

Fig. 1.1. Sampling and quantizing process

From Example 1.1, it is clear that the bandwidths of $u_1(t)$ and $u_2(t)$ are, respectively, 2π and 4π. Suppose we choose the sampling time to be $t_0 = 1$. Then we will have

$$u_1(nt_0) = u_2(nt_0) = \begin{cases} 1\,, & n = 0\,, \\ 0\,, & n = \pm 1\,, \pm 2\,, \cdots\,. \end{cases}$$

Hence, the two analog signals $u_1(t)$ and $u_2(t)$ are not distinguishable from the data set $\{u(nt_0)\} = \{\cdots, 0, 1, 0, \cdots\}$. In the following, we will see that in order to recover these two signals from their digitized samples, the sampling time t_0 for $u_1(t)$ must not exceed 1 and that for $u_2(t)$ must not exceed 1/2.

For a band-limited analog signal, the following celebrated result, commonly called the *sampling theorem*, provides us with a guideline for choosing the sampling time t_0.

Theorem 1.2. (Sampling Theorem)

Let $u(t)$ be a band-limited analog signal in $L^2(-\infty, \infty)$ with bandwidth $2\omega_0$, and let

$$0 < t_0 \leq \frac{\pi}{\omega_0}\,.$$

Then $u(t)$, $-\infty < t < \infty$, can be recovered from its digitized samples $u(nt_0)$, $n = \cdots, -1, 0, 1, \cdots$, by applying the formula

$$u(t) = \sum_{n=-\infty}^{\infty} u(nt_0) \frac{\sin \pi(t/t_0 - n)}{\pi(t/t_0 - n)}$$

where the convergence is uniform in t, $-\infty < t < \infty$.

Note that, in applications, we usually avoid the critical value $t_0 = \pi/\omega_0$, and the condition $0 < t_0 \le \pi/\omega_0$ is equivalent to $0 < 2\omega_0 \le 2\pi/t_0$, in which $2\omega_0$ is the bandwidth and $2\pi/t_0$ is the so-called *Nyquist frequency*.

To prove this theorem, we first note that since $\omega_0 \le \pi/t_0$, we may extend $\sigma(\omega)$ in (1.1), which is defined on $(-\omega_0, \omega_0)$, to a $2\pi/t_0$ periodic function $\tilde{\sigma}(\omega)$ by setting

$$\tilde{\sigma}(\omega) = \begin{cases} \sigma(\omega) & \text{for } -\omega_0 < \omega < \omega_0\,, \\ 0 & \text{for } -\pi/t_0 \le \omega \le -\omega_0 \quad \text{or} \quad \omega_0 \le \omega \le \pi/t_0\,, \end{cases}$$

and $\tilde{\sigma}(\omega + 2\pi k/t_0) = \tilde{\sigma}(\omega)$, where $-\pi/t_0 \le \omega \le \pi/t_0$, and $k = 0, \pm 1, \pm 2, \cdots$. Hence, $\tilde{\sigma}(\omega)$ has a Fourier series expansion

$$\tilde{\sigma}(\omega) \sim \sum_{k=-\infty}^{\infty} c_k e^{jkt_0\omega}\,, \tag{1.3}$$

where the Fourier coefficients c_k of $\tilde{\sigma}(\omega)$ are given by

$$c_k = \frac{t_0}{2\pi} \int_{-\pi/t_0}^{\pi/t_0} \tilde{\sigma}(\omega) e^{-jkt_0\omega} d\omega\,. \tag{1.4}$$

In particular, from (1.1), we have

$$u(t) = \int_{-\omega_0}^{\omega_0} \sigma(\omega) e^{j\omega t} d\omega = \int_{-\pi/t_0}^{\pi/t_0} \tilde{\sigma}(\omega) e^{j\omega t} d\omega \tag{1.5}$$

so that (1.4) and (1.5) give

$$c_k = \frac{t_0}{2\pi} u(-kt_0)\,. \tag{1.6}$$

Now, since $\sigma(\omega)$ is in $L^2(-\omega_0, \omega_0)$, it follows from (1.6) and the Parseval identity that

$$\begin{aligned} \sum_{n=-\infty}^{\infty} |u(nt_0)|^2 &= \sum_{k=-\infty}^{\infty} |u(-kt_0)|^2 \\ &= \left(\frac{2\pi}{t_0}\right)^2 \sum_{k=-\infty}^{\infty} |c_k|^2 \\ &= \frac{2\pi}{t_0} \int_{-\pi/t_0}^{\pi/t_0} |\tilde{\sigma}(\omega)|^2 d\omega \\ &= \frac{2\pi}{t_0} \int_{-\omega_0}^{\omega_0} |\sigma(\omega)|^2 d\omega < \infty\,. \end{aligned} \tag{1.7}$$

Hence, we may first conclude from (1.5), (1.3), and (1.6), consecutively, that

$$\begin{aligned} u(t) &= \int_{-\pi/t_0}^{\pi/t_0} \tilde{\sigma}(\omega) e^{j\omega t} d\omega \\ &= \frac{t_0}{2\pi} \sum_{k=-\infty}^{\infty} u(-kt_0) \int_{-\pi/t_0}^{\pi/t_0} e^{j\omega(t+kt_0)} d\omega \\ &= \frac{t_0}{2\pi} \sum_{k=-\infty}^{\infty} u(-kt_0) \frac{\sin \pi(t/t_0 + k)}{(t+kt_0)/2} \\ &= \sum_{n=-\infty}^{\infty} u(nt_0) \frac{\sin \pi(t/t_0 - n)}{\pi(t/t_0 - n)}, \end{aligned}$$

where the interchange of integration and summation is valid in view of (1.7). Finally, we may also conclude that the convergence of the infinite series is uniform for all t, $-\infty < t < \infty$, by using the Schwarz inequality and (1.7) again (Problem 1.5). This completes the proof of the sampling theorem.

Of course, most analog signals are not band-limited. However, by the Plancherel Theorem, which can be found in Rudin [1966], if $u(t)$ is in L^2 $(-\infty, \infty)$, so is its Fourier transform $\hat{u}(\omega)$. In addition, if $u(t) \in L^2(-\infty, \infty)$ $\cap L^1(-\infty, \infty)$, then $\hat{u}(\omega)$ is continuous on $(-\infty, \infty)$ so that $\hat{u}(\omega)$ decays to 0 as ω tends to $\pm\infty$, and $\hat{u}(\omega)$ can be well approximated by the truncated functions

$$\sigma_b(\omega) = \begin{cases} \frac{1}{2\pi} \hat{u}(\omega) & \text{for} b < \omega < b\,, \\ 0 & \text{otherwise}\,, \end{cases}$$

and this in turn implies that $u(t)$ can be well approximated by the band-limited signals

$$u_b(t) = \int_{-b}^{b} \sigma_b(\omega) e^{j\omega t} d\omega\,,$$

where $b \in (0, \infty)$ is usually determined by some practical criteria. For example, b may be chosen to be the so-called *equivalent rectangular bandwidth* defined by

$$b = \frac{1}{2} \frac{\int_{-\infty}^{\infty} |\hat{u}(\omega)|^2 d\omega}{\max_{-\infty<\omega<\infty} |\hat{u}(\omega)|^2},$$

which is formulated by matching the area under the curve $|\hat{u}(\omega)|^2$ with the area of the corresponding rectangle with height given by the maximum value of $|\hat{u}(\omega)|^2$ (see Fig. 1.2). Hence, the sampling theorem has very important applications to the study of analog and digital signals.

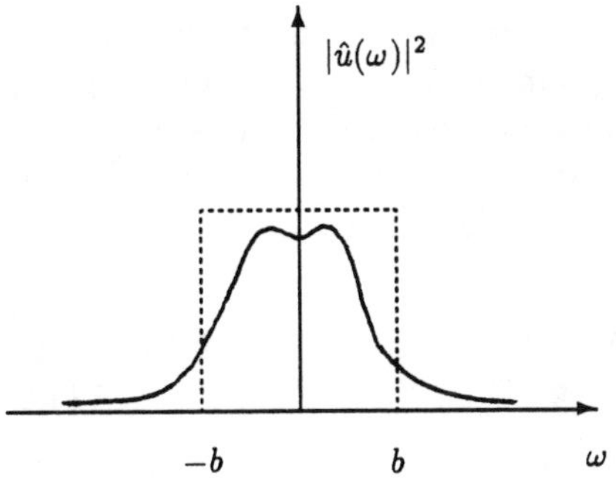

Fig. 1.2. Equivalent rectangular bandwidth

1.2 Time and Frequency Domains

We have already seen how digital signals are obtained and how analog signals are reconstructed from digital signals. With the recent rapid advance in digital technology (digital computers, digital processors, discrete-time Kalman filters, etc.), it is even more important than ever to work with digital signals. We have also seen the importance of studying analog signals in the so-called *frequency domain*, $-\infty < \omega < \infty$. It will be clear that it is equally important to be able to understand digital signals in the corresponding analogous frequency domain.

1.2.1 Fourier Transforms and Convolutions on Three Groups

To discuss the transitions between time and frequency domains, it is best to recall some basic concepts from harmonic analysis. In harmonic analysis, there are three basic *groups* of domains of definitions for signals: (1) the real line group $(-\infty, \infty)$, (2) the integer group $\{\cdots, -2, -1, 0, 1, 2, \cdots\}$, and (3) the circle group $\{e^{j\omega} : -\pi \le \omega < \pi\}$.

The Fourier transform takes analog signals $u(t)$ on the time domain $(-\infty, \infty)$, which is the real line group, to functions $\hat{u}(\omega)$ defined on the frequency domain $(-\infty, \infty)$, which is again the real line group via the Fourier transform. The reverse process is called the *inverse Fourier transform*. This pair, as we recall from Sect. 1.1, is given by

$$\begin{cases} \hat{u}(\omega) = \displaystyle\int_{-\infty}^{\infty} u(t) e^{-j\omega t} dt \,, & -\infty < \omega < \infty \,, \\ u(t) = \displaystyle\frac{1}{2\pi} \int_{-\infty}^{\infty} \hat{u}(\omega) e^{j\omega t} d\omega \,, & -\infty < t < \infty \,, \end{cases} \tag{1.8}$$

where the inverse Fourier transform exists provided that $\hat{u}(\omega) \in L^1(-\infty, \infty)$. Hence, we say that the real line group is *self-dual*; that is, the domain of $\hat{u}(\omega)$ is the "same" as that of $u(t)$.

While the time domain of analog signals is the real line group, the time domain of digital signals is the integer group. The dual of the integer group is the circle group, and conversely, the dual of the circle group is the integer group. The transition between a digital signal $\{u_n\}$ and its *"discrete Fourier transform"* $U^*(\omega)$ is given by the following pair of formulas

$$\begin{cases} U^*(\omega) = \displaystyle\sum_{n=-\infty}^{\infty} u_n e^{-jn\omega}, & -\pi \leq \omega < \pi, \\ u_n = \displaystyle\frac{1}{2\pi}\int_{-\pi}^{\pi} U^*(\omega) e^{jn\omega} d\omega, & n = 0, \pm 1, \pm 2, \cdots. \end{cases} \tag{1.9}$$

The first formula describes the digital signal in the frequency domain which is the circle group $\{e^{-j\omega} : -\pi \leq \omega < \pi\}$, and the second shows how the original digital signal is recovered from its discrete Fourier transform.

Note that by replacing ω with $-\omega$ in (1.9), $U^*(-\omega)$ becomes a Fourier series with Fourier coefficients u_n. Note also that the discrete Fourier transform of a discrete-time signal is 2π-periodic. If, in addition, $U^*(\omega) = 0$ for $\omega \in [-\pi, \omega_1) \cup (\omega_2, \pi)$, where $-\pi < \omega_1 < \omega_2 < \pi$, then the signal is said to be *periodically band-limited.* The "multiplicative operations" on the signals defined on these three groups of time and frequency domains are the appropriate *convolution* operations. We will use the notations $\overline{*}$, $*$, $\underline{*}$ for convolutions on the real-line, integer, and circle groups, respectively, as follows:

(1) *Convolution on the real line group:*

$$(f \overline{*} g)(x) = \int_{-\infty}^{\infty} f(y) g(x-y) dy.$$

(2) *Convolution on the integer group:*

$$\{w_n\} = \{u_n\} * \{v_n\},$$

defined by

$$w_n = \sum_{k=-\infty}^{\infty} u_k v_{n-k}, \quad n = \cdots, -2, -1, 0, 1, 2, \cdots.$$

(3) *Convolution on the circle group:*

$$(f \underline{*} g)(\omega) = \frac{1}{2\pi}\int_{-\pi}^{\pi} f(y) g(\omega - y) dy,$$

where $f(\omega)$ and $g(\omega)$ are 2π-periodic functions.

Example 1.4.

(1) Let

$$f(t) = g(t) = \begin{cases} 1 & \text{for} \quad -1/2 \le t \le 1/2\,, \\ 0 & \text{otherwise}\,. \end{cases}$$

Determine $(f\bar{*}g)(t)$.

(2) Consider the digital signals $\{u_n\}$ and $\{v_n\}$ defined by

$$u_n = \begin{cases} a^n & \text{for } \ 0 \le n \le N-1\,, \\ \ 0 & \text{otherwise}\,, \end{cases} \qquad \text{and} \quad v_n = \begin{cases} b^n & \text{for } \ n \le 0\,, \\ \ 0 & \text{for } \ n > 0\,. \end{cases}$$

Find $\{u_n\} * \{v_n\}$.

(3) Let

$$f(\omega) = g(\omega) = e^{j\omega}\,, \qquad -\pi + 2k\pi \le \omega < \pi + 2k\pi\,.$$

Compute $(f\underline{*}g)(\omega)$.

These convolutions can be found by direct computations as follows:

(1)

$$\begin{aligned} (f\bar{*}g)(t) &= \int_{-\infty}^{\infty} f(y)g(t-y)dy \\ &= \int_{-1/2}^{1/2} g(t-y)dy \\ &= \begin{cases} 1+t & \text{for } \ -1 \le t \le 0\,, \\ 1-t & \text{for } 0 \le t \le 1\,, \\ \quad 0 & \text{otherwise}\,. \end{cases} \end{aligned}$$

(2) $\{\omega_n\} = \{u_n\} * \{v_m\}$ with

$$\begin{aligned} \omega_n &= \sum_{k=-\infty}^{\infty} u_k v_{n-k} \\ &= \sum_{k=\max\{0,n\}}^{N-1} a^k b^{n-k} \\ &= \begin{cases} \dfrac{a^n b}{b-a}\left[1 - \left(\dfrac{a}{b}\right)^{N-n}\right] & \text{for } 0 \le n \le N-1\,, \\ \dfrac{b^{n+1}}{b-a}\left[1 - \left(\dfrac{a}{b}\right)^{N}\right] & \text{for } n < 0\,, \\ \qquad 0 & \text{otherwise}\,. \end{cases} \end{aligned}$$

(3)

$$\begin{aligned}(f\underline{*}g)(\omega) &= \frac{1}{2\pi}\int_{-\pi}^{\pi} f(y)g(\omega-y)dy \\ &= \frac{1}{2\pi}\int_{-\pi}^{\pi} e^{jy}e^{j(\omega-y)}dy \\ &= e^{j\omega}\,.\end{aligned}$$

It must be emphasized that all the three convolution operations are commutative, that is,

$$\begin{cases}(f\overline{*}g)(x) = (g\overline{*}f)(x) & \text{where } x = t \text{ or } \omega \\ \{u_n\} * \{v_n\} = \{v_n\} * \{u_n\} & \\ (f\underline{*}g)(\omega) = (g\underline{*}f)(\omega)\,, & \text{where } f(\omega) \text{ and } g(\omega) \text{ are } 2\pi\text{-periodic}.\end{cases}$$

This observation can be easily verified by a change of variables.

Now, let us denote by C_0^∞ and $\tilde{C}^\infty$, respectively, the classes of all infinitely differentiable (test) functions vanishing at $\pm\infty$ and all 2π-periodic infinitely differentiable (test) functions, namely,

$$C_0^\infty = \{f \in C^\infty : \;\; f(-\infty) = f(\infty) = 0\} \tag{1.10}$$

and

$$\tilde{C}^\infty = \{f \in C^\infty : \;\; f(x+2k\pi) = f(x)\,, \;\; k \text{ integer}\,, \;\; x \text{ real}\}\,. \tag{1.11}$$

The identities of these three convolution operations $\overline{*}$, $*$, $\underline{*}$ will be denoted by $\delta(x)$, $\{\delta_n\}$, and $\delta^*(\omega)$, respectively, as follows:

(1)

$$(f\overline{*}\delta)(x) = \int_{-\infty}^{\infty} f(y)\delta(x-y)dy = f(x)$$

for all $f(x)$ in C_0^∞.

(2) $\{u_n\} * \{\delta_n\} = \{u_n\}$, or

$$u_n = \sum_{k=-\infty}^{\infty} u_k\delta_{n-k}$$

for all n, where $n = \cdots, -2, -1, 0, 1, 2, \cdots$.

(3)

$$(f\underline{*}\delta^*)(\omega) = \frac{1}{2\pi}\int_{-\pi}^{\pi} f(y)\delta^*(\omega-y)dy = f(\omega)$$

for all $f(\omega)$ in $\tilde{C}^\infty$.

Our immediate need is to identify the identities $\delta(x)$, $\{\delta_n\}$, and $\delta^*(\omega)$. It is clear that $\delta(x)$ and $\delta^*(\omega)$ cannot be functions in the usual sense. More precisely, if we assume that $\delta(x)$ and $\delta^*(x)$ are continuous functions, then $\delta(x)$ must vanish for all $x \neq 0$ and $\delta^*(\omega) = 0$ for all $\omega \neq 2\pi k$, k any integer. Hence, $\delta(x)$ cannot have a finite value at $x = 0$, and $\delta^*(\omega)$ does not take on finite values at each $\omega = 2\pi k$, $k = 0, \pm 1, \cdots$ (Problem 1.13). The convolution identity $\{\delta_n\}$ for the operation $*$ is easy to identify, namely, for each integer k_0, $k_0 = \cdots, -2, -1, 0, 1, 2, \cdots, \delta_{n-k_0} = 1$ for $n = k_0$ and $\delta_{n-k_0} = 0$ for $n \neq k_0$. In other words, we have

$$\delta_n = \begin{cases} 1 & \text{for } n = 0 \\ 0 & \text{for } n \neq 0 \,. \end{cases}$$

The identities $\delta(x)$ and $\delta^*(\omega)$ are usually called *delta functions* or *delta distributions* and $\{\delta_n\}$ the sequence of *unit impulses* (Fig. 1.3).

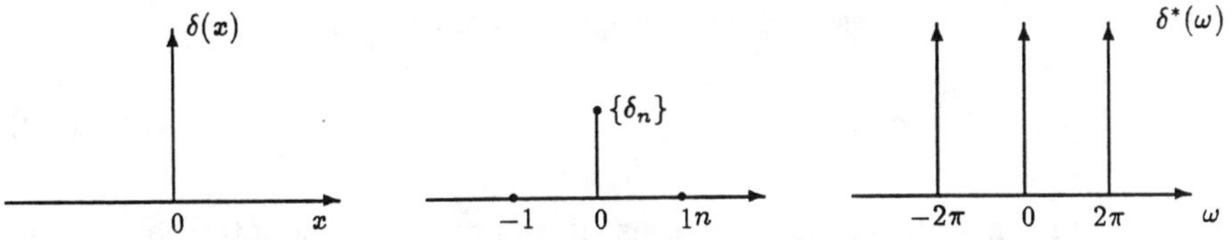

Fig. 1.3. Unit impulses

The most important property which is common to all convolution operations is that their Fourier transforms (that is, continuous Fourier transform for the real line group, discrete Fourier transform for the integer group, and inverse discrete Fourier transform for the circle group) take convolutions to actual algebraic multiplications:

(1) $\widehat{(f*g)}(\omega) = \hat{f}(\omega)\hat{g}(\omega)$

for all analog signals $f(t)$ and $g(t)$ in $L^1(-\infty, \infty)$;

(2) $W^*(\omega) = U^*(\omega)V^*(\omega)$,

where $\{w_n\} = \{u_n\} * \{v_n\}$; and

(3) $\displaystyle \frac{1}{2\pi}\int_{-\pi}^{\pi} (f \underline{*} g)(\omega) e^{-jn\omega} d\omega$

$$= \left(\frac{1}{2\pi}\int_{-\pi}^{\pi} f(\omega)\mathrm{e}^{-\mathrm{j}n\omega}d\omega\right)\left(\frac{1}{2\pi}\int_{-\pi}^{\pi} g(\omega)\mathrm{e}^{-\mathrm{j}n\omega}d\omega\right)$$

for all 2π-periodic analog signals $f(\omega)$ and $g(\omega)$ in $L^1(-\pi,\pi)$ (Problem 1.14).

1.2.2 Frequency Spectra of Digital Signals

Let us now pay more attention to the discrete Fourier transform of digital signals. Formally, for each digital signal $\{u_n\}$, we write

$$U^*(\omega) = \sum_{n=-\infty}^{\infty} u_n \mathrm{e}^{-\mathrm{j}n\omega}.$$

If $\{u_n\}$ is in l^1, and by this we mean

$$\sum_{n=-\infty}^{\infty} |u_n| < \infty,$$

then $U^*(\omega)$ is defined for all frequencies ω. However, even the constant signal $\{1\} = \{\cdots, 1, 1, \cdots\}$ does not produce a well-defined discrete Fourier transform. Nevertheless, we will accept the convention

$$\sum_{n=-\infty}^{\infty} \mathrm{e}^{-\mathrm{j}n\omega} = \delta^*(\omega). \tag{1.12}$$

This formula actually makes sense if we are allowed to interchange integration and summation. Indeed, for any test function $f(\omega)$ in $\tilde{C}^\infty$, we have, see (1.9),

$$f\underline{*}\left(\sum_{n=-\infty}^{\infty} \mathrm{e}^{-\mathrm{j}n\omega}\right) = \sum_{n=-\infty}^{\infty}\left(\frac{1}{2\pi}\int_{-\pi}^{\pi} f(y)\mathrm{e}^{\mathrm{j}ny}dy\right)e^{-jn\omega}$$
$$= f(\omega) = (f\underline{*}\delta^*)(\omega).$$

The discrete Fourier transform of a digital signal is called the *spectrum* (or more precisely, *frequency spectrum*) of the signal. We give a list of examples in the following table (Problem 1.16):

Table 1.1. Spectra of digital signals

Terminology	Digital signal $\{u_n\}$	Spectrum $U^*(\omega)$
unit impulse at $n = k_0$	$\{\delta_{n-k_0}\}$	$\mathrm{e}^{-\mathrm{j}k_0\omega}$
power pulse	$\{r^{\lvert n\rvert}\}, 0 \le r < 1$	$(1-r^2)/(1-2r\cos\omega + r^2)$
single frequency signal	$\{\mathrm{e}^{\mathrm{j}n\omega_0}\}$	$\delta^*(\omega-\omega_0)$
sinusoid	$\{\sin n\omega_0\}$	$\{\delta^*(\omega-\omega_0)$ $-\delta^*(\omega+\omega_0)\}/(2\mathrm{j})$
cosinusoid	$\{\cos n\omega_0\}$	$\{\delta^*(\omega-\omega_0)$ $+\delta^*(\omega+\omega_0)\}/2$

The spectrum of the unit impuse at the k_0th instant is a complex number with (constant) *magnitude* 1 and *phase* $-k_0\omega$. It is usually represented in the complex plane, as shown in Fig. 1.4.

On the other hand, the spectrum of the power pulse digital signal has zero phase and hence can be represented in the magnitude vs frequency plane, see Fig. 1.5. It is also called the *Poisson distribution.*

The last three examples in Table 1.1 are particularly interesting. First note that the (radian) frequency of the digital signal $\{e^{jn\omega_0}\}$ is ω_0, and its spectrum, being the delta distribution $\delta^*(\omega - \omega_0)$, actually "lives" only at this frequency. (Of course, the 2π-translates in both the time and frequency domains must be considered.) On the other hand, the spectra of the sinusoidal and cosinusoidal digital signals "live" at both frequencies ω_0 and $-\omega_0$ (and their 2π-translations), indicating that the digital signals $\{\sin n\omega_0\}$ and $\{\cos n\omega_0\}$ carry both positive and negative frequencies $\pm\omega_0$. The above observation will be important for a better understanding of digital filtering.

1.3 z-Transforms

The very important notion of z-transformation for digital signals will be introduced in this section. As will be seen from the context below, the z-transform of a digital signal is an analogue of the Fourier transform of a continuous (analog) signal, and plays a central role in digital signal analysis and digital filter design.

1.3.1 Properties of the z-Transform

In order to avoid the notion of distributions and be able to utilize complex variable techniques, it is more convenient to extend the frequency domain $-\infty < \omega < \infty$ to the complex plane by substituting $e^{j\omega}$ with

$$z = re^{j\omega}, \quad 0 \le r < \infty,$$

in the discrete Fourier transform. In doing so, we have introduced the notion of a *z-transform.* The z-transform of a digital signal $\{u_n\}$, which will be denoted by $Z\{u_n\} = U(z)$, and its inverse transform are given by the following pair of formulas

$$\begin{cases} U(z) = Z\{u_n\} = \displaystyle\sum_{n=-\infty}^{\infty} u_n z^{-n}, \\ \\ u_n = Z^{-1}\{U(z)\} = \dfrac{1}{2\pi j}\displaystyle\int_{|z|=1} U(z) z^{n-1} dz, \end{cases} \tag{1.13}$$

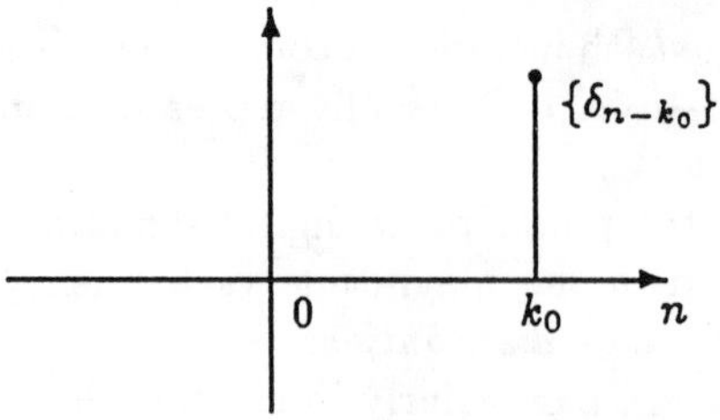

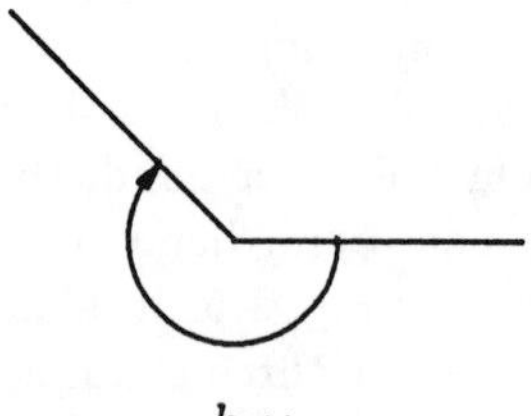

Fig. 1.4. Magnitude and phase of the unit impulse

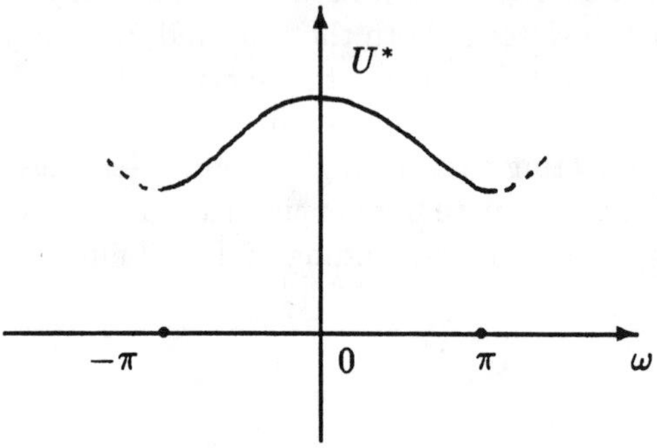

Fig. 1.5. Poisson distribution

where the contour integral is taken in the counterclockwise direction along the unit circle $|z| = 1$ on which the series for $U(z)$ converges. Note that the contour does not necessarily have to be a circle, and in fact it may be any appropriate (closed) Jordan curve. We remark that the first formula in (1.13) is only a formal series. However, if it converges uniformly on the contour, then $U(z)$ is continuous there, and the second formula is applicable. Hence, the contour of integration is usually chosen to be any Jordan curve that encloses the origin and lies in the region of uniform convergence of $U(z)$. In addition, the z-transform and discrete Fourier transform are related by

$$U(e^{j\omega}) = U^*(\omega), \quad -\infty < \omega < \infty .$$

Table 1.2. z−transforms of digital signals

Operation	Digital signal	z-transform
convolution	$\{u_n\}*\{v_n\}$	$U(z)V(z)$
delay	$\{u_{n-k_0}\}$	$z^{-k_0}U(z)$
linear combination	$\{au_n + bv_n\}$	$aU(z) + bV(z)$
differentiation	$\{nu_n\}$	$-z(dU(z)/dz)$
multiplication	$\{u_n v_n\}$	$\int_{\|w\|=1} U(w)V(z/w)w^{-1}dw/(2\pi j)$

Hence, z-transformation also changes convolution to algebraic multiplication. A list of properties of z-transformation is given in the Table 1.2 (Problem 1.17):

Example 1.5. Find the z-transform of the digital signal $\{u_n\}$, where

$$u_n = \begin{cases} a^n & \text{for} \quad n \geq 0\,, \\ b^n & \text{for} \quad n < 0\,, \end{cases}$$

with $0 < |a| < |b|$.

By definition, we have

$$\begin{aligned} Z\{u_n\} &= \sum_{n=0}^{\infty} a^n z^{-n} + \sum_{n=-\infty}^{-1} b^n z^{-n} \\ &= \frac{1}{1-az^{-1}} + \frac{1}{1-b^{-1}z} - 1 \\ &= \frac{(b-a)z}{(z-a)(b-z)}\,, \end{aligned}$$

which converges only on the region $|a| < |z| < |b|$.

Example 1.6. Find the inverse z-transform of

$$U(z) = \frac{1}{1-az^{-1}}\,, \qquad |a| < |z|\,,\ |a| < 1\,.$$

It follows from the definition that

$$\begin{aligned} u_n &= Z^{-1}\{U(z)\} \\ &= \frac{1}{2\pi \mathrm{j}} \int_{|z|=1} \frac{z^{n-1}}{1-az^{-1}} dz \\ &= \frac{1}{2\pi \mathrm{j}} \int_{|z|=1} \frac{z^n}{z-a} dz \\ &= \begin{cases} a^n & \text{for} \quad n \geq 0\,, \\ 0 & \text{for} \quad n < 0\,. \end{cases} \end{aligned}$$

From Table 1.2, it is interesting to note that the z-transformation and the *Laplace transformation* have very similar properties. For this and other reasons, the z-transform is also called the *discrete Laplace transform*. Since the Laplace-transform method is very useful in solving linear ordinary differential equations with initial conditions (usually called *initial value problems*), we should expect the z-transform to have analogous performance for linear difference equations. We will see next that indeed it does, and in fact, since discrete convolution is usually simpler than continuous convolution, this method is even more useful. To incorporate initial conditions, it is necessary to introduce the concept of *causality*.

1.3.2 Causal Digital Signals

A digital signal $\{u_n\}$ is said to be *causal* if $u_n = 0$ for each $n < 0$. In other words, a causal signal has the initial conditions $u_0, u_1, \cdots, u_k$ (which are not necessarily nonzero) for some positive integer k. The z-transform of a causal signal $\{u_n\}$ is given by

$$Z\{u_n\} = \sum_{n=0}^{\infty} u_n z^{-n}$$

and is usually called a *one-sided z-transform.*

Example 1.7. Let $\{u_n\}$ be a causal signal with initial conditions $u_0 = a$, $u_1 = b$, $u_2 = c$. Determine the z-transforms of $\{u_{n+k}\}$, for $k = 1, 2, 3$, in terms of the z-transform $U(z)$ of $\{u_n\}$ and the initial conditions.

Using the one-sided z-transform notation

$$U(z) = Z\{u_n\} = \sum_{n=0}^{\infty} u_n z^{-n},$$

we have

$$\begin{aligned} Z\{u_{n+k}\} &= \sum_{n=0}^{\infty} u_{n+k} z^{-n} = z^k \sum_{n=0}^{\infty} u_{n+k} z^{-(n+k)} \\ &= z^k \left(\sum_{n=0}^{\infty} u_n z^{-n} - u_0 - u_1 z^{-1} - \cdots - u_{k-1} z^{-(k-1)} \right). \end{aligned} \tag{1.14}$$

In particular,

$$Z\{u_{n+1}\} = z(U(z) - u_0) = zU(z) - az\,,$$

$$Z\{u_{n+2}\} = z^2(U(z) - u_0 - u_1 z^{-1}) = z^2 U(z) - az^2 - bz\,,$$

and

$$Z\{u_{n+3}\} = z^3(U(z) - u_0 - u_1 z^{-1} - u_2 z^{-2}) = z^3 U(z) - az^3 - bz^2 - cz\,.$$

1.3.3 Initial Value Problems

We now consider the solution of an initial value problem. A linear difference equation with unknowns u_n's and constant coefficients $a_0, \cdots, a_k$ is given by

$$\sum_{i=0}^{k} a_i u_{n-i} = b_{n-k}\,.$$

It is said to be of *order* k if $a_0 a_k \neq 0$. If $b_n = 0$ for all n, the equation is said to be *homogeneous.*

Suppose that we have a kth order linear difference equation with given constant coefficients $a_0, \cdots, a_k$, $a_0 a_k \neq 0$, and a "control" input sequence $\{b_{n-k}\}$. To guarantee uniqueness of the solution, we are usually given k initial values. This is called an *initial value problem.* More precisely, a kth order initial value problem can be stated as follows: Solve for the causal digital signal $\{u_n\}$ satisfying

$$\begin{cases} \sum_{i=0}^{k} a_i u_{n-i} = b_{n-k}\,, & n = k, k+1, \cdots, \\ u_0 = c_0, \ldots, u_{k-1} = c_{k-1}\,. \end{cases}$$

To utilize the z-transform, it is more convenient to make a change of indices from n to $n+k$. Hence, we have

$$a_0 u_{n+k} + a_1 u_{n+k-1} + \cdots + a_k u_n = b_n, \quad n = 0, 1, \cdots,$$

and an application of (1.14) yields

$$\begin{aligned} a_0 z^k \left[U(z) - \sum_{i=0}^{k-1} c_i z^{-i} \right] + a_1 z^{k-1} \left[U(z) - \sum_{i=0}^{k-2} c_i z^{-i} \right] \\ + \cdots + a_k U(z) = B(z)\,, \end{aligned} \tag{1.15}$$

where

$$U(z) = Z\{u_n\} = \sum_{n=0}^{\infty} u_n z^{-n}$$

and

$$B(z) = Z\{b_n\} = \sum_{n=0}^{\infty} b_n z^{-n}\,.$$

To be consistent, we also use the notation

$$A(z) = Z\{a_n\} = \sum_{n=0}^{k} a_n z^{-n}\,.$$

Then (1.15) may be written as

$$\begin{aligned} z^k A(z) U(z) = B(z) + \big[a_0 c_0 z^k + (a_0 c_1 + a_1 c_0) z^{k-1} + \cdots \\ + (a_0 c_{k-1} + \cdots + a_{k-1} c_0) z \big] \end{aligned}$$

or

$$U(z) = \frac{z^{-k}B(z)}{A(z)} + \frac{z^{-k}}{A(z)}\left[a_0 c_1 z^k + (a_0 c_1 + a_1 c_0)z^{k-1} + \cdots \right.$$
$$\left. + (a_0 c_{k-1} + \cdots + a_{k-1}c_0)z\right],$$

and via the inverse z-transform, we have

$$u_n = \frac{1}{2\pi \mathrm{j}} \int_{|z|=r} U(z) z^{n-1} dz .$$

Here, the contour integral is taken in the counterclockwise direction along the circle $|z| = r$ that encloses the origin and lies in the region of uniform convergence of the series $U(z)$. In solving for $\{u_n\}$, the residue theorem from complex analysis may be used.

Example 1.8. Solve the initial value problem

$$\begin{cases} u_{n+k} + \cdots + u_{n+1} + u_n = 1, & n = 0, 1, \cdots, \\ u_0 = u_1 = \cdots = u_{k-1} = 0 & \text{where } k \geq 1 . \end{cases}$$

Since

$$A(z) = \sum_{n=0}^{k} z^{-n} = z^{-k}(z^k + z^{k-1} + \cdots + z + 1)$$

and

$$B(z) = \sum_{n=0}^{\infty} z^{-n} = \frac{1}{1 - z^{-1}} = \frac{z}{z - 1},$$

where the series for $B(z)$ converges in $|z| > 1$, we have

$$U(z) = \frac{z^{-k}B(z)}{A(z)} = \frac{z}{z^{k+1} - 1} .$$

Hence, by using any value of $r > 1$, it follows that

$$u_n = \frac{1}{2\pi j} \int_{|z|=r} \left(\frac{z}{z^{k+1} - 1}\right) z^{n-1} dz .$$

The integrand

$$\left(\frac{z}{z^{k+1} - 1}\right) z^{n-1} = \frac{z^n}{z^{k+1} - 1}$$

has simple poles at each

$$z_\ell = e^{\mathrm{j}2\pi\ell/(k+1)}, \quad \ell = 1, \cdots, k + 1,$$

so that by the residue theorem, we have

$$
\begin{aligned}
u_n &= \sum_{\ell=1}^{k+1} \lim_{z\to z_\ell} (z - z_\ell) \frac{z^n}{z^{k+1}-1} \\
&= \sum_{\ell=1}^{k+1} z_\ell^n \Big/ \left(\lim_{z\to z_\ell} \frac{z^{k+1}-1}{z-z_\ell} \right) \\
&= \sum_{\ell=1}^{k+1} \frac{z_\ell^n}{(k+1) z_\ell^k} \\
&= \frac{1}{k+1} \sum_{\ell=1}^{k+1} e^{j2\pi\ell(n-k)/(k+1)} \\
&= \begin{cases} 1 & \text{for} \quad n = k,\ 2k+1,\ 3k+2,\ 4k+3, \cdots, \\ 0 & \text{otherwise}. \end{cases}
\end{aligned}
$$

1.3.4 Singular and Analytic Discrete Fourier Transforms

Since causal signals are most important, it is necessary to pay more attention to one-sided z-transforms. Let $\{u_n\}$ be any digital signal, and $U(z)$ its z-transform. Then we may write

$$U(z) = U_s(z) + U_a(z),$$

where

$$U_s(z) = \sum_{n=1}^{\infty} u_n z^{-n}$$

and

$$U_a(z) = \sum_{n=-\infty}^{0} u_n z^{-n} = \sum_{n=0}^{\infty} u_{-n} z^n$$

are called the *singular* and *analytic parts* of $U(z)$, respectively. Note that both $U_s(z)$ and $U_a(z)$ are one-sided z-transforms. By taking the radial limits (as $|z| \to 1$), we also have

$$U_s(e^{j\omega}) = U_s^*(\omega)$$

and

$$U_a(e^{j\omega}) = U_a^*(\omega),$$

which are also called the *singular* and *analytic discrete Fourier transforms* of the digital signal $\{u_n\}$, respectively, see (1.9). We may also decompose the convolution identity δ^* into

$$\delta^* = \delta_s^* + \delta_a^*$$

by using the definitions

$$(f \underline{*} \delta_s^*)(\omega) = \frac{1}{2\pi} \int_{-\pi}^{\pi} f(y) \delta_s^*(\omega - y) dy = f_s(\omega)$$

and

$$(f \underline{*} \delta_a^*)(\omega) = \frac{1}{2\pi} \int_{-\pi}^{\pi} f(y) \delta_a^*(\omega - y) dy = f_a(\omega)$$

for every test function

$$f(\omega) = f_s(\omega) + f_a(\omega)$$

in $\widetilde{C}^\infty$, with

$$f_s(\omega) = \sum_{n=1}^{\infty} \left(\frac{1}{2\pi} \int_{-\pi}^{\pi} f(y) \mathrm{e}^{\mathrm{j}ny} dy \right) \mathrm{e}^{-\mathrm{j}n\omega}$$

and

$$f_a(\omega) = \sum_{n=-\infty}^{0} \left(\frac{1}{2\pi} \int_{-\pi}^{\pi} f(y) \mathrm{e}^{\mathrm{j}ny} dy \right) \mathrm{e}^{-\mathrm{j}n\omega} .$$

In doing so, the convention

$$\sum_{n=1}^{\infty} \mathrm{e}^{-\mathrm{j}n\omega} = \delta_s^*(\omega)$$

and

$$\sum_{n=-\infty}^{0} \mathrm{e}^{-\mathrm{j}n\omega} = \delta_a^*(\omega) ,$$

see (1.12), may be used. Indeed, if we are allowed to interchange integration and summation, then for any test function f in $\widetilde{C}^\infty$, we have

$$\begin{aligned} f \underline{*} \left(\sum_{n=1}^{\infty} \mathrm{e}^{-\mathrm{j}n\omega} \right) &= \sum_{n=1}^{\infty} \left(\frac{1}{2\pi} \int_{-\pi}^{\pi} f(y) \mathrm{e}^{\mathrm{j}ny} dy \right) \mathrm{e}^{-\mathrm{j}n\omega} \\ &= f_s(\omega) = (f \underline{*} \delta_s^*)(\omega) \end{aligned}$$

and

$$\begin{aligned} f \underline{*} \left(\sum_{n=-\infty}^{0} \mathrm{e}^{-\mathrm{j}n\omega} \right) &= \sum_{n=-\infty}^{0} \left(\frac{1}{2\pi} \int_{-\pi}^{\pi} f(y) \mathrm{e}^{\mathrm{j}ny} dy \right) \mathrm{e}^{-\mathrm{j}n\omega} \\ &= f_a(\omega) = (f \underline{*} \delta_a^*)(\omega) . \end{aligned}$$

We conclude this section by saying that the z-transform is a very useful tool in the study of digital signals, and by taking the radial limit as $|z| \to 1$, the z-transform of a digital signal becomes its spectrum, which contains information on both the magnitude and the phase of the signal.

1.4 Digital Filters

A *digital filter* is a transformation that takes any digital signal $\{u_n\}$, called an *input* signal, to a digital signal $\{v_n\}$, called the corresponding *output* signal, as shown in Fig. 1.6:

Fig. 1.6. Digital filter

In practice, the desirable properties of a digital filter include the following:

(1) *Linearity.* If $\{v_n\}$ and $\{v'_n\}$ are outputs corresponding to the inputs $\{u_n\}$ and $\{u'_n\}$, respectively, and a, b are two arbitrary constants, then the output signal corresponding to the input $\{au_n+bu'_n\}$ is $\{av_n+bv'_n\}$. In notation, we write

$$\left.\begin{array}{l}\{u_n\} \to \{v_n\} \\ \{u'_n\} \to \{v'_n\}\end{array}\right\} \Longrightarrow \{au_n + bu'_n\} \to \{av_n + bv'_n\}.$$

(2) *Time-invariance.* If $\{v_n\}$ is the output signal corresponding to an input $\{u_n\}$ and k_0 is any fixed integer, then the output signal corresponding to the input $\{u_{n+k_0}\}$ is $\{v_{n+k_0}\}$. In notation, we write for any k_0,

$$\{u_n\} \to \{v_n\} \Longrightarrow \{u_{n+k_0}\} \to \{v_{n+k_0}\}.$$

(3) *Causality.* For any input signal $\{u_n\}$ its output signal $\{v_n\}$ at any time instant n_0 does not depend on u_n for $n > n_0$. In other words, for each n_0, v_{n_0} does not depend on the "future" information ($n > n_0$) of the input signal $\{u_n\}$.

(4) *Stability.* For every bounded input signal $\{u_n\}$, its output signal $\{v_n\}$ is also bounded. That is, if $|u_n| \leq M$ for all n and some $M < \infty$, then $|v_n| \leq N$ for all n and some $N < \infty$. In notation, we write

$$\{u_n\} \in l^\infty \Longrightarrow \{v_n\} \in l^\infty .$$

(For this reason, stability may also be called *Bounded-Input/Bounded-Output stability*, or BIBO *stability*.)

1.4.1 Basic Properties of Digital Filters

In the engineering and geophysics literature, a digital filter is usually defined by means of convolution with a sequence of complex numbers

$$h_0 \,,\; h_1 \,,\; h_2 \,,\; \cdots$$

in the sense that in the *filtering process* the output $\{v_n\}$ is obtained from the input $\{u_n\}$ by convolution with the "filter sequence" $\{h_n\}$ as

$$\{v_n\} = \{h_n\} * \{u_n\} ,$$

or equivalently,

$$v_n = \sum_{i=0}^{\infty} h_i u_{n-i} . \tag{1.16}$$

It is clear that a digital filter so defined satisfies the first three desirable properties, that is, linearity, time-invariance, and causality (Problem 1.21). In fact, as will be seen from Theorem 1.3 below, these three properties also guarantee that the digital filter is defined by convolution with some sequences $\{h_n\}$ as in (1.16). For this reason, this digital filter is also called a *causal linear time-invariant (LTI) digital filter.* We have the following theorem:

Theorem 1.3. (Characterization of Causal LTI Digital Filters)

A digital filter is linear, time-invariant, and causal if and only if there exists a sequence $\{h_n\}, n = 0, 1, \cdots$, of complex numbers such that the input-output relation is described by the convolution as given in (1.16).

Before we go into details, let us first make the following important observation: Suppose that we use the unit impulse $\{\delta_n\}$, where

$$\delta_n = \begin{cases} 1 & \text{for } n = 0 , \\ 0 & \text{for } n \neq 0 , \end{cases}$$

as the input signal to the digital filter defined by (1.16), then its corresponding output signal is

$$v_n = \sum_{i=0}^{\infty} h_i \delta_{n-i} = h_n \,.$$

That is, the sequence $\{h_n\}$, which is used to define the digital filter (1.16), is the output signal corresponding to the unit impulse input. Hence, this digital filter is also called a *unit impulse response digital filter*, and the defining sequence $\{h_n\}$, $n = 0, 1, \cdots$, is called the *unit impulse response sequence* of the filter, see Fig. 1.7.

Fig. 1.7. Unit impulse response sequence

To prove the nontrivial direction (necessity) of Theorem 1.3, let P denote the operator that governs the input-output relationship of the digital filter. That is, $P\{u_n\} = \{v_n\}$ where $\{u_n\}$ denotes the input and $\{v_n\}$ the output. As usual, the "standard basis" of any sequence space will be denoted by $\{e_n\}$, where

$$\begin{aligned} e_0 &= \{1, 0, 0, \cdots\}, \\ e_1 &= \{0, 1, 0, \cdots\}, \\ &\cdots \\ e_n &= \{0, \cdots, 0, 1, 0, \cdots\}, \\ &\cdots. \end{aligned}$$

Note that the sequence e_0 is the same as the sequence $\{\delta_n\}$ of the unit impulse. Hence, as suggested by the above discussion, we should define the *unit impulse response* sequence of the digital filter by

$$\{h_n\} = P\{\delta_n\} = Pe_0 \,.$$

Since P is time-invariant, we have

$$\begin{aligned} Pe_0 &= \{h_0, h_1, h_2, \cdots\}, \\ Pe_1 &= \{0, h_0, h_1, \cdots\}, \\ Pe_2 &= \{0, 0, h_0, \cdots\}, \\ &\cdots. \end{aligned}$$

To verify that $\{h_n\}$, indeed, governs the input-output relationship of the digital filter in the form (1.16), we introduce a sequence of "projection" operators P_m. Here, for each m, P_m which maps a sequence to a scalar is defined by

$$P_m\{u_n\} = v_m ,$$

where v_m is the mth term of the output sequence $\{v_n\} = P\{u_n\}$ of the digital filter P. Hence, if e_k is used as the input sequence, then we have

$$P_m e_k = \begin{cases} h_{m-k} & \text{for } m \geq k \\ 0 & \text{otherwise}. \end{cases}$$

In general, for any input sequence $\{u_m\}$, we may write

$$\{u_n\} = \sum_{k=0}^{\infty} u_k e_k ,$$

so that by the linearity of P_m (which is induced by that of P) and the above formula for $P_m e_k$, we have

$$v_m = P_m\{u_n\} = \sum_{k=0}^{\infty} u_k P_m e_k = \sum_{k=0}^{m} u_k h_{m-k} .$$

Here, the interchange of P_m with an infinite sum is permissible since

$$v_m = P_m\{u_0, u_1, \cdots, u_k, 0, \cdots\} = P_m\{u_0, u_1, \cdots, u_j, 0, \cdots\}$$

for all $j, k \geq m$, so that a standard limit argument applies. Finally, since $\{u_n\}$ is a causal signal in the sense that $u_{-1} = u_{-2} = \cdots = 0$, we have

$$v_m = \sum_{k=0}^{m} u_k h_{m-k} = \sum_{i=0}^{\infty} h_i u_{m-i} ,$$

as described by (1.16).

An advantage of using the convolution (1.16) for defining a digital filter is that the following stability criterion can be obtained immediately.

Theorem 1.4. (Stability Criterion for Convolution Filters)

A causal, linear, time-invariant digital filter with unit impulse-response sequence $\{h_n\}$ is stable if and only if $\{h_n\}$ is in l^1, that is,

$$\textit{stability} \Longleftrightarrow \sum_{n=0}^{\infty} |h_n| < \infty .$$

It is clear that if $\{h_n\}$ is in l^1, then for any input $\{u_n\}$ with $|u_n| \leq M$, we have

$$|v_n| \leq M \sum_{j=0}^{\infty} |h_j| .$$

The proof of the converse is left to the reader (Problem 1.22). Of course, the above stability criterion is always satisfied if $\{h_n\}$ is a finite sequence (i.e., $h_n = 0$ for all $n > M$, where M is some non-negative integer). Such a digital filter is called a *Finite Impulse Response* (FIR) digital filter. If infinitely many h_n are nonzero, the filter is called an *Infinite Impulse Response* (IIR) digital filter.

An FIR digital filter is easy to implement. Recall that since

$$v_n = \sum_{i=0}^{M} h_i u_{n-i} = h_0 u_n + \cdots + h_M u_{n-M} , \tag{1.17}$$

the only operations are scalar multiplication, summation, and delay which will be denoted by D, namely,

$$D u_n = u_{n-1} .$$

Since the operations of scalar multiplication and delay commute, an FIR digital filter can be implemented, as shown in Fig. 1.8.

Of course, the operations described here can be considered as a weighted average with weights $h_0, \cdots, h_M$. Since the same weights are used at any time instant n, an FIR filter is also called a *Moving-Average* (MA) digital filter.

On the other hand, an IIR digital filter described by (1.16) cannot be implemented in the same manner, simply because it is not possible to implement infinitely many scalar multiplications and delays!

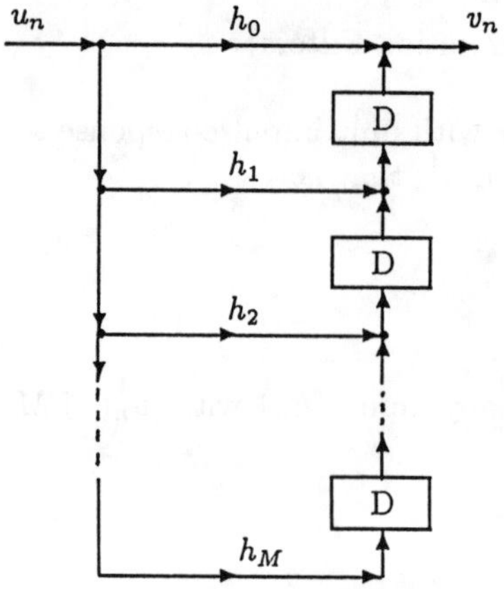

Fig. 1.8. Implementation of FIR filters

1.4.2 Transfer Functions and IIR Digital Filters

Before discussing how an IIR digital filter can be realized, it is necessary to study the (one-sided) z-transform of the filter equation (1.16). Recall from Table 1.2 that the z-transformation takes convolution to algebraic multiplication. In other words, we have

$$V(z) = H(z)U(z), \tag{1.18}$$

where

$$U(z) = \sum_{n=0}^{\infty} u_n z^{-n}, \quad V(z) = \sum_{n=0}^{\infty} v_n z^{-n},$$

and

$$H(z) = \sum_{n=0}^{\infty} h_n z^{-n}. \tag{1.19}$$

Note that in obtaining (1.18), we have set u_n, v_n, $h_n = 0$ for $n < 0$ (Problem 1.23). That is, the spectrum of the output signal is obtained by multiplying the spectrum of the input signal by $H(z)$. For this reason, $H(z)$, which is the z-transform of the unit impulse response sequence $\{h_n\}$ of the digital filter, is called the *transfer function* of the filter.

Now let us consider an example that demonstrates the importance of the transfer function. First, taking the radial limit as $|z| \to 1$ in (1.18), we have

$$V^*(\omega) = H^*(\omega)U^*(\omega),$$

where

$$H^*(\omega) = \sum_{n=0}^{\infty} h_n e^{-jn\omega}.$$

Example 1.9. Let

$$u_n = a e^{jn\omega_0} + b e^{jn\omega_1} ,$$

where a and b are positive constants, $0 < |\omega_1 - \omega_0| < 2\pi$, and $n = 1, 2, \cdots$. Also, set $u_n = 0$ for $n \leq 0$. This can be considered as a causal digital signal that carries two frequencies ω_0 and ω_1 with magnitudes a and b, respectively. Design a very simple FIR digital filter that passes the frequency ω_0 without changing its magnitude, but stops the frequency ω_1.

Since only two frequencies are being considered, the simplest FIR filter should have only two nonzero terms in $\{h_n\}$, and the filter must satisfy the design criterion

$$\{h_n\} * \{a e^{jn\omega_0} + b e^{jn\omega_1}\} = \{a e^{jn\omega_0}\} .$$

The "spectral" approach is to find $H(z)$ such that $V(z) = H(z)U(z)$, where

$$U(z) = a \sum_{n=1}^{\infty} e^{jn\omega_0} z^{-n} + b \sum_{n=1}^{\infty} e^{jn\omega_1} z^{-n}$$

and

$$V(z) = a \sum_{n=1}^{\infty} e^{jn\omega_0} z^{-n} ,$$

since $u_n = 0$ for $n \leq 0$. Recall from Sect. 1.3.4 that by taking the radial limit as $|z| \to 1$ we may write

$$U^*(\omega) = a\delta_s^*(\omega - \omega_0) + b\delta_s^*(\omega - \omega_1)$$

and

$$V^*(\omega) = a\delta_s^*(\omega - \omega_0) .$$

That is, the spectrum of the input signal "lives" at two frequencies ω_0 and ω_1. To pass the frequency ω_0 and stop the frequency ω_1, we must construct $H(z)$ such that

$$H^*(\omega_0) = 1 \quad \text{and} \quad H^*(\omega_1) = 0 .$$

This is easily satisfied by setting

$$H^*(\omega) = \frac{e^{-j\omega} - e^{-j\omega_1}}{e^{-j\omega_0} - e^{-j\omega_1}} ,$$

or equivalently,

$$H(z) = h_0 + h_1 z^{-1}$$

where

$$h_0 = \frac{-e^{-j\omega_1}}{e^{-j\omega_0} - e^{-j\omega_1}} \qquad \text{and} \qquad h_1 = \frac{1}{e^{-j\omega_0} - e^{-j\omega_1}}.$$

Note that the transfer function is only a linear polynomial in z^{-1}.

To verify that this FIR digital filter really works, we simply note that for $n \geq 1$,

$$\begin{aligned}
v_n &= \sum_{i=0}^{\infty} h_i u_{n-i} = h_0 u_n + h_1 u_{n-1} \\
&= \frac{-e^{-j\omega_1}}{e^{-j\omega_0} - e^{-j\omega_1}}(ae^{jn\omega_0} + be^{jn\omega_1}) \\
&\quad + \frac{1}{e^{-j\omega_0} - e^{-j\omega_1}}(ae^{j(n-1)\omega_0} + be^{j(n-1)\omega_1}) \\
&= a\frac{e^{jn\omega_0}(e^{-j\omega_0} - e^{-j\omega_1})}{e^{-j\omega_0} - e^{-j\omega_1}} + b\frac{e^{j(n-1)\omega_1} - e^{j(n-1)\omega_1}}{e^{-j\omega_0} - e^{-j\omega_1}} \\
&= ae^{jn\omega_0}.
\end{aligned}$$

Let us now return to the transfer function

$$H(z) = \sum_{n=0}^{\infty} h_n z^{-n}.$$

If the digital filter is an IIR filter, then $H(z)$ cannot be a polynomial in z^{-1}. The simplest non-polynomial function that is analytic at $z = \infty$ is a rational function in z^{-1}. We will see that indeed a rational function which is analytic at ∞ provides a realizable IIR digital filter. The general form of such a rational function is

$$H(z) = \frac{a_0 + a_1 z^{-1} + \cdots + a_M z^{-M}}{1 - b_1 z^{-1} - \cdots - b_N z^{-N}}, \tag{1.20}$$

where $a_0, \cdots, a_M$, $b_1, \cdots, b_N$ are complex numbers and M, N are non-negative integers. Note, in particular, that $a_0 = h_0$.

From (1.18) and (1.20), we may write

$$\left(1 - \sum_{n=1}^{N} b_n z^{-n}\right)\left(\sum_{n=0}^{\infty} v_n z^{-n}\right) = \left(\sum_{n=0}^{M} a_n z^{-n}\right)\left(\sum_{n=0}^{\infty} u_n z^{-n}\right)$$

and taking the inverse z-transform on both sides, we arrive at

$$v_n - \sum_{i=1}^{N} b_i v_{n-i} = \sum_{i=0}^{M} a_i u_{n-i}$$

or

$$v_n = \sum_{i=0}^{M} a_i u_{n-i} + \sum_{i=1}^{N} b_i v_{n-i}. \tag{1.21}$$

Observe that the filtered output v_{n-i}, $i = 1, \cdots, N$, is used again to give the output v_n. For this reason, the digital filter defined by (1.21) is also called a *recursive digital filter*, while an FIR filter is said to be *non-recursive*. Since there are only a finite number of parameters in (1.21), a rational function provides a realizable digital filter as shown in Fig. 1.9, where D is again the delay processor defined by $Du_n = u_{n-1}$.

The important question is if this filter is stable. The following result provides another stability criterion.

Theorem 1.5. (Stability Criterion for IIR Digital Filters)

An IIR digital filter with transfer function $H(z)$ given by (1.20) is stable if and only if all the poles of the rational function $H(z)$ lie in the open unit disk $|z| < 1$.

One direction of this result is clear. Indeed, if all the poles of $H(z)$ lie in $|z| < 1$, or equivalently $|z^{-1}| > 1$, then $H(z)$, as a function of z^{-1}, is analytic at $z^{-1} = 0$ and its Taylor series expansion

$$H(z) = \sum_{n=0}^{\infty} h_n z^{-n}$$

has radius of convergence greater than one, so that $\{h_n\} \in l^1$, and by Theorem 1.4, the digital filter is stable.

On the other hand, if the filter is stable, we have $\{h_n\} \in l^1$ by Theorem 1.4, so that the radius of convergence of the above power series is at least one, or all the poles of $H(z)$ lie on $|z^{-1}| \geq 1$, or equivalently, $|z| \leq 1$. To show that there could not be any poles on the unit circle $|z| = 1$ while the condition

$$\sum_{n=0}^{\infty} |h_n| < \infty$$

is still valid, one could use partial fractions. We leave this as an Problem to the reader (Problem 1.26).

Example 1.10. The IIR digital filter

$$v_n = u_n + u_{n-1} + \frac{5}{6} v_{n-1} - \frac{1}{6} v_{n-2}$$

is stable.

The poles of the transfer function

$$H(z) = \frac{1 + z^{-1}}{1 - \frac{5}{6} z^{-1} + \frac{1}{6} z^{-2}} = \frac{z(z+1)}{(z - \frac{1}{2})(z - \frac{1}{3})}$$

are $\frac{1}{2}$ and $\frac{1}{3}$, which lie in the open unit disk $|z| < 1$.

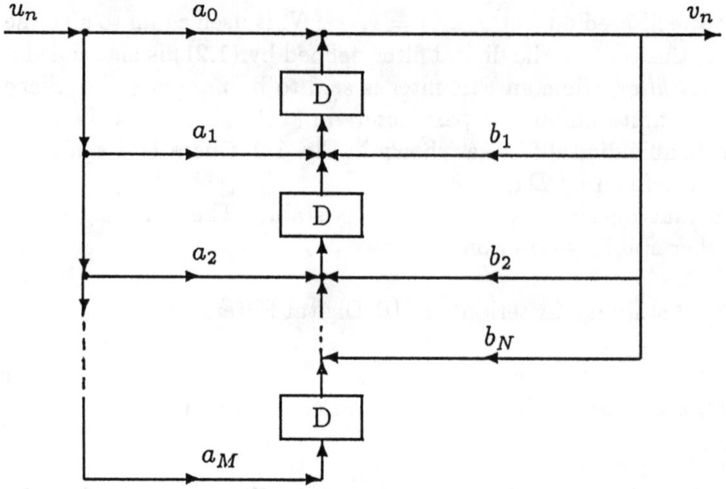

Fig. 1.9. Implementation of IIR filters

We close this section by introducing some commonly used terminologies. From Fig. 1.9, we see that it is natural to call the constants $a_0, \cdots, a_M$ *feedforward* parameters and $b_1, \cdots, b_N$ *feedback* parameters. Hence, an FIR digital filter does not have any feedback operation. On the other hand, the IIR filter defined by (1.21) takes advantage of the output for future regression. Such an operation is usually called *auto-regression.* Hence, the digital filter that takes advantage of both the input and past output to produce the present output as described by (1.21) is usually called an *Auto-Regressive Moving-Average* (ARMA) digital filter.

1.5 Optimal Digital Filter Design Criteria

This section will be devoted to the discussion of the central topic of optimal digital filters design. Two important research problems related to the optimal design of digital filters will also be posed. The second problem is the so-called Hankel-norm optimization problem that will be studied in some detail later in this book.

1.5.1 An Interpolation Method

We have learnt from Example 1.9 the important role the transfer function $H(z)$ of a digital filter plays in deciding what frequencies to pass and what frequencies to stop. If ω_0 is a certain frequency of interest and the transfer function is "designed" to take on a positive constant A at this frequency, i.e., $H^*(\omega_0) = H(e^{j\omega_0}) = A$, then the digital filter magnifies the input signal with this frequency by a factor of A.

We now consider a more general situation.

Example 1.11. Let $0 \leq \omega_0 < \cdots < \omega_k < 2\pi$; $a_0, \cdots, a_k > 0$; and $A_0, \cdots, A_k \geq 0$. Design an FIR digital filter which takes the input signal

$$u_n = a_0 e^{j\omega_0 n} + \cdots + a_k e^{j\omega_k n}$$

with $u_n = 0$ for $n \leq 0$, that carries (radian) frequencies $\omega_0, \cdots, \omega_k$ with amplitudes $a_0, \cdots, a_k$ respectively, to the output signal

$$v_n = a_0 A_0 e^{j\omega_0 n} + \cdots + a_k A_k e^{j\omega_k n}.$$

Determine the one that requires the minimum number of delays.

Note that if $A_i = 0$, then the frequency ω_i is stopped and if $A_i = 1$, it is passed without magnifying the amplitude.

To filter $(k+1)$ frequencies, we need to adjust $(k+1)$ parameters in determining the transfer function $H(z)$ (Example 1.9). For this reason, we consider

$$H(z) = h_0 + h_1 z^{-1} + \cdots + h_k z^{-k},$$

where the number of required delays is k. The spectrum is, of course,

$$H^*(\omega) = h_0 + h_1 e^{-j\omega} + \cdots + h_k e^{-jk\omega},$$

and following Example 1.9, we set

$$H^*(\omega_i) = A_i, \quad i = 0, \cdots, k.$$

That is, the system of linear equations

$$\begin{cases} h_0 + e^{-j\omega_0} h_1 + \cdots + e^{-jk\omega_0} h_k = A_0, \\ \cdots \\ h_0 + e^{-j\omega_k} h_1 + \cdots + e^{-jk\omega_k} h_k = A_k, \end{cases}$$

must be solved. In matrix form, we have

$$\begin{bmatrix} 1 & e^{-j\omega_0} & \cdots & e^{-jk\omega_0} \\ 1 & e^{-j\omega_1} & \cdots & e^{-jk\omega_1} \\ \vdots & \vdots & & \vdots \\ 1 & e^{-j\omega_k} & \cdots & e^{-jk\omega_k} \end{bmatrix} \begin{bmatrix} h_0 \\ h_1 \\ \vdots \\ h_k \end{bmatrix} = \begin{bmatrix} A_0 \\ A_1 \\ \vdots \\ A_k \end{bmatrix}.$$

The determinant of the coefficient matrix is (Problem 1.27)

$$\prod_{0 \leq \ell < m \leq k} (e^{-j\omega_m} - e^{j\omega_\ell}),$$

which is called the *Vandermonde determinant* of $\{e^{-j\omega_0}, \cdots, e^{-j\omega_k}\}$. Since $0 \leq \omega_\ell < \omega_m < 2\pi$ whenever $\ell < m$, each factor

$$e^{-j\omega_m} - e^{-j\omega_\ell} = e^{-j\omega_m}\left(1 - e^{j(\omega_m - \omega_\ell)}\right)$$

is nonzero. Hence, there is a unique solution for $h_0, \cdots, h_k$. In general, for suitable choices of $A_0, \cdots, A_k$, the unit impulse responses $h_0, \cdots, h_k$ are all non-zero. Hence, the minimum number of delays is k.

1.5.2 Ideal Filter Characteristics

In practice, a whole range of frequencies must be filtered. For instance, if we wish to stop all frequencies ω in the range $a < \omega < b$ and pass those frequencies ω in the range $c < \omega < d$, where the intervals (a, b) and (c, d) lie in $(0, \pi)$ but do not overlap, then we require

$$H^*(\omega) = \begin{cases} 0 & \text{for } a < \omega < b\,, \\ 1 & \text{for } c < \omega < d\,. \end{cases}$$

The intervals (a, b) and (c, d) are called the *stopband* and *passband* of the digital filter, respectively. Since there are infinitely many frequencies in a stopband or passband, the interpolation method discussed in the previous section fails to give a transfer function that satisfies the design criterion exactly.

In a digital filter design problem, the design criterion is usually specified on the magnitude spectrum $|H^*(\omega)|$ of the transfer function $H(z)$. Hence, the design criterion is also called the *amplitude filter characteristic.* If a stopband is required, for example, the transfer function must have zero radial limit on an arc of the unit circle. No causal digital filter can satisfy this property, since any function analytic in $|z| < 1$ that has zero radial limit on an arc of $|z| = 1$ must be identically zero (Problem 1.28). In other words, a transfer function $H(z)$ with a nontrivial stopband cannot be represented as a power series

$$H(z) = \sum_{n=0}^{\infty} h_n z^{-n}\,.$$

Hence, in general, a prescribed amplitude filter characteristic cannot be achieved, and must be called an *ideal amplitude filter characteristic.* To remind ourselves of this important fact, we will use the subscript I. Some typical ideal amplitude filter characteristics are given in the following:

(1) Ideal low-pass filter
(2) Ideal band-pass filter
(3) Ideal differentiator
(4) Ideal notch filter

In each of the above four figures, we only show the graph for $0 \le \omega \le \pi$. Since negative frequencies are not considered here, we will use even extensions. That is, we will set

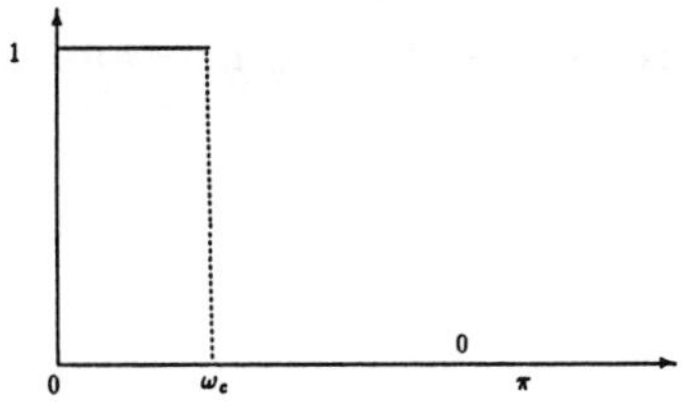

Fig. 1.10. $|H^*_{\ell,I}(\omega)|$

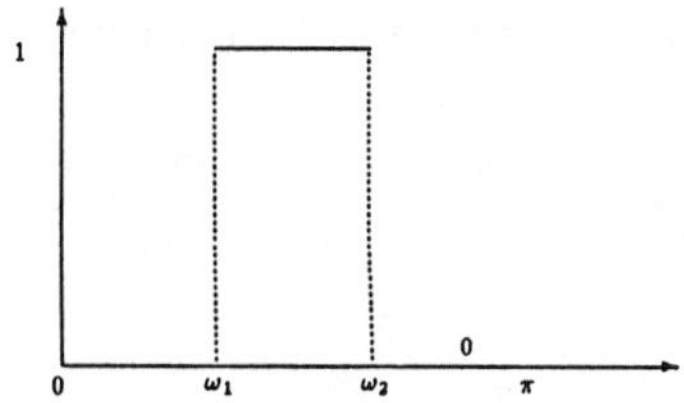

Fig. 1.11. $|H^*_{b,I}(\omega)|$

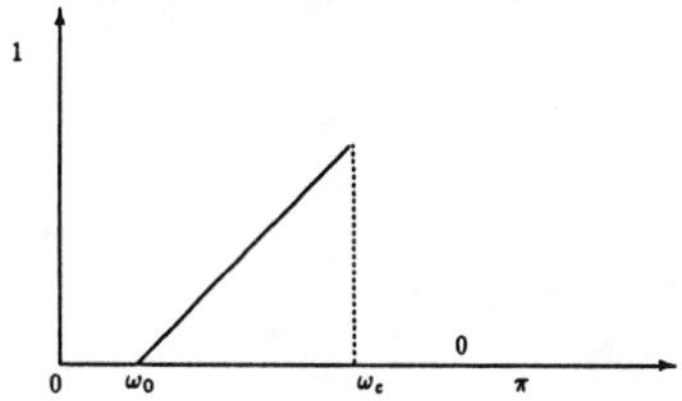

Fig. 1.12. $|H^*_{d,I}(\omega)|$

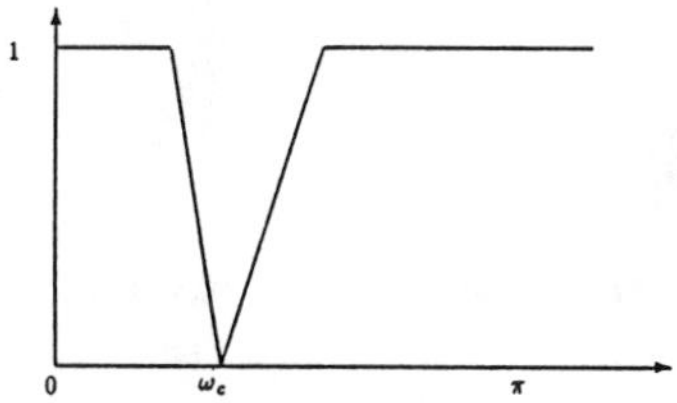

Fig. 1.13. $|H^*_{n,I}(\omega)|$

$$|H_I^*(-\omega)| = |H_I^*(\omega)|$$

so that if we consider its Fourier expansion, writing

$$|H_I^*(\omega)| = \sum_{n=-\infty}^{\infty} c_n \mathrm{e}^{-\mathrm{j}n\omega},$$

where

$$c_n = \frac{1}{2\pi} \int_{-\pi}^{\pi} |H_I^*(\omega)| \mathrm{e}^{\mathrm{j}n\omega} d\omega,$$

then we have $c_{-n} = c_n$ for each $n = 1, 2, \cdots$ (Problem 1.29). In other words, if an even extension is used, then

$$|H_I^*(\omega)| = c_0 + \sum_{n=1}^{\infty} c_n z^{-n} + \sum_{n=1}^{\infty} c_n z^n$$

where $z = e^{j\omega}$. Unfortunately, this is not a causal representation of the transfer function of the digital filter. In fact, as we mentioned above, any ideal amplitude filter characteristic $|H_I^*(\omega)|$ which has a stopband (consisting of at least one interval) cannot have a causal representation.

Hence, to obtain a causal representation, the stopband must be raised. Let ε be a small positive number. Then we can raise the stopband by ε by defining

$$|H_\varepsilon^*(\omega)| = \max(|H_I^*(\omega)|, \varepsilon),$$

where the maximum is taken at each frequency ω (see the Fig. 1.14, where a low-pass filter is considered).

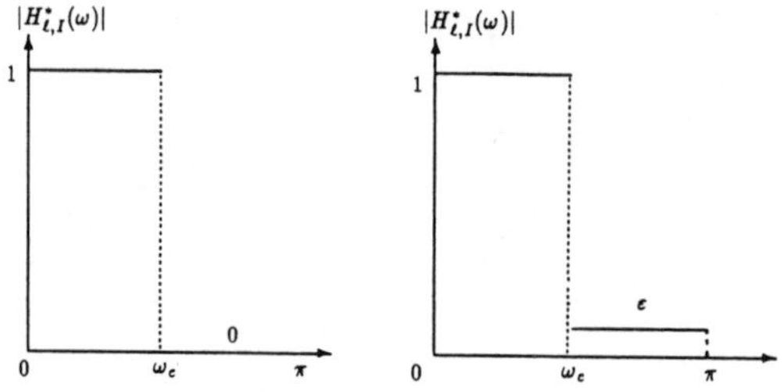

Fig. 1.14. Causal low-pass filter

In applications, since there is always some tolerance for the stopband, this transformation is certainly permissible. Now, we are able to take the natural logarithm of $|H_\varepsilon^*(\omega)|$ and construct the transfer function

$$\hat{H}_\varepsilon(z) := \exp\left(\frac{1}{2\pi}\int_{-\pi}^{\pi} \frac{e^{jt} + z^{-1}}{e^{jt} - z^{-1}} \ln |H_\varepsilon^*(t)| dt\right),$$

which is certainly analytic in $|z| > 1$, and hence has the causal representation

$$\hat{H}_\varepsilon(z) = \sum_{n=0}^{\infty} h_n(\varepsilon) z^{-n} \tag{1.22}$$

for some constants $h_n(\varepsilon)$ on the open unit disk $|z| < 1$. In addition, since

$$\mathrm{Re}\left\{\frac{e^{jt} + z^{-1}}{e^{jt} - z^{-1}}\right\} = \frac{1 - |z|^{-2}}{1 - 2|z|^{-1}\cos(\omega - t) + |z|^{-2}},$$

where $z = |z|e^{j\omega}$ and $-\pi \le \omega \le \pi$, is the Poisson kernel, it follows that

$$\lim_{r\uparrow 1} |\hat{H}_\varepsilon(re^{j\omega})| = |H_\varepsilon^*(\omega)|$$

at each ω where $|H_\varepsilon^*(\omega)|$ is continuous (Problem 1.32). That is, the one-sided z-transform (1.22) is indeed the causal representation of the "raised" ideal amplitude filter characteristic.

To determine the unit impulse responses $h_n(\varepsilon)$, we consider

$$a_k := \frac{1}{\pi}\int_{-\pi}^{\pi} e^{jk\omega} \ln |H_\varepsilon^*(\omega)| d\omega \,,$$

which can be computed for instance by the Fast Fourier Transform (FFT). Then we have (Problem 1.33)

$$h_0(\varepsilon) = e^{a_0/2}$$

and

$$h_n(\varepsilon) = h_0(\varepsilon)\sum_{k=1}^{n}\frac{1}{k!}\sum_{i_1+\cdots+i_k=n} a_{i_1}\cdots a_{i_k}\,.$$

1.5.3 Optimal IIR Filter Design Criteria

Although there are many digital-filter design methods available in the literature, see, for example, Bose [1985], we will pose two optimum design criteria in this section. At this writing, no complete solution is available to any of these extremal problems. To facilitate our discussion, we first need some notation.

Denote by $\|\cdot\|_{L^p}$ the usual L^p norm on $(-\pi,\pi)$ defined by

$$\|f(\omega)\|_{L^p} = \left(\frac{1}{2\pi}\int_{-\pi}^{\pi} |f(\omega)|^p d\omega\right)^{1/p}$$

if $1 \le p < \infty$, and

$$\|f(\omega)\|_{L^\infty} = ess \sup_{\omega\in(-\pi,\pi)} |f(\omega)|$$

if $p = \infty$. Let R_{MN} be the class of all rational functions of the form (1.20), namely,

$$R_{MN} = \left\{ H_{MN}(z) :\ H_{MN}(z) = \frac{a_0 + a_1 z^{-1} + \cdots + a_M z^{-M}}{1 - b_1 z^{-1} - \cdots - b_N z^{-N}} \right\}, \quad (1.23)$$

where all the poles of $H_{MN}(z)$ lie in the unit disk $|z| < 1$.

Research Problem 1. Let $|H_I^*(\omega)|$ be an ideal digital filter characteristic. For $1 \le p \le \infty$, determine $H_{MN}(z)$ in R_{MN} such that

$$\|\, |H_I^*(\omega)| - |H_{MN}(e^{j\omega})|\, \|_{L^p} = \min_{G\in R_{MN}} \|\, |H_I^*(\omega)| - |G(e^{j\omega})|\, \|_{L^p}\,. \quad (1.24)$$

The next problem is concerned with approximating a causal representation. Hence, in some situations, it is necessary to consider a "raised" ideal digital filter characteristic. We will use the usual notation H^p, $1 \le p \le \infty$, for the *Hardy space* of all functions $f(z)$ analytic in $|z| < 1$ such that

$$\|f\|_{H^p} = \sup_{0 \le r < 1} \left(\frac{1}{2\pi} \int_{-\pi}^{\pi} |f(re^{j\omega})|^p \right)^{1/p} < \infty \tag{1.25}$$

for $1 \le p < \infty$, and

$$\|f\|_{H^\infty} = ess \sup_{0 \le r < 1} \max_{\omega} |f(re^{j\omega})| < \infty \tag{1.26}$$

for $p = \infty$. For more details on Hardy spaces, the reader is referred to Chap. 3. In addition, if

$$H(z) = \sum_{n=0}^{\infty} h_n z^{-n}$$

is analytic at infinity, the *Hankel matrix* corresponding to $H(z)$ is the infinite matrix $\Gamma_H = \left[h_{\ell+m-1}\right]_{\ell,m=1,2,\cdots}$, namely,

$$\Gamma_H = \begin{bmatrix} h_1 & h_2 & h_3 & \cdots \\ h_2 & h_3 & \cdots & \cdots \\ h_3 & \cdots & \cdots & \cdots \\ \cdots & \cdots & \cdots & \cdots \end{bmatrix}. \tag{1.27}$$

Note that the "analytic part" h_0 of $H(z)$ is not considered here. Γ_H can be viewed as an operator on the sequence space l^2. Then the operator-norm of Γ_H is given by

$$\|\Gamma_H\|_s := \sup_{\|\mathbf{x}\|_{l^2}=1} \|\Gamma_H \, \mathbf{x}\|_{l^2}, \tag{1.28}$$

and we define the *Hankel (semi-)norm* of $H(z)$ by

$$\|H(z)\|_\Gamma = \|\Gamma_H\|_s. \tag{1.29}$$

If h_0 is zero, then $\|H(z)\|_\Gamma = 0$ implies that $H(z)$ is identically zero, and the semi-norm becomes an actual norm. The subscript s in the operator norm indicates that this norm is actually the spectral radius of the operator Γ_H when Γ_H is real. In this case, it is also called the *spectral norm* of the operator.

Research Problem 2. Let

$$H(z) = \sum_{n=0}^{\infty} h_n z^{-n} .$$

For each p, $1 \le p \le \infty$, determine $H_{MN}(z)$ in R_{MN}, such that

$$\|H(z^{-1}) - H_{MN}(z^{-1})\|_{H^p} = \min_{G \in R_{MN}} \|H(z^{-1}) - G(z^{-1})\|_{H^p} . \tag{1.30}$$

Also, determine $\tilde{H}_{MN}(z)$ in R_{MN}, such that

$$\|H(z) - \tilde{H}_{MN}(z)\|_{\Gamma} = \min_{G \in R_{MN}} \|H(z) - G(z)\|_{\Gamma} . \tag{1.31}$$

Recently, Hankel norm approximation has been quite popular in systems theory and has also gained some attention as a possible tool in digital filter design. A detailed study of this topic will be given in Chaps. 4–6.

Problems

Problem 1.1. Determine the bandwidth of the analog signal

$$u(t) = \frac{\sin \omega_0 t}{\omega_0 t} + 2\frac{\sin \omega_1 t}{\omega_1 t} , \qquad -\infty < t < \infty ,$$

where $0 < \omega_0 < \omega_1$.

Problem 1.2. Show that a continuous-time signal cannot be both time-limited and band-limited unless it is zero almost everywhere. (This is also true for discrete-time signals.)

Problem 1.3. Consider the system shown in Fig. 1.15, in which the two continuous-time signals $u_1(t)$ and $u_2(t)$ are multiplied and the product $v(t)$ is sampled by a periodic impulse train. Suppose that both $u_1(t)$ and $u_2(t)$ are band-limited with

$$\hat{u}_1(\omega) = 0 \qquad |\omega| > \omega_1 ,$$
$$\hat{u}_2(\omega) = 0 \qquad |\omega| > \omega_2 .$$

Determine the maximum sampling time t_0 such that the signal $v(t)$ can be recovered from the sampling data by using an ideal low-pass filter, see Problem 1.30 below.

Problem 1.4. In Fig. 1.16, a system in which the sampling signal is an impulse train with alternating sign is shown. Find the maximum sampling

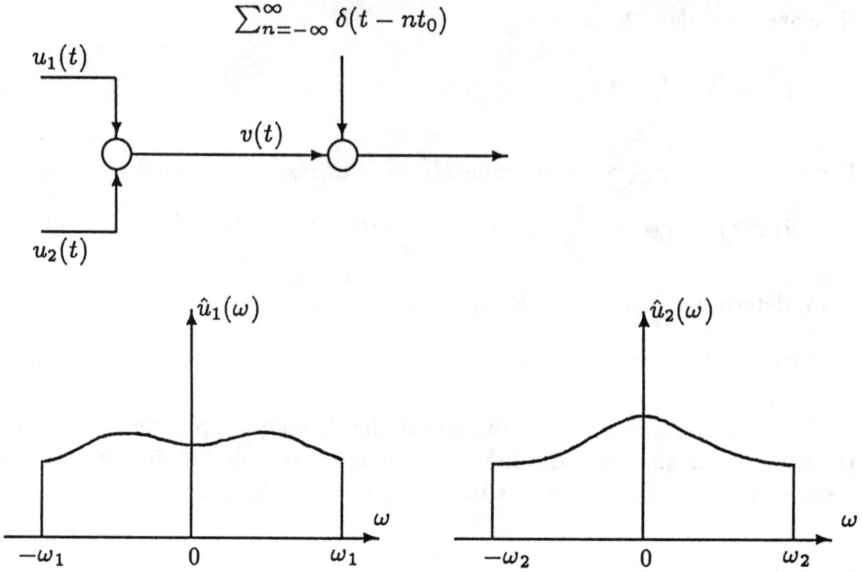

Fig. 1.15. Sampling process

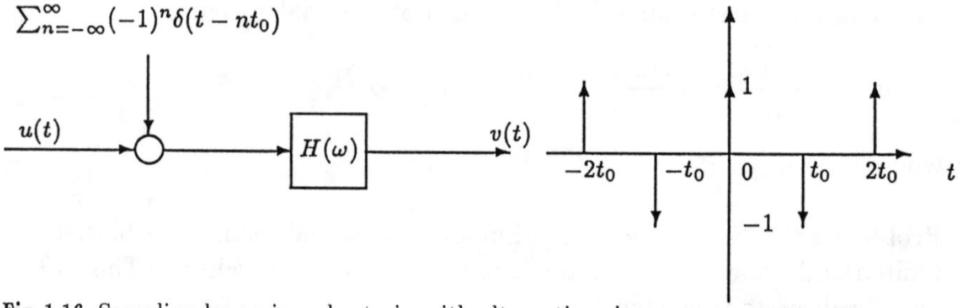

Fig. 1.16. Sampling by an impulse train with alternating sign

time t_0 such that the input signal $u(t)$ can be recovered from the output $v(t)$.

Problem 1.5. Use the Schwarz inequality and (1.7) to demonstrate that the convergence of the infinite series

$$u(t) = \sum_{n=-\infty}^{\infty} u(nt_0) \frac{\sin \pi(t/t_0 - n)}{\pi(t/t_0 - n)}$$

is uniform for all t, $-\infty < t < \infty$. (This completes the proof of the sampling theorem.)

Problem 1.6. Let $u(t)$ be a band-limited analog signal in $L^2(-\infty,\infty)$ with bandwidth ω_0, and let $0 < t_0 < \pi/\omega_0$. Moreover, let

$$\hat{u}(\omega) = \int_{-\infty}^{\infty} u(t)e^{-j\omega t}dt$$

and

$$U^*(\omega) = \sum_{n=-\infty}^{\infty} u(nt_0)e^{-j\omega n t_0}$$

be the continuous and discrete Fourier transforms of $u(t)$ and $\{u_n(nt_0)\}$, respectively. Show that

$$U^*(\omega) = \frac{1}{t_0} \sum_{n=-\infty}^{\infty} \hat{u}\big(\omega + 2n\pi/t_0\big)\,.$$

Problem 1.7. Consider the infinite series

$$u(t) = \sum_{n=-\infty}^{\infty} u(nt_0)\phi_n(t)\,,$$

where

$$\phi_n(t) := \frac{\sin \pi(t/t_0 - n)}{\pi(t/t_0 - n)}\,.$$

Is it true that the system $\{\phi_n(t)\}$ is orthogonal in the sense that

$$\int_{-\infty}^{\infty} \phi_n(t)\phi_m(t)dt = 0 \quad \text{for} \quad n \neq m\,?$$

If so, then the system would be automatically linearly independent. If not, is the system still linearly independent?

Problem 1.8. In the sampling theorem, all sampling data values are equally spaced. However, a band-limited signal with bandwidth $2\omega_0$ can also be recovered from non-equally spaced samples as long as the average sampling rate is ω_0/π samples per a unit of time. More precisely, let $u(t)$ be a band-limited signal with

$$\hat{u}(\omega) = 0 \qquad \text{for} \qquad |\omega| > \omega_0\,,$$

and the sampling impulse train be displayed in Fig. 1.17. Then, the system depicted in Fig. 1.18 can be used to recover the unknown input signal $u(t)$, where $s(t)$ is a $2\pi/\omega_0$ -periodic signal with $s(0) = a$ and $s(t_0) = b$ for some real constants a and b, $H_1(\omega)$ is a phase shifter with

$$H_1(\omega) = \begin{cases} \mathrm{j} & \text{for} \quad \omega > 0 \\ -\mathrm{j} & \text{for} \quad \omega < 0\,, \end{cases}$$

and $H_2(\omega)$ is an ideal low-pass filter given by

$$H_2(\omega) = \begin{cases} c + \mathrm{j}d & \text{for} \quad 0 < \omega < \omega_0 \\ c - \mathrm{j}d & \text{for} \quad -\omega_0 < \omega < 0 \\ 0 & \text{for} \quad |\omega| < \omega_0\,, \end{cases}$$

where c and d are two real constants. Find the four real constants $a, b, c,$ and d such that the system output $v(t)$ recovers the signal $u(t)$ in the sense that $v(t) = u(t)$ for all $t \in (-\infty, \infty)$.

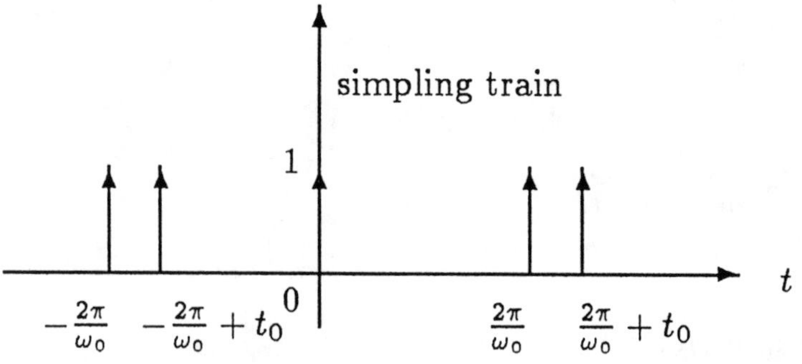

Fig. 1.17. A sampling impulse train

Problem 1.9. Let $\{u_n\}$ be a digital signal of the so-called *unit-sample responses* given by

$$u_n = \begin{cases} 1 & \text{for} \quad 0 \le n \le N-1\,, \\ 0 & \text{otherwise}\,. \end{cases}$$

Find the discrete Fourier transform $U^*(\omega)$ of $\{u_n\}$.

Problem 1.10. Compute the convolution $f \bar{*} g$ in the real line group of the analog signals f and g given by

$$f(t) = g(t) = \begin{cases} 1 - t & \text{for} \quad 0 \le t \le 1\,, \\ 1 + t & \text{for} \quad -1 \le t \le 0\,, \\ 0 & \text{otherwise}\,. \end{cases}$$

Problem 1.11. Determine the convolution $\{u_n\} * \{v_n\}$ over the integer group of the digital signals $\{u_n\}$ and $\{v_n\}$ given by

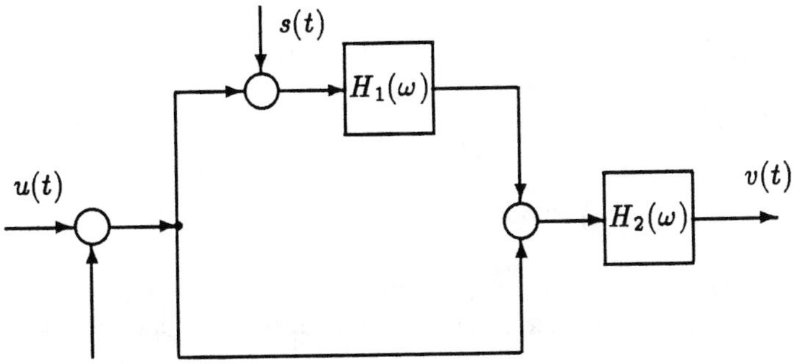

Fig. 1.18. Recovering the unknown signal

$$u_n = \begin{cases} a^n, & 0 \le n \le N-1, \\ 0, & \text{otherwise}, \end{cases} \qquad \text{and} \qquad v_n = \begin{cases} b^{n-\ell} & \text{for } \ell \le n, \\ 0 & \text{for } n < \ell, \end{cases}$$

where N and ℓ are fixed integers.

Problem 1.12. Compute the convolution $f \underline{*} g$ in the circle group of the analog signals f and g given respectively by

$$f(\omega) = e^{j\omega}, \qquad -\pi + 2k\pi \le \omega < \pi + 2k\pi,$$

and

$$g(\omega) = \cos\omega, \qquad -\pi + 2k\pi \le \omega < \pi + 2k\pi.$$

Problem 1.13. Let $\delta(x)$ and $\delta^*(\omega)$ be the convolution identities in the real line and circle groups, respectively (Sect. 1.2.1). Convince yourself that $\delta(x)$ and $\delta^*(\omega)$ cannot be functions in the usual sense.

Problem 1.14. Show that the Fourier transformation takes convolutions to algebraic multiplications, in the sense that

$$\begin{aligned} &(1) \quad (\widehat{f \overline{*} g})(\omega) = \hat{f}(\omega)\hat{g}(\omega), && f, g \in L^1(-\infty, \infty); \\ &(2) \quad W^*(\omega) = U^*(\omega)V^*(\omega), && \text{where } \{w_n\} = \{u_n\} * \{v_n\}; \end{aligned}$$

and

(3)
$$\frac{1}{2\pi}\int_{-\pi}^{\pi}(f\underline{*}g)(\omega)e^{-jn\omega}d\omega = \left(\frac{1}{2\pi}\int_{-\pi}^{\pi}f(\omega)e^{-jn\omega}d\omega\right)\left(\frac{1}{2\pi}\int_{-\pi}^{\pi}g(\omega)e^{-jn\omega}d\omega\right),$$

where $f, g \in L^1(-\pi,\pi)$ are 2π-periodic.

Problem 1.15. Consider the digital filter shown in Fig. 1.19.
(a) Determine the transfer function $H(z)$ of this filter and indicate the region of convergence.
(b) If the input signal is $u_n = (2/3)^n$, find the corresponding output signal v_n.

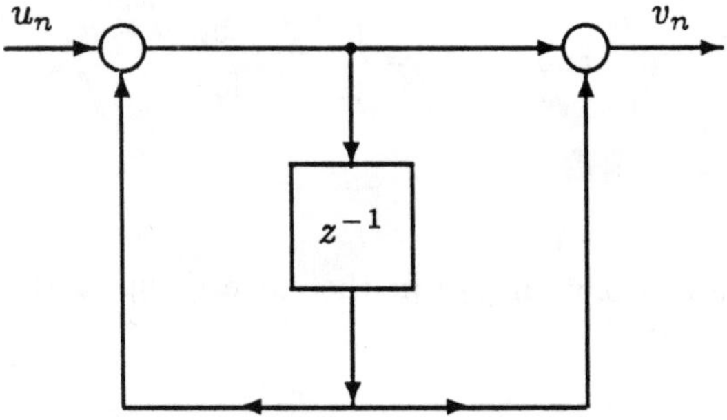

Fig. 1.19. A digital filter

Problem 1.16. Verify the digital-signal and spectrum relationships given in Table 1.1.

Problem 1.17. Verify the digital-signal and z-transform relationships given in Table 1.2.

Problem 1.18. Find the z-transform of the digital signal $\{u_n\}$ where

$$u_n = \begin{cases} (-1)^{n+1}a^n/n & \text{for} \quad n \geq 1, \\ 0 & \text{for} \quad n \leq 0. \end{cases}$$

Problem 1.19. Find the inverse z-transform of

$$U(z) = \frac{1}{1-az^{-1}}, \qquad |z| < |a|.$$

Problem 1.20. Solve the initial value problem

$$\begin{cases} u_{n+k} - u_{n+k-1} + \cdots + (-1)^{k-1} u_{n+1} + (-1)^k u_n = (-1)^n \,, \\ u_0 = u_1 = \cdots = u_{k-1} = 0 \,, \end{cases}$$

$n = 0, 1, \cdots$.

Problem 1.21. (a) Verify that the digital filter considered in problem 1.15 is linear, time-invariant, and causal. (b) Verify that the digital filter defined by (1.16) is linear, time-invariant, and causal.

Problem 1.22. Supply a proof of Theorem 1.4.

Problem 1.23. Show that the one-sided z-transform of the input-output relationship (1.16) is given by

$$V(z) = H(z)U(z) \,,$$

where

$$U(z) = \sum_{n=0}^{\infty} u_n z^{-n} \,, \quad V(z) = \sum_{n=0}^{\infty} v_n z^{-n} \,, \quad H(z) = \sum_{n=0}^{\infty} h_n z^{-n} \,.$$

Problem 1.24. Determine the transfer function $H(z)$ of the digital filter defined by

$$v_n = \sum_{i=0}^{\infty} \left(\frac{a^{i+1} - b^{i+1}}{a - b} \right) u_{n-i} \,,$$

where $a \neq b$.

Problem 1.25. Let

$$u_n = a_0 e^{jn\omega_0} + a_1 e^{jn\omega_1} + \cdots + a_k e^{jn\omega_k}$$

be an input signal, where the a_i's are positive constants and $0 \le \omega_0 < \cdots < \omega_k < 2\pi$. Design a simple FIR digital filter that passes the frequency ω_0 without changing its magnitude, but stops the frequencies $\omega_1, \cdots, \omega_k$.

Problem 1.26. Consider the transfer function

$$H(z) = \sum_{n=0}^{\infty} h_n z^{-n}$$

of a (stable) ARMA digital filter. By using partial fractions, demonstrate that the radius of convergence of this power series is strictly less than 1. This completes the proof of Theorem 1.5.

Problem 1.27. By using mathematical induction, demonstrate that the determinant of the matrix

$$\begin{bmatrix} 1 & x_0 & x_0^2 & \cdots & x_0^k \\ 1 & x_1 & x_1^2 & \cdots & x_1^k \\ \vdots & \vdots & \vdots & & \vdots \\ 1 & x_k & x_k^2 & \cdots & x_k^k \end{bmatrix}$$

is given by

$$\prod_{0 \le \ell < m \le k} (x_m - x_\ell) .$$

Problem 1.28. Let $f(z) = H(z^{-1})$. Prove that any function $f(z)$ analytic in $|z| < 1$ that has zero radial limit on an arc of $|z| = 1$ must be identically zero. [Hint: Assume without loss of generality that $f(0) \neq 0$ and consider $F(z) = f(z)f(ze^{j\omega_0}) \cdots f(ze^{jn\omega_0})$ for some appropriate choices of ω_0 and n so that $F(z)$ has zero radial limit on all of the unit circle.]

Problem 1.29. Let $f(\omega)$ be a 2π-periodic, even, and non-negative real-valued function with Fourier expansion

$$f(\omega) = \sum_{n=-\infty}^{\infty} c_n e^{-jn\omega} ,$$

where

$$c_n = \frac{1}{2\pi} \int_{-\pi}^{\pi} f(\omega) e^{jn\omega} d\omega .$$

Verify that $c_{-n} = c_n$ for each $n = 1, 2, \cdots$.

Problem 1.30. Consider the system shown in Fig. 1.15, Problem 1.3. Design an ideal low-pass filter to recover a signal $v(t)$ from the sampling data obtained with the maximum sampling time t_0.

Problem 1.31. The system shown in Fig. 1.20 is widely used to convert a high-pass filter from a low-pass filter, and vice versa.

(a) If $H(\omega)$ is an ideal low-pass filter with cutoff frequency ω_c, show that the overall system is an ideal high-pass filter. Find its corresponding cutoff frequency ω_c'.

(b) If $H(\omega)$ is an ideal high-pass filter with cutoff frequency ω_c', show that the overall system is an ideal low-pass filter. Find its corresponding cutoff frequency ω_c.

Problem 1.32. Let

$$\hat{f}(z) = \exp\left(\frac{1}{2\pi}\int_{-\pi}^{\pi} \frac{e^{jt} + z^{-1}}{e^{jt} - z^{-1}} \ln|f(t)|dt\right),$$

where $f(t)$ is a real-valued continuous function which does not vanish anywhere on $[-\pi, \pi]$. Show that

$$\lim_{r\uparrow 1} |\hat{f}(re^{j\omega})| = |f(\omega)|$$

at each ω. [Hint: Recall the solution of the Dirichlet problem on the unit disk.]

Problem 1.33. Let

$$h_n = \frac{1}{2\pi}\int_{|z|=1} F(z)z^{n-1}dz,$$

where

$$F(z) = \exp\left(\frac{1}{2\pi}\int_{-\pi}^{\pi} \frac{e^{jt} + z^{-1}}{e^{jt} - z^{-1}} \ln|f(t)|dt\right)$$

and define

$$a_k = \frac{1}{\pi}\int_{-\pi}^{\pi} e^{jk\omega} \ln|f(\omega)|d\omega.$$

Verify that

$$h_0 = e^{a_0/2}$$

and

$$h_n = h_0 \sum_{k=1}^{n} \frac{1}{k!} \sum_{i_1+\cdots+i_k=n} a_{i_1}\cdots a_{i_k}.$$

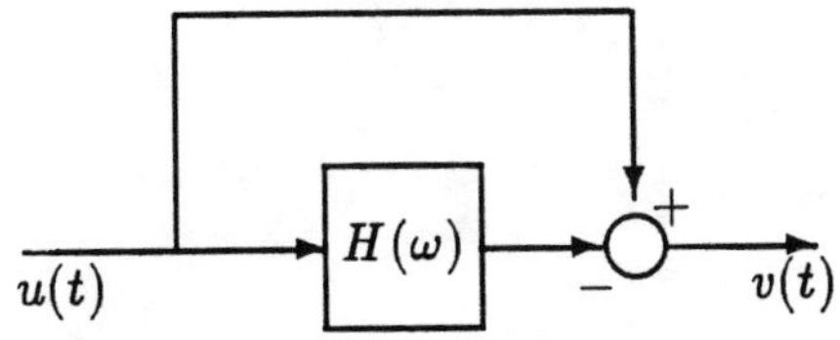

Fig. 1.20. Converting a high-pass filter to a low-pass filter or vice versa

Problem 1.34. Determine which of the following functions belong to H^1, H^2, or H^∞ in $|z| < 1$:

(1) $\sum_{n=1}^{\infty} z^n/\sqrt{n}$;
(2) $\sum_{n=1}^{\infty} z^n/(\sqrt{n}\ln(n+1))$;
(3) $\sum_{n=1}^{\infty} z^n/n^\alpha, \quad \alpha > 1$.

Problem 1.35. Find the Hankel matrices corresponding to the following functions:

(1) $\sin z/z$;
(2) $z^2/((z-a)(z-b)), \qquad |z| > \max\{|a|, |b|\}$;
(3) $\ln(1 + az^{-1}), \qquad |z| > |a|$.

2. Linear Systems

There is a very important common ground for digital signal processing and linear systems theory; and this is rational approximation of the transfer function. Indeed, the frequency domain approach for time-invariant linear systems amounts to certain minimization problems such as system reduction in linear control systems optimization, robust stabilization, disturbance rejection, and sensitivity minimization. This chapter will be devoted to the study of state-space descriptions, transfer functions, and stability theorems for general linear control systems; Kronecker's theorem and its application to the system reduction problem for SISO systems; and general MIMO linear feedback systems and their sensitivity minimization problems.

2.1 State-Space Descriptions

As mentioned in the previous chapter, a digital signal may also be considered as a *digital control sequence* input to a linear system. To introduce this viewpoint and the notion of a *digital control system*, we consider the following example.

2.1.1 An Example of Flying Objects

Let $\mathbf{x}(t)$, $0 \leq t < \infty$, denote the trajectory in 3-space of a flying object, where t denotes the time variable as shown in Fig. 2.1. This vector-valued function is discretized by sampling and quantizing with sampling time $t_0 > 0$ to yield

$$\mathbf{x}_n \doteq \mathbf{x}(nt_0), \quad n = 0, 1, \cdots .$$

For convenience, $\mathbf{x}_n$ will be treated as a 3×1 column vector. For practical purposes, $\mathbf{x}(t)$ can be assumed to have continuous first and second order derivatives, denoted by $\dot{\mathbf{x}}(t)$ and $\ddot{\mathbf{x}}(t)$, respectively, so that for small values of t_0 with $\dot{\mathbf{x}}_n \doteq \dot{\mathbf{x}}(nt_0)$ and $\ddot{\mathbf{x}}_n \doteq \ddot{\mathbf{x}}(nt_0)$, the position, velocity, and acceleration vectors $\mathbf{x}_n$, $\dot{\mathbf{x}}_n$ and $\ddot{\mathbf{x}}_n$ are governed by the equations

$$\begin{cases} \mathbf{x}_{n+1} = \mathbf{x}_n + t_0 \dot{\mathbf{x}}_n + \dfrac{1}{2} t_0^2 \ddot{\mathbf{x}}_n \,, \\ \dot{\mathbf{x}}_{n+1} = \dot{\mathbf{x}}_n + t_0 \ddot{\mathbf{x}}_n \,, \quad n = 0, 1, \cdots . \end{cases}$$

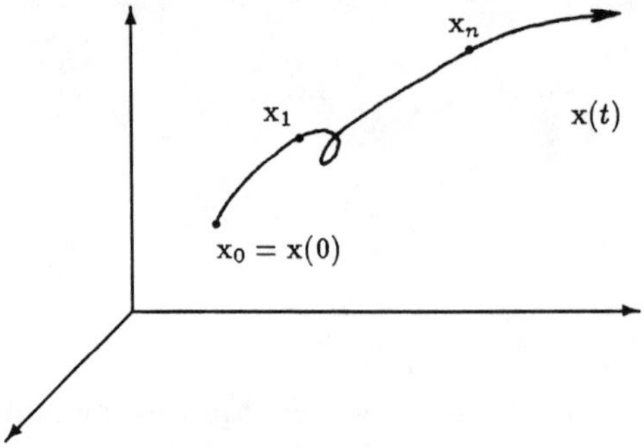

Fig. 2.1. Trajectory of a flying object

These equations are, of course, not exact since they are second and first order Taylor approximations, respectively, but the approximation improves with smaller values of the sampling time t_0. Suppose that we wish to guide this flying object by controlling its acceleration $\ddot{\mathbf{x}}_n$ with a vector-valued digital sequence $\{\mathbf{u}_n\}$, as described by the rule

$$\ddot{\mathbf{x}}_{n+1} = \ddot{\mathbf{x}}_n + \mathbf{u}_n ,$$

where again, for each n, $\mathbf{u}_n$ is treated as a 3×1 column vector. Then by setting

$$\bar{\mathbf{x}}_n = \begin{bmatrix} \mathbf{x}_n \\ \dot{\mathbf{x}}_n \\ \ddot{\mathbf{x}}_n \end{bmatrix}, \quad A = \begin{bmatrix} I_3 & t_0 I_3 & \frac{1}{2}t_0^2 I_3 \\ 0_3 & I_3 & t_0 I_3 \\ 0_3 & 0_3 & I_3 \end{bmatrix}, \quad B = \begin{bmatrix} 0_3 \\ 0_3 \\ I_3 \end{bmatrix},$$

where 0_3 and I_3 are 3×3 zero and identity blocks, respectively, $\bar{\mathbf{x}}_n$ is a 9×1 column vector, A a 9×9 matrix, and B a 9×3 matrix, the above information may be collected as a single matrix formulation

$$\bar{\mathbf{x}}_{n+1} = A\bar{\mathbf{x}}_n + B\mathbf{u}_n . \tag{2.1}$$

In addition, to facilitate the control process, the position (vector) of the flying object may be observed (or measured) at each time instant in the sense that

$$\mathbf{v}_n = C\bar{\mathbf{x}}_n , \tag{2.2}$$

where $C = [I_3 \; 0_3 \; 0_3]$ is a 3×9 matrix. In (2.1), the vector $\bar{\mathbf{x}}_n$, which contains all of the important information on the motion of the flying object, is called the *state vector*, the matrix A which governs the state is called the *system matrix*, and the matrix B that dictates the input control sequence is called the *control matrix*. Also, (2.1) itself is called the *system equation* of the flying object. In (2.2), the matrix C that describes how the state is measured is called the *observation* (or *measurement*) *matrix*, and the vector $\mathbf{v}_n$ the measurement *data vector*, while (2.2) itself is called the *observation* (or *measurement*) *equation*. The two equations together give a *state-space description* of the flying object. Note that we may treat $\{\mathbf{u}_n\}$ as an input sequence with corresponding output $\{\mathbf{v}_n\}$. The transformation that takes $\{\mathbf{u}_n\}$ to its output $\{\mathbf{v}_n\}$ is called a *system*. In other words, the "system" in this example is the flying object governed by the state-space description (2.1) and (2.2).

2.1.2 Properties of Linear Time-Invariant Systems

In general, let us consider a system described by the input-output relations $(\{\mathbf{u}_n\},\{\mathbf{v}_n\})$ as shown in Fig. 2.2.

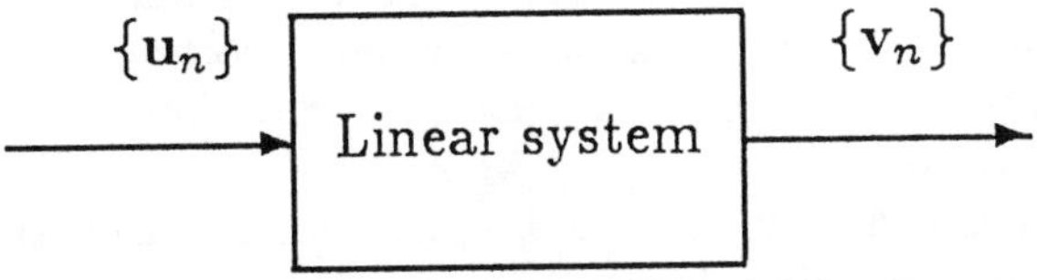

Fig. 2.2. Input-output relation in the time domain

Here, $\mathbf{u}_n$ and $\mathbf{v}_n$ are $p\times 1$ and $q\times 1$ column vectors, respectively. Hence, the system may also be called a *Multi-Input/Multi-Output* (MIMO) system. In the special case when both p and q are equal to 1, it is called a *Single-Input/Single-Output* (SISO) system. Of course, the digital filter we studied in Chap. 1 is an SISO system. Analogous to a digital filter, the desirable properties of a system can also be stated as follows:

(1) *Linearity*. If $\{\mathbf{u}_n^1\}\to\{\mathbf{v}_n^1\}$ and $\{\mathbf{u}_n^2\}\to\{\mathbf{v}_n^2\}$ and a, b are constants, then $\{a\mathbf{u}_n^1+b\mathbf{u}_n^2\}\to\{a\mathbf{v}_n^1+b\mathbf{v}_n^2\}$.
(2) *Time-invariance*. For each $k_0\geq 0$, if $\{\mathbf{u}_n\}\to\{\mathbf{v}_n\}$, then $\{\mathbf{u}_{n+k_0}\}\to\{\mathbf{v}_{n+k_0}\}$.
(3) *Causality*. If $\{\mathbf{u}_n\}\to\{\mathbf{v}_n\}$, then for each n_0, $\mathbf{v}_{n_0}$ does not depend on $\mathbf{u}_{n_0+1},\mathbf{u}_{n_0+2},\cdots$.
(4) *Stability* [or *Bounded Input/Bounded Output* (*BIBO*) *stability*]. If $\{\mathbf{u}_n\}\to\{\mathbf{v}_n\}$ where $\{|\mathbf{u}_n|\}\in l^\infty$, then $\{|\mathbf{v}_n|\}\in l^\infty$.

Here and throughout, if $\mathbf{x}$ is an m-vector, say $\mathbf{x}=[x_1\cdots x_m]^\top$, then $|\mathbf{x}|$ denotes its *length*, namely,

$$|\mathbf{x}|=\sqrt{x_1^2+\cdots+x_m^2}\,.$$

In view of the above flying object example, we will study MIMO systems which are governed by the pair of equations

$$\begin{cases} \mathbf{x}_{n+1} = A\mathbf{x}_n + B\mathbf{u}_n \,, \\ \quad \mathbf{v}_n = C\mathbf{x}_n + D\mathbf{u}_n \,, \end{cases} \tag{2.3}$$

where A, B, C, D are $m \times m$, $m \times p$, $q \times m$, and $q \times p$ matrices independent of n and, usually, $1 \le p, q \le m$. The same terminology introduced in the flying object example will be used. In particular, (2.3) will be called a *state-space description* of the system. The observation equation in (2.3), however, is more general than the one discussed in the above example, since when the "measurement" is made, the input (or control sequence) is sometimes observed as well. The state vector $\mathbf{x}_n$, which is an $m \times 1$ column vector, is introduced to facilitate the description of the MIMO system. It usually carries a lot of information on the system. In the above example, it represents the position, velocity, and acceleration in 3-space, and it traces the trajectory of the flying object. In the general setting, we simply say, for convenience, that the state vector $\mathbf{x}_n$ at the nth time instant is a *position vector* in the m-dimensional space, and the path traced by $\mathbf{x}_n$, $n = 1, 2, \cdots$, will be called the *trajectory* of the state vector.

The system which is governed by the state-space description (2.3) with the initial position $\mathbf{x}_0 = 0$ is certainly linear, time-invariant, and causal (Problem 2.1). Hence, we will say that it is a *Linear Time-Invariant* (*LTI*) *system*. When one of the matrices A, B, C, and D in (2.3) depends on the time variable n, then the linear system becomes *time-varying*. Since we will only be concerned with frequency-domain methods, time-varying systems will not be considered in this text.

2.1.3 Properties of State-Space Descriptions

Let us now study the important aspects of a state-space description.

(1) *State transition*. To go from the position $\mathbf{x}_k$ to the position $\mathbf{x}_n$, where $k < n$, we have the relationship

$$\mathbf{x}_n = A^{n-k}\mathbf{x}_k + \sum_{\ell=k+1}^{n} A^{n-\ell}B\mathbf{u}_{\ell-1} \tag{2.4}$$

(Problem 2.2). This equation is usually called the *state-transition equation* of the linear system. In particular, from the initial position $\mathbf{x}_0 = 0$ to $\mathbf{x}_n$, we have

$$\mathbf{x}_n = \sum_{\ell=1}^{n} A^{n-\ell}B\mathbf{u}_{\ell-1} \,. \tag{2.5}$$

(2) *Input-output relation*. From the state-space description (2.3), it is possible to obtain the input-output relation. By choosing the initial state to be 0 (in order to ensure the linearity of the relationship) and applying (2.5), we have

$$\mathbf{v}_n = D\mathbf{u}_n + \sum_{\ell=1}^{n} CA^{n-\ell}B\mathbf{u}_{\ell-1}\,. \tag{2.6}$$

(3) *ARMA model.* Suppose that an MIMO linear time-invariant system is given by the ARMA model

$$\mathbf{v}_n = \sum_{k=1}^{N} B_k\mathbf{v}_{n-k} + \sum_{k=0}^{M} A_k\mathbf{u}_{n-k}\,,$$

where the $q \times q$ matrices $B_1, \cdots, B_N$ and the $q \times p$ matrices $A_0, \cdots, A_M$ are independent of the time variable n, then the linear system has a state-space description given by (2.3) or

$$\begin{cases} \mathbf{x}_{n+1} = A\mathbf{x}_n + B\mathbf{u}_n\,, \\ \quad \mathbf{v}_n = C\mathbf{x}_n + D\mathbf{u}_n\,, \end{cases} \tag{2.7}$$

with

$$A = \begin{bmatrix} B_1 & I & 0 & \cdots & 0 \\ B_2 & 0 & \ddots & \ddots & \vdots \\ \vdots & \vdots & & \ddots & 0 \\ B_{N-1} & 0 & \cdots & \cdots & I \\ B_N & 0 & \cdots & \cdots & 0 \end{bmatrix}, \qquad B = \begin{bmatrix} A_1 + B_1A_0 \\ A_2 + B_2A_0 \\ \vdots \\ A_M + B_MA_0 \\ B_{M+1}A_0 \\ \vdots \\ B_NA_0 \end{bmatrix},$$

$$C = [\,I \quad 0 \quad \cdots \quad 0\,] \qquad \text{and} \qquad D = A_0\,,$$

where we have assumed $M \le N$ (Problem 2.3).

(4) *Controllability.* A linear system with state-space description (2.3) is said to be *controllable* (or *completely controllable*) if starting from any position $\mathbf{y}_0$ in $\mathbf{R}^m$, and any initial time $n = n_0$, the state vector sequence $\{\mathbf{x}_n\}$ can be brought to any desired position $\mathbf{y}_1$ in $\mathbf{R}^m$ by applying a certain digital control sequence $\{\mathbf{u}_n\}$ in a finite amount of time. In other words, from (2.4), the linear system is controllable if there is a sequence $\{\mathbf{u}_n\}$, $n = n_0, \cdots, N$, such that

$$A^{N-n_0}\mathbf{y}_0 + \sum_{\ell=n_0+1}^{N} A^{N-\ell}B\mathbf{u}_{\ell-1} = \mathbf{y}_1\,.$$

The following well-known controllability criterion for time-invariant systems is very useful (Problem 2.4).

Theorem 2.1. (Controllability Criterion)

A linear time-invariant system with state-space description given by (2.3) is controllable if and only if the $m \times pm$ matrix

$$M_{AB} = [B \quad AB \quad \cdots \quad A^{m-1}B] \tag{2.8}$$

has full rank.

The matrix M_{AB} defined in (2.8) is called the *controllability matrix* of the linear system described by (2.3).

(5) *Observability.* In many applications, it is necessary to determine an unknown state vector $\mathbf{x}_{n_0}$ at any time instant $n = n_0$ from the input-output information. Of course, once $\mathbf{x}_{n_0}$ is determined, then any position $\mathbf{x}_n$ for $n > n_0$ can be located by using the state-transition equation (2.4). A linear system with state-space description (2.3) is said to be *observable* (or *completely observable*) if for each n_0, there exists an $N > n_0$, such that $\mathbf{x}_{n_0}$ is uniquely determined by the input-output information $\{\mathbf{u}_n, \mathbf{v}_n\}$ where $n = n_0, \cdots, N$. The following well-known observability criterion is important (Problem 2.5).

Theorem 2.2. (Observability Criterion)

A time-invariant linear system with state-space description given by (2.3) is observable if and only if the $qm \times m$ matrix

$$N_{CA} = \begin{bmatrix} C \\ CA \\ \vdots \\ CA^{m-1} \end{bmatrix} \tag{2.9}$$

has full rank.

The matrix N_{CA} defined in (2.9) is called the *observability matrix* of the linear system described by (2.3).

For instance, the state-space description of the flying object example in Sect. 2.1.1 is both controllable and observable (Problem 2.6).

We have already discussed five important aspects of the state-space description of a linear time-invariant system. The input-output relationship in (2.6) is particularly important for the study of stability. For any digital control sequence $\{\mathbf{u}_n\}$, we always set $\mathbf{u}_{-1} = \mathbf{u}_{-2} = \cdots = 0$. Hence, if we define the $q \times p$ matrices h_n by

$$h_0 = D \qquad \text{and} \qquad h_n = CA^{n-1}B\,, \tag{2.10}$$

where $n = 1, 2, \cdots$, with $A^0 = I_m$, then (2.6) becomes

$$\mathbf{v}_n = \sum_{\ell=0}^{n} h_\ell \mathbf{u}_{n-\ell} = \sum_{\ell=0}^{\infty} h_\ell \mathbf{u}_{n-\ell}. \tag{2.11}$$

This is the vector-valued analog of the convolution version of a digital filter. For this reason, we will also call the $q \times p$ matrices h_n the *unit impulse responses* of the linear system. Similar to the proof of the stability criterion for convolution digital filters (Theorem 1.4), we have the following result (Problem 2.8).

Theorem 2.3. (First Stability Criterion)

A linear time-invariant system with state-space description given by (2.3) is (BIBO) stable if and only if there is a positive number M such that

$$\sum_{n=0}^{\infty} \|CA^n B\| \le M < \infty .$$

Here and throughout, for an $r \times s$ matrix G, we use the operator norm defined by

$$\|G\| = \max_{|\mathbf{x}|=1} |G\mathbf{x}| ,$$

where again $|\mathbf{x}|$, $|G\mathbf{x}|$ denote the lengths of the vectors $\mathbf{x}$, $G\mathbf{x}$ in $\mathbf{R}^s$, $\mathbf{R}^r$, respectively.

It is clear that a sufficient condition for the BIBO stability of a linear time-invariant system is $\|A\| < 1$, since the condition implies

$$\sum_{n=0}^{\infty} \|CA^n B\| \le \|C\| \; \|B\| \sum_{n=0}^{\infty} \|A\|^n = \frac{\|C\| \; \|B\|}{1 - \|A\|} < \infty .$$

Example 2.1. Apply the state-space description (2.7) and the stability criterion in Theorem 2.3 to determine values of b so that the ARMA model

$$v_n = b v_{n-2} + a_0 u_n + a_1 u_{n-1}$$

is stable.

A corresponding state-space description is given via (2.7) by

$$\begin{cases} \mathbf{x}_{n+1} = \begin{bmatrix} 0 & 1 \\ b & 0 \end{bmatrix} \mathbf{x}_n + \begin{bmatrix} a_1 \\ a_0 b \end{bmatrix} u_n , \\ v_n = [1 \;\; 0]\mathbf{x}_n + a_0 u_n . \end{cases}$$

Since

$$CA^nB = \begin{cases} a_0 b^{(n+1)/2}, & n = 1, 3, \cdots, \\ a_1 b^{n/2}, & n = 0, 2, 4, \cdots, \end{cases}$$

we have

$$\sum_{n=0}^{\infty} \|CA^nB\| = (|a_1| + |a_0 b|) \sum_{\ell=0}^{\infty} |b|^{\ell}.$$

Hence, the ARMA model under consideration is stable if and only if $|b| < 1$.

We conclude this section with the following remark: If a linear time-invariant system is governed by the state-space description (2.3), then its important properties on *controllability*, *observability*, and *stability* are completely determined by the three matrices A, B, C as can be seen from the above three theorems. In other words, the matrix D does not play any role in these considerations. Hence, for conciseness, if a linear system has a state-space description given by (2.3), we simply say that the linear system is described by the triple $\{A, B, C\}$. Here, A is an $m \times m$ square matrix, B an $m \times p$ matrix, and C a $q \times m$ matrix with $1 \le p, q \le m$. The positive integer m is called the *dimension* of the linear system. This dimension is also the dimension of its state vector $\mathbf{x}$.

2.2 Transfer Matrices and Minimal Realization

The first half of this section is devoted to the study of the *transfer matrix* of a linear time-invariant system described by the triple $\{A, B, C\}$, or more precisely, having the state-space description given in (2.3). One important application of the transfer matrix is that it contains the stability information of the linear system. It will be seen that the transfer matrix can be easily computed from the matrices A, B, C, and D. On the other hand, if the transfer matrix of a linear system is given, it is important to be able to find a state-space description of the system. It will be clear, however, that there is no unique way to describe the system. A very important problem is to find a system with the *minimum dimension*. This is called the problem of *minimal* (*state-space*) *realization*, a topic that will be discussed in Sects. 2.2.2 and 2.3.2.

2.2.1 Transfer Matrices of Linear Time-Invariant Systems

We first need the notion of the one-sided z-transform of a sequence of $r \times s$ matrices, $E_0, E_1, E_2, \cdots$. If the (i,k)th entry of E_n is denoted by e^n_{ik}, or

$$E_n = [e^n_{ik}]_{r \times s},$$

then the (one-sided) z-transform of $\{E_n\}$, $n = 0, 1, \cdots$, is an $r \times s$ matrix

$$\sum_{n=0}^{\infty} E_n z^{-n} = E_0 + E_1 z^{-1} + \cdots$$

whose (i,k)th entry is the (scalar-valued) one-sided z-transform

$$\sum_{n=0}^{\infty} e_{ik}^n z^{-n} .$$

It is clear that the z-transform is a linear operation, and this means that

$$\sum_{n=0}^{\infty}(C_1 E_n + C_2 F_n) z^{-n} = C_1 \sum_{n=0}^{\infty} E_n z^{-n} + C_2 \sum_{n=0}^{\infty} F_n z^{-n}$$

for all $r \times s$ matrices E_n, F_n and $t \times r$ matrices C_1 and C_2. Hence, taking the z-transforms of both sides of the two equations in the state-space description (2.3) with zero initial state, we have

$$\begin{cases} zX = AX + BU , \\ V = CX + DU , \end{cases}$$

where

$$X = \sum_{n=0}^{\infty} \mathbf{x}_n z^{-n} , \quad U = \sum_{n=0}^{\infty} \mathbf{u}_n z^{-n} \quad \text{and} \quad V = \sum_{n=0}^{\infty} \mathbf{v}_n z^{-n} .$$

Now, for large values of $|z|$, the $m \times m$ matrix $(zI_m - A)$ is diagonal dominant, and is therefore nonsingular. It follows that by defining

$$H(z) = C(zI_m - A)^{-1} B + D \tag{2.12}$$

we have the input-output relationship

$$V = H(z) U . \tag{2.13}$$

For this reason, the $q \times p$ matrix $H(z)$ defined in (2.12) is called the *transfer matrix* of the linear system.

Example 2.2. It is easy to see that the state-space description discussed in Example 2.1 has the (1×1) transfer matrix given by

$$H(z) = [1 \quad 0] \begin{bmatrix} z & -1 \\ -b & z \end{bmatrix}^{-1} \begin{bmatrix} a_1 \\ a_0 b \end{bmatrix} + a_0 = \frac{a_1 z + a_0 b^2}{z^2 - b} + a_0 .$$

Two remarks are in order. First, by using the Neumann series expansion, we may write

$$\begin{aligned}H(z) &= z^{-1}C(I_m - z^{-1}A)^{-1}B + D\\ &= z^{-1}C[I_m + Az^{-1} + A^2z^{-2} + \cdots]B + D\\ &= D + CBz^{-1} + CABz^{-2} + CA^2Bz^{-3} + \cdots\\ &= h_0 + h_1z^{-1} + h_2z^{-2} + h_3z^{-3} + \cdots\\ &= \sum_{n=0}^{\infty} h_n z^{-n}\,,\end{aligned} \tag{2.14}$$

where $h_0, h_1, \cdots$ have already been defined in (2.10). This shows that the transfer matrix $H(z)$ is a one-sided z-transform of the unit impulse response sequence $\{h_n\}$ of the linear system. In addition, (2.13) may be viewed as the z-transform of (2.11), so that even in the matrix-valued situation, the z-transform of the convolution of two sequences is the product of their z-transforms. Next, let us recall the following adjoint formulation of the inverse of a nonsingular matrix. The adjoint of an $m \times m$ matrix E, denoted by $\mathrm{adj}(E)$, is an $m \times m$ matrix whose (i,k)th entry is $(-1)^{i+k}\det(E_{ki})$ where E_{ki} is obtained from E by deleting its kth row and ith column. The adjoint formulation of the inverse of a nonsingular $m \times m$ matrix E is then given by

$$E^{-1} = \frac{1}{\det(E)}\mathrm{adj}(E)\,.$$

Hence, from (2.12), the transfer matrix of the linear system becomes

$$H(z) = \frac{1}{\det(zI_m - A)}C[\mathrm{adj}(zI_m - A)]B + D\,. \tag{2.15}$$

Since $\det(zI_m - A)$ is a polynomial of degree m with leading coefficient 1, it is clear that each entry of the $q \times p$ matrix $H(z) - D$ is a (strictly) *proper rational function*. Consequently, each entry of the transfer matrix $H(z)$ is a rational function whose numerator has degree no greater than that of the denominator. In view of Theorem 2.3, the following result analogous to Theorem 1.4 can be obtained (Problem 2.10).

Theorem 2.4. (Second Stability Criterion)

A linear time-invariant system described by the triple $\{A, B, C\}$ is stable if and only if each entry of its $q \times p$ transfer matrix $H(z)$ in (2.15) has all its poles lying in the unit disk $|z| < 1$. In particular, if all the zeros of the mth degree polynomial $\det(zI_m - A)$ lie in $|z| < 1$, then the linear system is stable.

Example 2.3. Discuss the stability of the ARMA model

$$v_n = bv_{n-2} + a_0u_n + a_1u_{n-1}$$

given in Example 2.1 using the second stability criterion.

As has been pointed out in Example 2.1, its corresponding state-space description is given by

$$\begin{cases} \mathbf{x}_{n+1} = \begin{bmatrix} 0 & 1 \\ b & 0 \end{bmatrix} \mathbf{x}_n + \begin{bmatrix} a_1 \\ a_0 b \end{bmatrix} u_n \,, \\ v_n = [1 \quad 0]\mathbf{x}_n + a_0 u_n \,, \end{cases}$$

and by (2.15) its transfer matrix is

$$H(z) = \frac{a_1 z + a_0 b^2}{z^2 - b} + a_0 \,.$$

Hence, by the second stability criterion, this linear system is stable if and only if $|b| < 1$.

We remark that in many situations the second stability criterion is more convenient to apply than the first one, as can be seen from Examples 2.1 and 2.3.

Of course, due to the possibility of cancellation of poles and zeros, there are stable linear time-invariant systems with some zeros of

$$\det(zI_m - A) \tag{2.16}$$

lying in $|z| \geq 1$ (Problem 2.15). However, the stability of controllable and observable linear systems is determined by the location of the zeros of the polynomial in (2.16), which are the eigenvalues of the system matrix A. To state this result more precisely, we need the following terminology: The transfer function $H(z)$ in (2.15) is said to have *no pole-zero cancellation* if none of the zeros of the denominator $\det(zI_m - A)$ in (2.15) disappears by all possible cancellations with the zeros of the numerator (matrix of polynomials), although there might be certain reduction of multiplicities. The following result can be found, for example, in the monographs of Chui and Chen [1989] and Kailath [1980].

Theorem 2.5. The transfer matrix $H(z)$ in (2.15) of a controllable and observable linear time-invariant system has no pole-zero cancellation.

As an immediate consequence, the following stability criterion is obtained.

Corollary 2.1. (Third stability criterion)

Let A be the system matrix of a controllable and observable linear time-invariant system. Then the system is stable if and only if all the eigenvalues of the system matrix A are of magnitude less than 1.

2.2.2 Minimal Realization of Linear Systems

We next study the inverse problem, namely, from the transfer matrix, we want to find a state-space description of the linear system. In many applications, a state-space description with very small dimension is required. The problem of finding one with the smallest dimension is called the problem of *minimal realization*, or more precisely, minimal state-space realization. This problem can be stated as follows:

Let $H(z)$ be a $q \times p$ matrix of the form

$$H(z) = \frac{1}{a_m(z)} \begin{bmatrix} b_{11}(z) & \cdots & b_{1p}(z) \\ \vdots & & \vdots \\ b_{q1}(z) & \cdots & b_{qp}(z) \end{bmatrix}, \tag{2.17}$$

where $a_m(z)$ is a polynomial of degree m with leading coefficient 1, and each $b_{ik}(z)$ is a polynomial of degree no greater than m. The problem is to determine a state-space description

$$\begin{cases} \mathbf{x}_{n+1} = A\mathbf{x}_n + B\mathbf{u}_n\,, \\ \mathbf{v}_n = C\mathbf{x}_n + D\mathbf{u}_n\,, \end{cases}$$

where A is an $\hat{m} \times \hat{m}$ matrix, and B, C, D are $\hat{m} \times p$, $q \times \hat{m}$, and $q \times p$ matrices, respectively, such that $\hat{m}$ is the smallest possible, and that

$$C(zI_{\hat{m}} - A)^{-1}B + D = H(z)\,. \tag{2.18}$$

This (smallest) $\hat{m}$ is called the *dimension of minimal realization* of the linear system.

By using the adjoint formulation of the left-hand side of (2.18), it is clear that

$$0 < \hat{m} \leq m.$$

Example 2.4. Verify that the state-space description S_1 described by the triple $\{A_1, B_1, C_1\} = \{[1], [1], [1]\}$ and the state-space description S_2 with

$$\{A_2, B_2, C_2\} = \left\{ \begin{bmatrix} 1 & 1 \\ 0 & 1 \end{bmatrix}, \begin{bmatrix} 1 \\ 0 \end{bmatrix}, [1 \ \ 0] \right\}$$

have the same transfer matrix if and only if they have the same D matrix.

It is easy to verify that the transfer matrix of the state-space description S_2 is

$$H_2(z) = C_2(zI_2 - A)^{-1}B + D = (z-1)^{-1} + D.$$

Hence, $H_2(z)$ is the same as the transfer function $H_1(z)$ of S_1 if and only if they have the same D.

From the above example, we observe that the matrix D of a linear system is unique but the triple $\{A, B, C\}$ is not. To find D, we simply divide each $b_{ik}(z)$ in (2.17) by $a_m(z)$, where $1 \leq i \leq q$ and $1 \leq j \leq p$, yielding

$$b_{ik}(z) = d_{ik} a_m(z) + c_{ik}(z),$$

where each d_{ik} is some constant (which may be zero), and $c_{ik}(z)$ is a polynomial of degree at most $(m-1)$. That is, we have now determined the constant matrix D, namely,

$$D = \begin{bmatrix} d_{11} & \cdots & d_{1p} \\ \vdots & & \vdots \\ d_{q1} & \cdots & d_{qp} \end{bmatrix}.$$

Finding a triple $\{A, B, C\}$ is much harder. Nevertheless, we can establish a relationship between a triple $\{A, B, C\}$ and its corresponding transfer matrix $H(z)$ as follows. First, we note that

$$H(z) = \frac{1}{a_m(z)} \begin{bmatrix} c_{11}(z) & \cdots & c_{1p}(z) \\ \vdots & & \vdots \\ c_{q1}(z) & \cdots & c_{qp}(z) \end{bmatrix} + D\,, \tag{2.19}$$

where each $c_{ik}(z)/a_m(z)$ is a (strictly) proper rational function, and hence may be written as

$$\frac{c_{ik}(z)}{a_m(z)} = h_1(i,k)z^{-1} + h_2(i,k)z^{-2} + \cdots$$

for some sequence of complex numbers $\{h_n(i,k)\}$. Hence, (2.19) may also be expressed as a one-sided z-transform, namely,

$$H(z) = D + h_1 z^{-1} + h_2 z^{-2} + \cdots,$$

where

$$h_n = \begin{bmatrix} h_n(1,1) & \cdots & h_n(1,p) \\ \vdots & & \vdots \\ h_n(q,1) & \cdots & h_n(q,p) \end{bmatrix}$$

are $q \times p$ matrices. We now form the infinite block-Hankel matrix $\Gamma_H = [h_{k+\ell-1}]$, $k, \ell = 1, 2, \cdots$, or

$$\Gamma_H = \begin{bmatrix} h_1 & h_2 & h_3 & \cdots \\ h_2 & h_3 & \cdots & \cdots \\ h_3 & \cdots & \cdots & \cdots \\ \cdots & \cdots & \cdots & \cdots \end{bmatrix}.$$

This matrix contains a lot of information on any triple $\{A, B, C\}$ which provides a realization of the linear system. For instance, the following relationship will be useful in determining the dimension of minimal realizations:

$$\Gamma_H = \begin{bmatrix} C \\ CA \\ CA^2 \\ \vdots \end{bmatrix} [B \quad AB \quad A^2B \quad \cdots] \,. \tag{2.20}$$

It must be emphasized that this identity holds for all triples $\{A, B, C\}$ that satisfy (2.12), as long as the dimension of A does not exceed m [since the corresponding $H(z)$ shown in (2.19) has been assumed to be (strictly) proper, see Problem 2.16]. Hence, it does not solve the problem of minimal realization directly. The following result, which can be proved by methods from linear algebra and can be found from, for example, Kailath [1980], gives a characterization of minimal realization.

Theorem 2.6. (Criterion for Minimal Realization)

Let A, B, C be $\tilde{m} \times \tilde{m}$, $\tilde{m} \times p$, $q \times \tilde{m}$ matrices, respectively, where $\tilde{m} \leq m$, such that the triple $\{A, B, C\}$ provides a realization of a linear system with transfer matrix given by (2.17). Then the realization is a minimal realization if and only if both the controllability and observability matrices

$$M_{AB} = [B \quad AB \quad \cdots \quad A^{\tilde{m}-1}B] \qquad \text{and} \qquad N_{CA} = \begin{bmatrix} C \\ CA \\ \vdots \\ CA^{\tilde{m}-1} \end{bmatrix}$$

have full rank.

Hence, if the triple $\{A, B, C\}$ provides a realization in the sense that

$$C(zI_{\tilde{m}} - A)^{-1}B = H(z) - D \,,$$

and M_{AB}, N_{CA} are of full rank, then the dimension $\tilde{m}$ of the matrix A must be the dimension $\hat{m}$ of minimal realizations. To determine $\hat{m}$, the following result provides a useful tool.

Theorem 2.7. (Dimension of Minimal Realization)

The dimension of minimal realization of a linear system with transfer matrix $H(z)$ given by (2.17) is the rank of the corresponding infinite block-Hankel matrix Γ_H.

This result can be proved by using the identity (2.20), see Kailath [1980, p.442]. It may be considered as a generalization of the classical Kronecker theorem, which we will study in some detail in the next section.

Example 2.5. Discuss the minimal realization of the two state-space descriptions S_1 and S_2 in Example 2.4.

It is clear that, by Theorems 2.1 and 2.2, S_1 is both controllable and observable, and S_2 is observable but not controllable. Hence, by Theorem 2.6, S_1 is a minimal realization of the linear system but S_2 is not. Note also that S_1 and S_2 have the same infinite Hankel matrix

$$\Gamma_H = \begin{bmatrix} 1 & 1 & 1 & \cdots \\ 1 & 1 & \cdots & \cdots \\ 1 & \cdots & \cdots & \cdots \\ \cdots & \cdots & \cdots & \cdots \end{bmatrix},$$

which has rank 1. Hence, by Theorem 2.7 the dimension of the minimal realization is 1.

2.3 SISO Linear Systems

We have seen that both a digital filter and a discrete-time linear system, which are time-invariant, causal, and has single input and single output, are characterized by the sequence of unit impulse responses. Hence, to unify the two subjects and to consider a more general theory, we consider the following "convolution" system, which will be called an SISO system for short [see (1.16) and Theorem 1.3]. That is, given any sequence of complex numbers,

$$h_0, h_1, h_2, \cdots,$$

an SISO system with unit impulse responses $\{h_n\}$ is defined by the following input-output relationship:

$$v_n = \sum_{i=0}^{n} h_i u_{n-i} \tag{2.21}$$

for any input sequence $\{u_n\}$, $n = 0, 1, \cdots$. The sequence $\{v_n\}$ is called the output sequence corresponding to the input sequence $\{u_n\}$. In the study of SISO systems, we only consider sequences with non-negative indices, so that we may use the notation

$$h_{-1} = h_{-2} = \cdots = 0$$

and

$$u_{-1} = u_{-2} = \cdots = 0 .$$

Such sequences are called *causal sequences* (Sect. 1.3.2). Hence, for causal sequences, (2.21) may be written as

$$v_n = \sum_{i=0}^{\infty} h_i u_{n-i} = \sum_{i=-\infty}^{\infty} h_i u_{n-i} , \tag{2.22}$$

or equivalently, $\{v_n\} = \{h_n\} * \{u_n\}$. In Chapter 1, $\{u_n\}$ is also called a digital signal, and in this chapter it is called a digital control sequence, or simply a sequence.

It should be remarked that we have not put any assumption on the unit impulse response sequence $\{h_n\}$ that defines the SISO system. We recall from Theorems 1.3 and 2.3, however, that the SISO *system is stable if and only if the sequence $\{h_n\}$ is in l^1, namely, if and only if*

$$\sum_{n=0}^{\infty} |h_n| < \infty .$$

2.3.1 Kronecker's Theorem

Two important concepts associated with $\{h_n\}$ have been introduced. The first one is the transfer function of an SISO system, and in this general setting, it is a formal power series

$$H(z) = \sum_{n=0}^{\infty} h_n z^{-n} .$$

Under very mild restrictions on $\{h_n\}$, $H(z)$ will be an analytic function, at least for large values of $|z|$. The second important concept is the infinite Hankel matrix $\Gamma_H = [h_{i+\ell-1}]$, $i, \ell = 1, 2, \cdots$, or

$$\Gamma_H = \begin{bmatrix} h_1 & h_2 & h_3 & \cdots \\ h_2 & h_3 & \cdots & \cdots \\ h_3 & \cdots & \cdots & \cdots \\ \cdots & \cdots & \cdots & \cdots \end{bmatrix} . \tag{2.23}$$

These two entities are closely related as can be seen from the following beautiful result of Kronecker, see, for example, Gantmacher [1966]. As before (Sect. 1.3.4), we consider the singular and analytic parts of $H(z)$, namely,

$$H_s(z) = \sum_{n=1}^{\infty} h_n z^{-n} \tag{2.24}$$

and

$$H_a(z) = h_0 ,$$

so that $H(z) = H_s(z) + H_a(z)$.

Theorem 2.8. (Kronecker's Theorem)

The infinite Hankel matrix Γ_H in (2.23) has finite rank if and only if the singular part $H_s(z)$ in (2.24) is a (strictly) proper rational function in z. Furthermore, in this situation, the rank of Γ_H agrees with the number of poles of $H_s(z)$, with multiplicities taken into consideration.

We will give a constructive proof of Kronecker's theorem below. To facilitate our discussion, we first establish the following lemma, where $\underline{\gamma}_i$ will denote the ith column of the infinite matrix Γ_H; that is, $\underline{\gamma}_i = [h_i \quad h_{i+1} \quad \cdots]^\top$.

Lemma 2.1. Γ_H has finite rank k if and only if the column vectors $\underline{\gamma}_1, \cdots, \underline{\gamma}_k$ are linearly independent and

$$\underline{\gamma}_{k+1} = \sum_{i=1}^{k} c_{k-i+1} \underline{\gamma}_{k-i+1} \tag{2.25}$$

for some constants $c_1, \cdots, c_k$.

From the structure of an infinite Hankel matrix, it is clear that (2.25) is equivalent to

$$\underline{\gamma}_{n+1} = \sum_{i=1}^{k} c_{k-i+1} \underline{\gamma}_{n-i+1}, \qquad n = k, k+1, \cdots. \tag{2.26}$$

The sufficiency of the condition that guarantees rank $\Gamma_H = k$ is obvious. To show that this condition is necessary, we first note that if rank $\Gamma_H = k$, then the set $\{\underline{\gamma}_1, \cdots, \underline{\gamma}_{k+1}\}$ must be linearly dependent. Let $\tilde{k}$ be the largest integer with $0 < \tilde{k} \leq k$ such that $\{\underline{\gamma}_1, \cdots, \underline{\gamma}_{\tilde{k}}\}$ is a linearly independent set. Then $\underline{\gamma}_{\tilde{k}+1}$ is in the linear span of this set, or

$$\underline{\gamma}_{\tilde{k}+1} = \sum_{i=1}^{\tilde{k}} \tilde{c}_{\tilde{k}-i+1} \underline{\gamma}_{\tilde{k}-i+1}$$

for some constants $\tilde{c}_1, \cdots, \tilde{c}_{\tilde{k}}$. Thus, as we have seen from the sufficiency condition,

$$\underline{\gamma}_{n+1} = \sum_{i=1}^{\tilde{k}} c_{\tilde{k}-i+1} \underline{\gamma}_{n-i+1}, \quad n = \tilde{k}, \tilde{k}+1, \cdots,$$

so that the rank of Γ_H is $\tilde{k}$, or $\tilde{k} = k$, completing the proof of the lemma.

It is important to note that if Γ_H has rank k, then the matrix

$$\Gamma_H^k = \begin{bmatrix} h_1 & \cdots & h_k \\ \vdots & & \vdots \\ h_k & \cdots & h_{2k-1} \end{bmatrix}$$

which is the principal minor of order k in Γ_H, must be nonsingular, and the constants $c_1, \cdots, c_k$ in (2.26) can be uniquely determined by solving the matrix equation

$$\Gamma_H^k \begin{bmatrix} c_1 \\ \vdots \\ c_k \end{bmatrix} = \begin{bmatrix} h_{k+1} \\ \vdots \\ h_{2k} \end{bmatrix}. \tag{2.27}$$

This observation can be justified by using (2.26) (Problem 2.17). It should be noted, however, that the analogous statement for a finite Hankel matrix is false (Problem 2.18).

We are now ready to prove Theorem 2.8. Suppose that $H(z)$ is a rational function in z with k poles, counting multiplicities. Then we may write its singular part as $H_s(z) = P(z)/Q(z)$ where

$$P(z) = p_1 z^{k-1} + \cdots + p_k \quad \text{and} \quad Q(z) = z^k + q_1 z^{k-1} + \cdots + q_k$$

are *coprime* (or *relatively prime*). Hence, by multiplying out the left-hand side of

$$(z^k + q_1 z^{k-1} + \cdots + q_k)(h_1 z^{-1} + h_2 z^{-2} + \cdots) = p_1 z^{k-1} + \cdots + p_k$$

and equating coefficients of equal powers of z, we arrive at the following two sets of equations:

$$\begin{cases} p_1 = h_1 \,, \\ p_2 = h_2 + h_1 q_1 \,, \\ \cdots \\ p_k = h_k + h_{k-1} q_1 + \cdots + h_1 q_{k-1} \,, \end{cases} \tag{2.28}$$

and

$$\begin{cases} h_{k+1} + \sum_{i=1}^{k} q_i h_{k-i+1} = 0 \,, \\ h_{k+2} + \sum_{i=1}^{k} q_i h_{k-i+2} = 0 \,, \\ \cdots \,. \end{cases} \tag{2.29}$$

It is clear that (2.29) is equivalent to (2.25) with $q_i = -c_{k-i+1}, i = 1, \cdots, k$. Since $P(z)$ and $Q(z)$ are coprime, the integer k in (2.29) or (2.25) cannot be reduced. Hence, $\underline{\gamma}_1, \cdots, \underline{\gamma}_k$ are linearly independent, and it follows from Lemma 2.1 that rank $\Gamma_H = k$. Conversely, if rank $\Gamma_H = k$, then (2.25) is satisfied where k is the smallest possible integer. By setting $q_i = -c_{k-i+1}, i = 1, \cdots, k$, and $p_1, \cdots, p_k$ as described by (2.28), we have shown that $H_s(z) = P(z)/Q(z)$. Since k is the smallest, $P(z)$ and $Q(z)$ are coprime, so that $H_s(z)$ has exactly k poles, counting multiplicities. This completes the proof of Kronecker's theorem.

The proof of Kronecker's theorem given here is constructive. In fact, if the rank of Γ_H is known, then the rational function $P(z)/Q(z)$ can be computed easily. We summarize this observation as follows.

Theorem 2.9. Suppose that the infinite Hankel matrix $\Gamma_H = [h_{i+\ell-1}]$, $i, \ell = 1, 2, \cdots$, has finite rank k. Then

$$H(z) = \sum_{n=0}^{\infty} h_n z^{-n} = h_0 + \frac{p_1 z^{k-1} + \cdots + p_k}{z^k - c_k z^{k-1} - \cdots - c_1}, \tag{2.30}$$

where $c_1, \cdots, c_k$ satisfy (2.27) and $p_1, \cdots, p_k$ satisfy (2.28). Furthermore, the (strictly) proper rational function in (2.30) is in irreducible form.

In matrix form, we may also write

$$\begin{aligned} \begin{bmatrix} c_1 \\ \vdots \\ c_k \end{bmatrix} &= \begin{bmatrix} h_1 & \cdots & h_k \\ \vdots & & \vdots \\ h_k & \cdots & h_{2k-1} \end{bmatrix}^{-1} \begin{bmatrix} h_{k+1} \\ \vdots \\ h_{2k} \end{bmatrix}, \\ \begin{bmatrix} p_1 \\ \vdots \\ p_k \end{bmatrix} &= \begin{bmatrix} 1 & 0 & \cdots & \cdots & 0 \\ -c_k & 1 & \ddots & & \vdots \\ -c_{k-1} & -c_k & 1 & \ddots & \vdots \\ \vdots & \ddots & \ddots & \ddots & 0 \\ -c_2 & \cdots & -c_{k-1} & -c_k & 1 \end{bmatrix} \begin{bmatrix} h_1 \\ \vdots \\ h_k \end{bmatrix}. \end{aligned} \tag{2.31}$$

Example 2.6. Suppose that

$$H(z) = \sum_{n=0}^{k} h_n z^{-n}, \qquad h_k \neq 0\,.$$

Then

$$\Gamma_H = \begin{bmatrix} h_1 & \cdots & h_k & 0 & \cdots \\ \cdots & \cdots & 0 & \cdots & \cdots \\ h_k & 0 & \cdots & \cdots & \cdots \\ 0 & \cdots & \cdots & \cdots & \cdots \\ \cdots & \cdots & \cdots & \cdots & \cdots \end{bmatrix}$$

has rank k. Since $h_{k+1} = h_{k+2} = \cdots = 0$, we have, from (2.31), that $c_1 = \cdots = c_k = 0$, so that $p_i = h_i$ for $i = 1, \cdots, k$. That is,

$$H(z) = h_0 + \frac{h_1 z^{k-1} + \cdots + h_k}{z^k}$$

a polynomial of degree k in z^{-1}, as it should be.

Example 2.7. Let $H(z) = 1 + rz^{-1} + r^2 z^{-2} + \cdots$ with $r \neq 0$. Then

$$\Gamma_H = \begin{bmatrix} r & r^2 & \cdots \\ r^2 & \cdots & \cdots \\ \cdots & \cdots & \cdots \end{bmatrix}$$

obviously has rank 1, and $\Gamma_H^1 = [r]$. Hence, we have

$$c_1 = \frac{1}{r} h_2 = r$$

and

$$p_1 = h_1 = r$$

by (2.31), so that

$$H(z) = h_0 + \frac{p_1}{z - c_1} = 1 + \frac{r}{z - r} = \frac{z}{z - r},$$

which is exactly the sum of the given geometric series.

2.3.2 Minimal Realization of SISO Linear Systems

We now discuss the problem of minimal realization. Again, suppose that rank $\Gamma_H = k$. Then by Theorem 2.9, we have

$$H(z) = \frac{h_0 + (p_1 - c_k h_0) z^{-1} + \cdots + (p_k - c_1 h_0) z^{-k}}{1 - c_k z^{-1} - \cdots - c_1 z^{-k}}.$$

Hence, in the time domain the input-output relationship of the SISO linear system can be described by the ARMA model

$$v_n = \sum_{i=1}^{k} c_{k-i+1} v_{n-i} + \sum_{i=0}^{k} a_i u_{n-i} \tag{2.32}$$

where

$$a_0 = h_0 \qquad \text{and} \qquad a_i = p_i - c_{k-i+1} h_0, \qquad i = 1, \cdots, k.$$

It follows from (2.7) that this linear system has a state-space description given by

$$\begin{cases} \mathbf{x}_{n+1} = A\mathbf{x}_n + Bu_n \,, \\ v_n = C\mathbf{x}_n + Du_n \,, \end{cases} \tag{2.33}$$

where

$$A = \begin{bmatrix} c_k & 1 & 0 & \cdots & 0 \\ c_{k-1} & 0 & 1 & \ddots & \vdots \\ \vdots & \vdots & \ddots & \ddots & 0 \\ c_2 & 0 & \cdots & 0 & 1 \\ c_1 & 0 & \cdots & \cdots & 0 \end{bmatrix}, \quad B = \begin{bmatrix} p_1 \\ \vdots \\ p_k \end{bmatrix},$$

$$C = [\,1 \quad 0 \quad \cdots \quad 0\,] \quad \text{and} \quad D = [h_0]$$

are $k \times k$, $k \times 1$, $1 \times k$, and 1×1 matrices, respectively. Here, the constants $c_1, \cdots, c_k$ and $p_1, \cdots, p_k$ can be computed by using (2.31). Since the system matrix A has dimension k, which is the rank of Γ_H, then according to Theorem 2.7 it is clear that the above state-space description indeed provides a *minimal* realization of the SISO linear system.

Example 2.8. Suppose that

$$H(z) = \frac{\frac{1}{2} + z^{-1} - \frac{1}{3}z^{-2}}{1 - z^{-1} - \frac{1}{6}z^{-2}} \,.$$

Then

$$h_0 = \frac{1}{2}, \quad c_1 = \frac{1}{6}, \quad c_2 = 1 \,, \quad a_1 = 1 \,, \quad \text{and} \quad a_2 = -\frac{1}{3} \,.$$

It follows that

$$p_1 = a_1 + c_2 h_0 = \frac{3}{2} \qquad \text{and} \qquad p_2 = a_2 + c_1 h_0 = -\frac{1}{4} \,,$$

so that

$$A = \begin{bmatrix} 1 & 1 \\ 1/6 & 0 \end{bmatrix}, \quad B = \begin{bmatrix} 3/2 \\ -1/4 \end{bmatrix}, \quad C = [1 \;\; 0] \,, \quad \text{and} \quad D = [1/2]$$

together provide a minimal realization of $H(z)$.

We have now derived an ARMA realization (2.32) and a minimal state-space realization (2.33) of an SISO linear system from the sequence of unit impulse responses $\{h_n\}$. Both realizations depend on the knowledge of the rank of the infinite Hankel matrix $\Gamma_H = [h_{i+\ell-1}]$, $i, \ell = 1, 2, \cdots$. In practice, it is difficult to obtain such information, and even if it is known, the rank is usually too large for physical feasibility. In fact, a small perturbation to the unit impulse responses h_i would change a finite rank Hankel matrix to one with infinite rank (Problem 2.20). Hence, the first step usually requires replacing Γ_H by another Hankel matrix with preassigned, and certainly smaller, rank. This procedure is called *system reduction*, and it is a very important problem in systems engineering.

2.3.3 System Reduction

The problem of system reduction for linear systems is posed in this section. We will study the SISO systems in detail here and delay the discussion of MIMO systems to Chap. 6. From the information on the sequence of unit impulse responses $\{h_n\}$ of an MIMO linear system, where each h_n is a $q \times p$ matrix, the first and most difficult step in solving the problem of system reduction is to replace this sequence by another sequence of $q \times p$ matrices $\{\tilde{h}_n\}$, such that the rank of the corresponding block-Hankel matrix

$$\tilde{\Gamma} = \begin{bmatrix} \tilde{h}_1 & \tilde{h}_2 & \tilde{h}_3 & \cdots \\ \tilde{h}_2 & \tilde{h}_3 & \cdots & \cdots \\ \tilde{h}_3 & \cdots & \cdots & \cdots \\ \cdots & \cdots & \cdots & \cdots \end{bmatrix}$$

does not exceed a prescribed integral value m. A novel approach to this problem is based on the work of Adamjan, Arov, and Krein (or AAK) [1971] via *best approximation* in the Hilbert space operator norm. Details will be given in Chaps. 4–6. A block-Hankel infinite matrix is considered as a linear operator on the sequence space l^2 with the norm defined by (1.28), namely,

$$\|\Gamma\|_s = \sup_{\|\mathbf{x}\|_{l^2}=1} \|\Gamma \mathbf{x}\|_{l^2} .$$

This is also called the spectral norm as discussed in Section 1.5.3. Let

$$\mathcal{H}^{pq} = \{ \Gamma : \ \Gamma \ \ p \times q \ \text{ block} - \text{Hankel matrix with } \|\Gamma\|_s < \infty \} \tag{2.34}$$

and for a positive integer m, let

$$\mathcal{H}^{pq}_m = \{ \Gamma \in \mathcal{H}^{pq} : \ \ \text{rank } \Gamma \le m \} . \tag{2.35}$$

Our problem in system reduction can be stated as follows:

Research Problem 3. Let $\Gamma \in \mathcal{H}^{pq}$ and m be a positive integer. Determine a $\hat{\Gamma}_m$ in $\mathcal{H}^{pq}_m$ such that

$$\|\Gamma - \hat{\Gamma}_m\|_s = \inf_{\Gamma_m \in \mathcal{H}^{pq}_m} \|\Gamma - \Gamma_m\|_s .$$

It is important to note that $\mathcal{H}_m^{pq}$ is not a linear manifold (Problem 2.21). In the special case where $p = q = 1$ (that is, when SISO linear systems are considered), we will use the notation

$$\mathcal{H} = \mathcal{H}^{11}, \qquad \mathcal{H}_k = \mathcal{H}_k^{11}.$$

In this special case, the above problem is contained in Problem 2 discussed in Sect. 1.5.3. Moreover, in this SISO setting once a rank-m Hankel matrix $\hat{\Gamma}_m \in \mathcal{H}_m$ has been obtained, an ARMA realization is readily available by applying (2.31), and a minimal (state-space) realization of $\hat{\Gamma}_m$ can be derived by using (2.33). In Chapt. 4, we will discuss a solution of the special case when the given Hankel matrix $\Gamma \in \mathcal{H}$ has finite-rank. Chapter 6 will be devoted to the study of finite-rank block-Hankel matrices $\Gamma \in \mathcal{H}^{pq}$. This is the so-called *AAK approach*. In fact, even the situation when Γ has infinite rank is not excluded in the AAK paper [1971]. However, at this writing, no algorithm or computational method is available to give a best finite-rank Hankel approximant of an infinite-rank bounded Hankel matrix operator exactly. Approximation procedures using truncations of Γ are available in Chui, Li and Ward [1989,1991]; Glover, Curtain, and Partington [1988]; and Hayashi, Trefethen, and Gutknecht [1990].

Of course, there are other system reduction methods available in the literature. These approaches are usually based on the Padé method. However, although Padé approximants are very easy to compute, they are usually unstable rational functions (or equivalently, the corresponding finite-rank Hankel matrix operators are no longer bounded), even if the original infinite Hankel matrices are bounded operators on l^2. To overcome this serious drawback of the Padé method, we may force stability on the rational approximant by choosing a denominator whose zeros lie in $|z| < 1$. There are again many methods in the literature for this purpose. The most well-known ones are probably the least-squares inverse procedure which we will discuss in some detail in Sect. 3.2.2, the Routh method and the Schwarz method, see Bultheel [1987]. The numerator can be determined by the "first half" of the set of Padé equations, or more preferably, by applying a second stage least-squares method considered in Chui and Chan [1982]. On the other hand, if a direct method based on an optimization criterion such as the AAK approach is applied to yield a bounded Hankel matrix in $\mathcal{H}_m$, then the corresponding rational function, which as a consequence of Kronecker's theorem has m poles, can be formulated by first solving a set of linear equations analogous to (2.29) and then applying another set of formulas analogous to (2.28), to obtain the denominator and numerator, respectively.

2.4 Sensitivity and Feedback Systems

Consider a discrete-time time-invariant linear system with state-space description

$$\begin{cases} \mathbf{x}_{n+1} = A\mathbf{x}_n + B\mathbf{u}_n\,, \\ \mathbf{v}_n = C\mathbf{x}_n + D\mathbf{u}_n\,, \end{cases}$$

where A, B, C, D are $m \times m$, $m \times p$, $q \times m$, $q \times p$ constant matrices, respectively. As we have already seen, the matrix D does not play any role in the study of stability, and for convenience we will assume that $D = 0$ in this section. Then the transfer matrix of this system is given by

$$H(z) = C(zI - A)^{-1}B\,,$$

and the input-output relationship in the frequency domain is $V(z) = H(z)U(z)$, as shown in Fig. 2.3 which is usually called an *open-loop configuration.* In the following, we will see that an open-loop system is in general sensitive to perturbation of system parameters and a standard technique to take care of this problem is to introduce a feedback yielding a so-called *closed-loop system.*

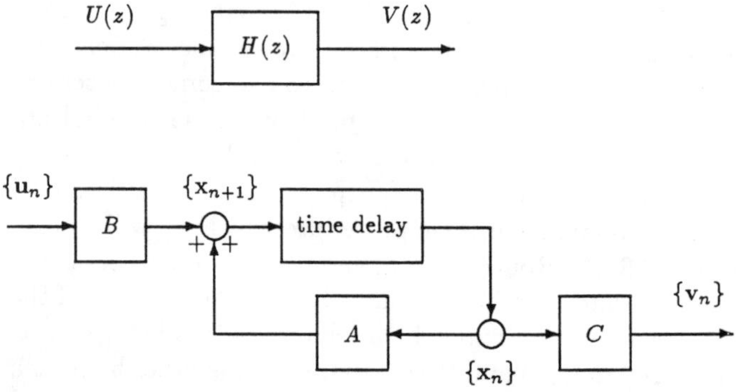

Fig. 2.3. Input-output relation in the frequency domain

2.4.1 Plant Sensitivity

In general, an open loop system is very sensitive to perturbation of the system parameters (or entries of the system matrix A). This means that a very small change in any parameter could easily transform a stable system to an unstable one. Since the system matrix is the dominating component of the *plant* of the system, this phenomenon is called *plant sensitivity.* In general, plant sensitivity may be a consequence of overheating, vibration, etc. The following example illustrates that the higher the order of the system, the more likely it is to be subject to plant sensitivity.

Example 2.9. Let the system matrix be given by

$$A = \begin{bmatrix} r & 1 & 0 & \cdots & 0 \\ 0 & \ddots & \ddots & \ddots & \vdots \\ \vdots & \ddots & \ddots & \ddots & 0 \\ \vdots & & \ddots & \ddots & 1 \\ 0 & \cdots & \cdots & 0 & r \end{bmatrix},$$

which is an $m \times m$ Toeplitz matrix with only two nonzero diagonals. For simplicity, set $B = C = I_m$, the identity matrix of order m. Then this system is stable if and only if $|r| < 1$. Suppose that $0 < r < 1$ and $\varepsilon > 0$ is a small perturbation of A at its lower left-hand corner [or the $(m, 1)$ entry], that is,

$$A_\varepsilon = A + \begin{bmatrix} 0 & \cdots & \cdots & 0 \\ \vdots & & & \vdots \\ 0 & \cdots & \cdots & 0 \\ \varepsilon & 0 & \cdots & 0 \end{bmatrix}.$$

Then the eigenvalues of A_ε are given by

$$z_k = r + \sqrt[m]{\varepsilon}\left(\cos\frac{2\pi k}{m} + \mathrm{j}\sin\frac{2\pi k}{m}\right),$$

$k = 0, 1, \cdots, m-1$ (Problem 2.24). In particular, the eigenvalue z_0 of A_ε satisfies

$$z_0 = r + \sqrt[m]{\varepsilon} \geq 1$$

whenever $\varepsilon \geq (1-r)^m$. For instance, if $r = 1/2$, then a perturbation of A by $\varepsilon \geq 2^{-m}$ transforms a very stable system to an unstable one. Note that for a fairly high order linear system, the perturbation of ε may be a small number. For example, if $m = 7$, then $\varepsilon = 0.01 > 2^{-7}$ gives an unstable system with the system matrix $A_\varepsilon = A_{0.01}$.

2.4.2 Feedback Systems and Output Sensitivity

A standard technique to take care of plant sensitivity is to introduce a *feedback matrix* F. This procedure may even stabilize an unstable system. In this consideration, F will be a $p \times m$ constant matrix. With this feedback, the original open-loop configuration now becomes a closed-loop system as shown in Fig. 2.4. Note that the input-output relationship becomes

$$V = C[zI - (A - BF)]^{-1}BU \tag{2.36}$$

and this means that the system matrix is changed from A to $A - BF$ (Problem 2.25). The following result can be found from Wonham [1979] which guarantees that a controllable linear system can be made less sensitive. More precisely, if the controllability condition stated in the following theorem is satisfied, then a suitable feedback can be designed such that all the poles of the resultant closed-loop system are located inside the unit circle and, in fact, far away from the boundary.

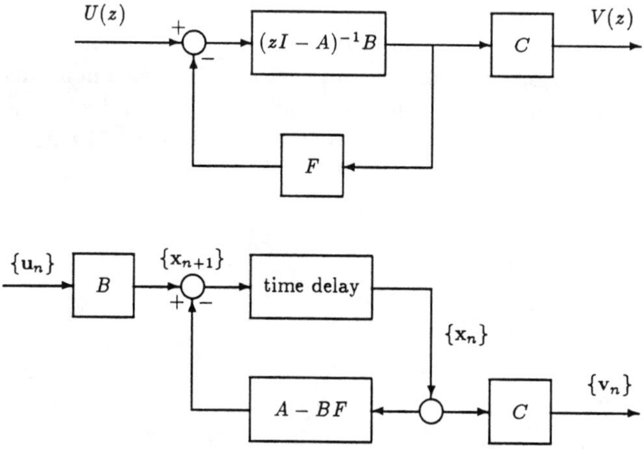

Fig. 2.4. Closed-loop system

Theorem 2.10. (Pole-Assignment Condition)

There exists a feedback matrix F such that the eigenvalues of the matrix $A - BF$ can be arbitrarily assigned if and only if the controllability matrix $M_{AB} = [\, B \quad AB \quad \cdots \quad A^{m-1}B \,]$ has full rank.

Let us now turn to the general setting. Denote by $\mathcal{M}$ the collection of all finite rectangular matrices M of rational functions such that for each $M \in \mathcal{M}$ there exists a non-negative integer n and a polynomial $q_n(z) = z^n + q_1 z^{n-1} + \cdots + q_n$, so that

$$M = M(z) = \frac{1}{q_n(z)} N_n(z) \tag{2.37}$$

where each entry of the matrix $N_n(z)$ is a polynomial of degree $\leq n$, that is,

$$\mathcal{M} = \Big\{ \frac{1}{z^n + q_1 z^{-1} + \cdots + q_n} N_n(z): \quad N_n(z) \text{ a finite matrix of polynomials with degree } \leq n;\ q_1, \cdots, q_n \text{ complex; and } n = 0, 1, \cdots \Big\}. \tag{2.38}$$

Also, define

$$\mathcal{M}^s = \{\ M \in \mathcal{M}: \quad M \ \text{ analytic on } \ |z| \geq 1\ \}. \tag{2.39}$$

Hence, all the poles of each $M \in \mathcal{M}^s$ lie in $|z| < 1$ so that all such matrices are stable. Note that two matrices in $\mathcal{M}^s$ may have different dimensions and their entries may be (stable) rational functions of different degrees.

Consider an open-loop system describing an MIMO linear control system in the frequency domain in terms of two matrices P and W_2 in $\mathcal{M}^s$, which are called the *plant* and *weight matrices* of the system, respectively. The plant plays the dominant role in the control system while the weight matrix W_2 is sometimes considered as a filtering device. The input-output relationship is shown in Fig. 2.5, where $V = W_2PU$.

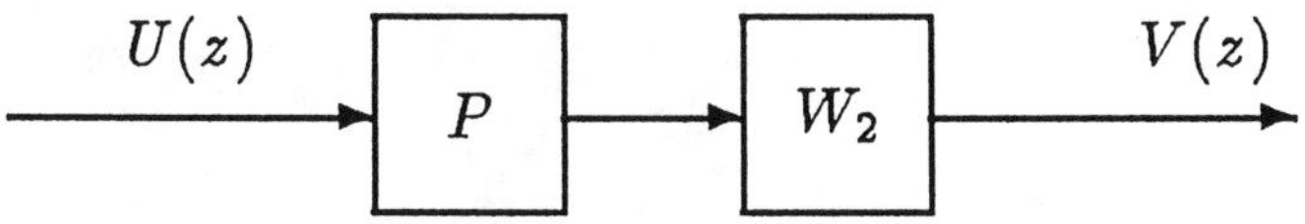

Fig. 2.5. Open-loop input-output relation in the frequency domain

As we have seen in the above example, a feedback matrix F may be introduced to take care of the plant sensitivity. In the general setting, F is a (not necessarily constant) matrix of appropriate dimension chosen from $\mathcal{M}^s$ such that the matrix $(I + PF)^{-1}$ is also in $\mathcal{M}^s$. The feedback matrix F is sometimes called a *compensator*, and with this compensator we have a closed-loop system as shown in Fig. 2.6.

The input-output relationship is now given by (Problem 2.26)

$$V = W_2(I + PF)^{-1}PU\,. \tag{2.40}$$

Unfortunately, by introducing a compensator to change an open-loop system to a closed-loop one, we have usually created another sensitivity phenomenon which occurs even if the input is zero, as shown in Fig. 2.7.

The reason is that there is the possibility of "noise" within the system, although a filtering device W_1, also called a *weight matrix*, chosen from $\mathcal{M}^s$ has been added. This is called *output sensitivity*. Let us use the notation N

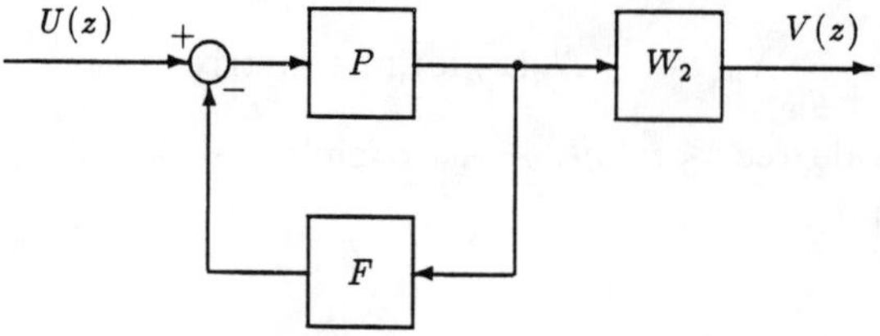

Fig. 2.6. Closed-loop system in the frequency domain

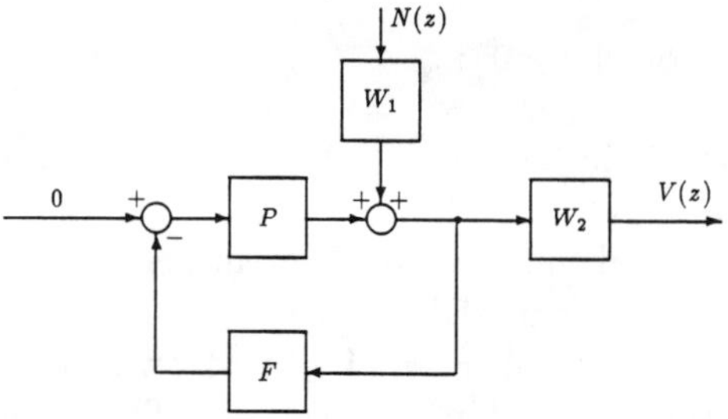

Fig. 2.7. Closed-loop system with disturbance input

for the noise input when the compensator is added and V_N for the corresponding output induced by the noise input alone. Then this input-output relationship is given by (Problem 2.27)

$$V_N = H_N N\,, \tag{2.41}$$

where

$$H_N = W_2(I + PF)^{-1}W_1\,. \tag{2.42}$$

Hence, it is important to minimize (a norm measurement of) V_N in (2.41) among all possible noise input N with certain maximum noise level (say the norm of N does not exceed 1).

2.4.3 Sensitivity Minimization

As we have seen from the input-output relationship (2.41), in order to reduce the output sensitivity due to possible noise input, it is important to minimize (a norm measurement of) the transfer matrix H_N. To be more precise, we consider N to belong to a certain Banach space $\mathcal{N}$ with norm $\|\ \|_{\mathcal{N}}$ and H_N defined by (2.42) as an operator mapping $\mathcal{N}$ into a Banach space $\mathcal{V}$ with norm $\|\ \|_{\mathcal{V}}$ by means of matrix multiplication. For convenience, we will simply use $\|H_N\|$ to denote the operator norm of H_N which is considered as an operator mapping $\mathcal{N}$ to $\mathcal{V}$. The problem is to minimize $\|V_N\|_{\mathcal{V}}$ subject to the constraint $\|N\|_{\mathcal{N}} \leq 1$ or, equivalently, to minimize the operator norm $\|H_N\|$ of H_N.

Let $\mathcal{F}$ denote the class of all admissible compensators defined by

$$\mathcal{F} = \{\, F \in \mathcal{M}^s : \quad W_2(I+PF)^{-1}W_1 \in \mathcal{M}^s \,\} . \tag{2.43}$$

Then we may pose the following problem:

Research Problem 4. Let $P \in \mathcal{M}$ and $W_1, W_2 \in \mathcal{M}^s$ of appropriate dimensions be given. Determine an $\hat{F} \in \mathcal{F}$ such that

$$\|W_2(I+P\hat{F})^{-1}W_1\| = \inf_{F \in \mathcal{F}} \|W_2(I+PF)^{-1}W_1\| . \tag{2.44}$$

To specify the operator norm in (2.44), we must decide what norms should be used for the (noise) input and (system) output spaces $\mathcal{N}$ and $\mathcal{V}$, respectively. First, note that N, V_N, and the transfer matrix H_N are not necessarily scalars; and hence, the Euclidean norm $|\ \ |_2$ (i.e. length) on N and V_N and the spectral norm $\|\ \ \|_s$ on H_N must be used. To be more precise, the spectral norm of H_N is

$$\|H_N(z)\|_s = \|\bar{H}_N^\top(z) H_N(z)\|_s^{1/2} , \tag{2.45}$$

where $\bar{H}$ is the complex conjugate of H, as usual. After applying these vector and matrix norms, we have scalar-valued functions on $|z| = 1$ and we may use the L^p norm on $[0, 2\pi]$. Here, it should be noted that all transfer matrices are assumed to have no poles on the unit circle $|z| = 1$. To simplify the notation, we define

$$\begin{cases} \|N\|_{H^p} := \|\ |N(e^{j\theta})|_2\ \|_{L^p[0,2\pi]} , \\ \|V_N\|_{H^p} := \|\ |V_N(e^{j\theta})|_2\ \|_{L^p[0,2\pi]} , \\ \|H_N\|_{H^p} := \|\|\ H_N(e^{j\theta})\|_s\ \|_{L^p[0,2\pi]} , \end{cases} \tag{2.46}$$

for $1 \le p \le \infty$. The reason for using the subscript H^p in the above norms is that if N, V_N, or H_N is scalar-valued, then the above corresponding definition will coincide with the Hardy space with the exception that z must be changed to z^{-1}. Hardy spaces have been introduced briefly in Sect. 1.5.3 and will be studied in detail in the next chapter. For N and V_N, the most important norms in applications are the H^2 and H^∞ norms defined in (2.46). If the H^2 norm is used, then the operator norm in (2.45) of H_N, considered as an operator on H^2, is (Problem 2.28)

$$\|H_N\| = \|H_N\|_{H^\infty} . \tag{2.47}$$

When disturbances must be uniformly rejected, we should use the H^∞-norm for both N and V_N. In this situation, it can be shown that the operator norm of H_N becomes the l^1-norm of the sequence of Fourier coefficients of $\|H_N(e^{j\theta})\|_s = \|\bar{H}_N^\top(e^{j\theta})H_N(e^{j\theta})\|_s^{1/2}$ (Problem 2.29). We denote this norm by $\|H_N\|_A$; and hence, the operator norm for uniform rejection of disturbances is

$$\|H_N\| = \|H_N\|_A := \sum_{i=-\infty}^{\infty} |c_i|, \tag{2.48}$$

where $\{c_i\}$ is the sequence of Fourier coefficients of $\|H_N(e^{j\theta})\|_s$.

Let us return to Research Problem 4. Two norms are to be used:

(i) To optimally reject disturbances measured in the least-squares norm, we may consider

$$\min_{F\in\mathcal{F}} \|W_2(I+PF)^{-1}W_1\|_{H^\infty}. \tag{2.49}$$

(ii) To optimally reject disturbances uniformly, we consider

$$\min_{F\in\mathcal{F}} \|W_2(I+PF)^{-1}W_1\|_A. \tag{2.50}$$

The A-norm is very difficult to handle and so far not much is known, see Vidyasagar [1985]. In addition, it is inconvenient to estimate the inverse operator $(I+PF)^{-1}$. Hence, in many cases, we may reformulate Research Problem 4 by making the following transformation:

$$F = G(I-PG)^{-1}. \tag{2.51}$$

Since $I = (I-PG)(I-PG)^{-1} = (I-PG)^{-1} - PF$, it follows that

$$(I+PF)^{-1} = I - PG.$$

Consider the class

$$\mathcal{G} = \{\, G \in \mathcal{M}^s : \quad W_2(I-PG)W_1 \in \mathcal{M}^s \,\}$$

of matrices in $\mathcal{M}^s$. Then, Research Problem 4 can be reformulated as follows (Problem 2.30):

Research Problem 5. Let $P \in \mathcal{M}$ and $W_1, W_2 \in \mathcal{M}^s$. Determine a $\hat{G}$ in $\mathcal{G}$ such that

$$\|W_2(I-P\hat{G})W_1\| = \inf_{G\in\mathcal{G}} \|W_2(I-PG)W_1\|. \tag{2.52}$$

Note that, in doing so, the inverse of the objective matrix is removed, so that the minimization procedure becomes much simpler. In many applications, a minimization is directly formulated in the form of (2.52) and the above reformulation is not needed. Here again, if the least-squares norm is used to measure disturbances, then the H^∞-norm must be used in (2.52), and if the disturbances must be rejected uniformly then the A-norm has to be used in (2.52).

It should also be noted that, for the scalar-valued setting with $W_2W_1 = 1$, (2.52) becomes the problem of "inverse approximation" to be discussed in the next chapter.

Finally, we remark that once Research Problem 5 has been solved, we have an optimal $\hat{G} \in \mathcal{G}$. To obtain $\hat{F}$ from $\hat{G}$, we may simply implement a subsystem as shown in Fig. 2.8, where the two subsystems are equivalent. This can be easily verified as follows: For any input-output pair (u, v) such that $v = F(u)$, we have, from the connection of the second subsystem, that

$$\begin{cases} w = u + P(y), \\ y = \hat{G}(w), \\ v = \hat{G}(w). \end{cases}$$

Hence,

$$w = u + P\hat{G}(w)$$

or

$$w = (I - P\hat{G})^{-1}(u),$$

so that

$$v = \hat{G}(I - P\hat{G})^{-1}(u) = \hat{F}(u).$$

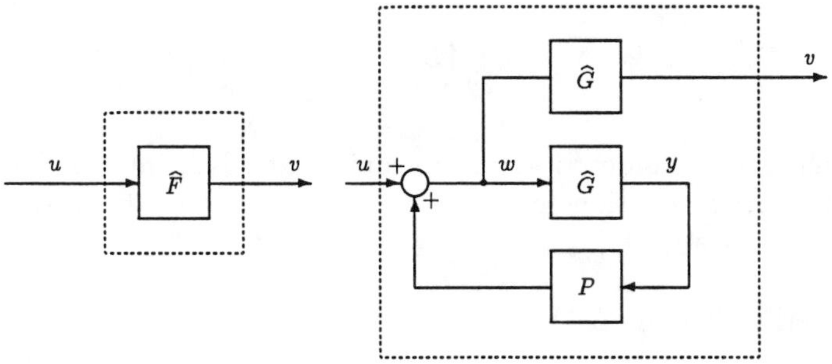

Fig. 2.8. Two equivalent systems

Problems

Problem 2.1. Verify that the MIMO system with the state-space description

$$\begin{cases} \mathbf{x}_{n+1} = A\mathbf{x}_n + B\mathbf{u}_n, \quad \mathbf{x}_0 = 0, \\ \mathbf{v}_n = C\mathbf{x}_n + D\mathbf{u}_n, \end{cases}$$

is linear (in the sense that, for any fixed $\mathbf{u}_n$, the system output is linear in $\mathbf{x}_n$), time-invariant, and causal. What can we say if $\mathbf{x}_0 \neq 0$?

Problem 2.2. (a) Derive the state-transition equation

$$\mathbf{x}_n = A^{n-k}\mathbf{x}_k + \sum_{\ell=k+1}^{n} A^{n-\ell} B\mathbf{u}_{\ell-1}$$

for the MIMO linear system in Problem 2.1.

(b) Suppose that a state vector is moved first from the position $\mathbf{x}_{k_0}$ to position $\mathbf{x}_{k_1}$ and then from $\mathbf{x}_{k_1}$ to $\mathbf{x}_{k_2}$. According to part (a), we have either

$$\begin{cases} \mathbf{x}_{k_1} = A^{k_1-k_0}\mathbf{x}_{k_0} + \displaystyle\sum_{\ell=k_0+1}^{k_1} A^{k_1-\ell} B\mathbf{u}_{\ell-1}\,, \\ \mathbf{x}_{k_2} = A^{k_2-k_1}\mathbf{x}_{k_1} + \displaystyle\sum_{\ell=k_1+1}^{k_2} A^{k_2-\ell} B\mathbf{u}_{\ell-1}\,, \end{cases}$$

or more directly

$$\mathbf{x}_{k_2} = A^{k_2-k_0}\mathbf{x}_{k_0} + \sum_{\ell=k_0+1}^{k_2} A^{k_2-\ell} B\mathbf{u}_{\ell-1}\,.$$

Verify that these two descriptions are equivalent. [This implies that a state-transition equation depends only on the initial and final state vectors of the system.]

Problem 2.3. Verify that the ARMA model

$$\mathbf{v}_n = \sum_{k=1}^{N} B_k \mathbf{v}_{n-k} + \sum_{k=0}^{M} A_k \mathbf{u}_{n-k}$$

(with $M > N$) can be reformulated as the state-space description (2.7). [Hint: Set $\mathbf{x}_n^1 = \mathbf{v}_n - A_0\mathbf{u}_n$ and

$$\mathbf{x}_n = \begin{bmatrix} \mathbf{x}_n^1 \\ \vdots \\ \mathbf{x}_n^N \end{bmatrix}.]$$

Problem 2.4. Prove Theorem 2.1. (See Chui and Chen [1989, Theorem 3.4])

Problem 2.5. Prove Theorem 2.2. (See Chui and Chen [1989, Theorem 4.2])

Problem 2.6. Verify that the state-space description of the system in the example described by (2.1) is both controllable and observable.

Problem 2.7. Construct a linear system which is controllable but not observable, one which is observable but not controllable, and one which is neither controllable nor observable.

Problem 2.8. Supply a proof for Theorem 2.3.

Problem 2.9. Consider the simple ARMA model

$$\mathbf{v}_n = B\mathbf{v}_{n-1} + \mathbf{u}_{n-1},$$

where $\mathbf{v}_n, \mathbf{u}_n \in \mathbf{R}^p$. Show that this linear system is stable if and only if the operator norm $\|B\|$ of B is less than 1.

Problem 2.10. Supply a proof for Theorem 2.4.

Problem 2.11. Consider the causal discrete-time linear system shown in Fig. 2.9a, where the number 4 indicates simply multiplication by 4.

(a) Verify that this system is not stable.
(b) Design a feedback by means of determining the maximum range of the constant c indicated in Fig. 2.9b such that the overall system becomes stable.

Problem 2.12. Consider the discrete-time linear feedback system shown in Fig. 2.10, where the system function $H(z)$ is given by

$$H(z) = \frac{1}{(z-1)(z+\frac{1}{2})}.$$

(a) Find the transfer function of the overall feedback system. Is this feedback system stable?
(b) Show that this system can track the unit step signal

$$x_n = s_n = \begin{cases} 1, & n = 0, 1, 2, \cdots, \\ 0, & \text{otherwise}, \end{cases}$$

in the sense that if $u_n = s_n$ then

$$\lim_{n\to\infty} e_n = 0.$$

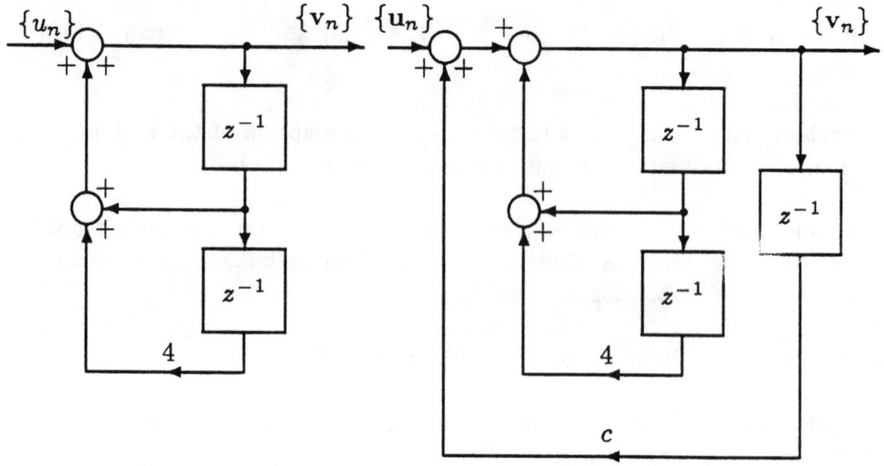

Fig. 2.9. (a) Discrete-time linear system (b) Discrete-time linear system

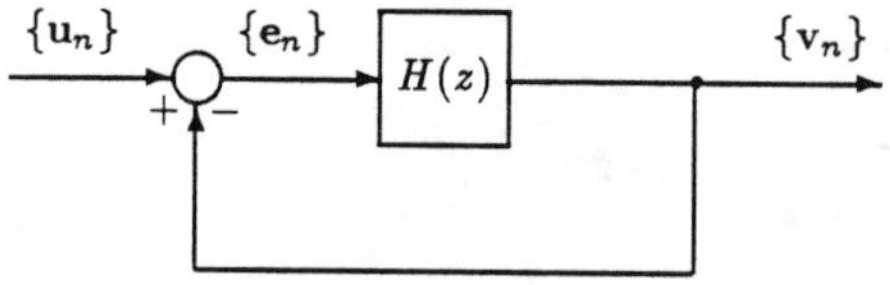

Fig. 2.10. Unity feedback system

Problem 2.13. Consider the discrete-time linear feedback system shown in Fig. 2.11, where the system function $H(z)$ is given by

$$H(z) = \frac{1}{1 - \frac{1}{2}z^{-1}}\,.$$

(a) Find the transfer function of the overall system, and determine the maximum range of the constant c for which the overall system is stable.

(b) Show that the overall feedback system is causal except for the value of c for which the closed-loop pole is at $z = \infty$.

Problem 2.14. Consider the discrete-time linear feedback system shown in Fig. 2.12, with

$$H(z) = \frac{z+1}{z^2 + z + \frac{1}{4}} \quad \text{and} \quad C(z) = \frac{c}{z-1}\,.$$

(a) Find the transfer function of the overall system, and determine the maximum range of the constant c for which the overall system is stable.

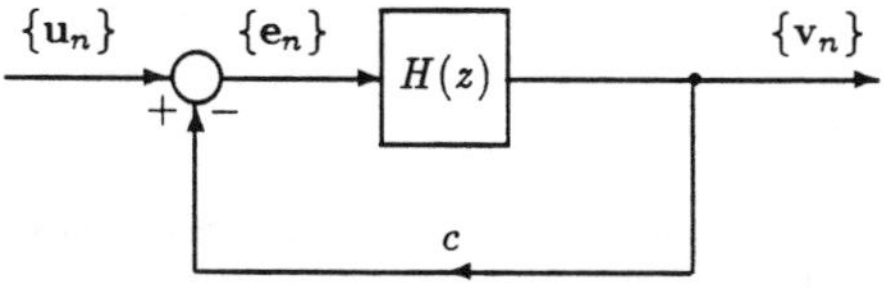

Fig. 2.11. Constant feedback system

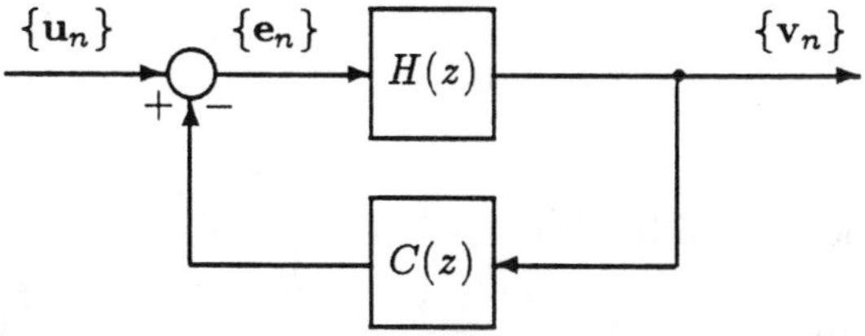

Fig. 2.12. Feedback system described in both time/frequency domains

(b) Show that the sum of the closed-loop poles is independent of the constant c.

Problem 2.15. Give an example to show that the denominator of the transfer matrix of a stable time-invariant linear system may happen to have some zeros in $|z| \geq 1$. Of course, this linear system cannot be both controllable and observable. Why?

Problem 2.16. Let $H(z)$ be a (strictly) proper transfer matrix given by (2.19). Give an example to demonstrate that the identity (2.20) may not hold if the dimension of A is larger than the degree of the denominator of $H(z)$.

Problem 2.17. Let Γ_H be a rank-k infinite Hankel matrix and set $\Gamma_H = [\,\underline{\gamma}_1\ \underline{\gamma}_2\ \cdots\,]$. Show that in the relation

$$\underline{\gamma}_{n+1} = \sum_{i=1}^{k} c_{k-i+1}\underline{\gamma}_{n-i+1}\,, \qquad n = k, k+1, \cdots,$$

in (2.26) the constants c_i's are uniquely determined by

$$\begin{bmatrix} c_1 \\ \vdots \\ c_k \end{bmatrix} = \begin{bmatrix} h_1 & \cdots & h_k \\ \vdots & & \vdots \\ h_k & \cdots & h_{2k-1} \end{bmatrix}^{-1} \begin{bmatrix} h_{k+1} \\ \vdots \\ h_{2k} \end{bmatrix}$$

in (2.27).

Problem 2.18. Give an example to show that for finite Hankel matrices, a formula analogous to Problem 2.17 is not valid.

Problem 2.19. Let

$$H(z) = 1 + az^{-1} + bz^{-2} + az^{-3} + bz^{-4} + \cdots$$

with $a \neq b$ and $a, b \neq 0$. Use (2.31) to find the c_i's and p_i's. Verify the result by summing two geometric series directly as follows:

$$H(z) = 1 + a \sum_{k=1}^{\infty} z^{-(2k-1)} + b \sum_{k=1}^{\infty} z^{-2k} .$$

Problem 2.20. Verify that the infinite Hankel matrix $[h_{i+\ell-1}]$ with $h_i \equiv 1$ for all i has rank 1, and the infinite matrix $[\tilde{h}_{k+\ell-1}]$ with $\tilde{h}_i = h_i + 1/i = 1 + 1/i$ for all i has infinite rank.

Problem 2.21. Convince yourself that the set $\mathcal{H}_m^{pq}$ defined in (2.35) is not a linear manifold.

Problem 2.22. Find a minimal state-space realization of the linear system with transfer function given by

$$H(z) = \frac{1 - \frac{1}{2}z^{-1}}{1 + z^{-1} + \frac{1}{3}z^{-2}} .$$

Problem 2.23. Use the algorithm given in Sect. 2.4.2 to compute the coprime form of

$$H(z) = \frac{-z^4 + z^3 + z^2 + z + 2}{z^4 + 3z^3 + 3z^2 + 3z + 2} .$$

Problem 2.24. Show that the eigenvalues of the $m \times m$ matrix

$$A_\epsilon = \begin{bmatrix} r & 1 & & \\ & \ddots & \ddots & \\ & & \ddots & 1 \\ & & & r \end{bmatrix} + \begin{bmatrix} 0 & 0 & \cdots & 0 \\ \vdots & \vdots & & \vdots \\ 0 & 0 & \cdots & 0 \\ \varepsilon & 0 & \cdots & 0 \end{bmatrix}$$

are given by

$$z_k = r + \sqrt[m]{\varepsilon}\left(\cos\frac{2\pi k}{m} + \mathrm{j}\sin\frac{2\pi k}{m} \right), \quad k = 0, 1, \ldots, m-1 .$$

Problem 2.25. Verify that the transfer matrix of the closed-loop system shown in Fig. 2.4 is given by

$$H(z) = C[zI - (A - BF)]^{-1}B\,.$$

Problem 2.26. Verify that the transfer matrix of the closed-loop system shown in Fig. 2.6 is given by

$$H(z) = W_2(I + PF)^{-1}P\,.$$

Problem 2.27. Verify that the transfer matrix of the closed-loop system shown in Fig. 2.7 is given by

$$H_N(z) = W_2(I + PF)^{-1}W_1\,.$$

Problem 2.28. Let $F(z)$ be a matrix-valued rational function with norm $\|\;\|_{H^\infty}$ defined by (2.46). Show that when F is considered as an operator on H^2, its operator norm, defined by (2.45), is given by

$$\|F\| = \|F\|_{H^\infty}\,.$$

Problem 2.29. Let $F(z)$ be a matrix-valued rational function with the norm $\|\;\|_{H^p}$ defined in (2.46), and let the sequence of Fourier coefficients of $\|F(e^{j\theta})\|_s$ be $\{c_i\}_{i=-\infty}^{\infty}$. Show that the operator norm of F as an operator on H^∞ is given by

$$\|F\| = \|\{c_i\}\|_{l^1} = \sum_{i=-\infty}^{\infty} |c_i|\,.$$

Problem 2.30. Verify, for SISO systems, that under the transformation (2.51), Research Problem 4 in Sect. 2.5.3 reduces to Research Problem 5 in the same section. (See Zames [1981])

3. Approximation in Hardy Spaces

Since the transfer function (or matrix) of a stable linear system is analytic in $|z| \geq 1$, the transformation $z \to z^{-1}$ guarantees its analyticity in the unit disk $|z| \leq 1$. In this chapter we will study approximation by polynomials and "stable" rational functions in $|z| < 1$, using the Hardy spaces H^2 and H^∞ norms. Here, due to the reciprocal transformation, "stability" requires location of poles in $|z| > 1$. Approximation in the Hankel norm will be considered in the next chapter. In order to achieve a firm understanding of the rudiments of these important mathematical tools for optimal system reduction, filter design, model matching, feedback control, and disturbance rejection, etc., we will first consider SISO systems. Standard methods are sometimes applicable to generalize our discussions to MIMO systems which will be studied in Chap. 6. Here and throughout, $\mathcal{P}_n$ will always denote the collection of all polynomials of degree no greater than n and R_n the collection of all rational functions of degree n. More precisely,

$$\mathcal{P}_n = \{p_0 z^n + \cdots + p_{n-1} z + p_n \;:\; p_0, \cdots, p_n \in \mathcal{C}\} \tag{3.1}$$

and

$$R_n = \left\{ \frac{p_0 z^n + \cdots + p_n}{z^n + q_1 z + \cdots + q_n} \;:\; p_0, \cdots, p_n, q_1, \cdots, q_n \in \mathcal{C} \right\}. \tag{3.2}$$

Of course, since some of the coefficients $p_n, p_{n-1}, \cdots$ and $q_n, q_{n-1}, \cdots$ may be zero, multiplying both the numerator and denominator of any $r(z) \in R_n$ by z^{-n} gives a rational function in z^{-1} of the form (1.20) in the ARMA realization of an arbitrary (finite rank but not necessarily stable) SISO linear system (1.21), which may represent a digital filter as has been discussed in Chap. 1.

3.1 Hardy Space Preliminaries

In this section we give a brief description of the Hardy spaces H^p, factorization of H^p functions, and some useful inequalities. Since this is a well-established subject in mathematics, we do not go into details but only refer the reader to Duren [1970] and Hoffman [1962].

3.1.1 Definition of Hardy Space Norms

For an analytic function $f(z)$ in the open unit disk $|z| < 1$, set

$$M_p(r, f) = \begin{cases} \left(\dfrac{1}{2\pi} \displaystyle\int_0^{2\pi} |f(re^{j\theta})|^p d\theta \right)^{1/p} & \text{for } 0 < p < \infty\,, \\ \sup\limits_{|z|=r} |f(z)| & \text{for } p = \infty\,, \end{cases} \tag{3.3}$$

where $0 \le r < 1$. Then $M_p(r, f)$ is a nondecreasing function of r (Problem 3.1), and in fact, if $f(z)$ is not the zero function, then $\ln M_p(r, f)$ is a convex function of $\ln r$, $0 < r < 1$. It is also well known, by a result usually attributed to Fatou who was actually only responsible for the case $p = \infty$, see Hoffman [1962], that if

$$\sup_{0 \le r < 1} M_p(r, f) < \infty\,,$$

then the nontangential limit of $f(z)$ exists almost everywhere as $r \to 1$, and the limit with $z = e^{j\theta}$ is an $L^p[0, 2\pi]$ function, which we will also denote, for convenience, by $f(e^{j\theta})$. In addition, if $f(z)$ is not the zero function, then $\ln |f(e^{j\theta})|$ is in $L^1[0, 2\pi]$.

Hence, by setting

$$\|f\|_{H^p} = \sup_{0 \le r < 1} M_p(r, f)\,, \tag{3.4}$$

we have

$$\begin{aligned} \|f\|_{H^p} &= \lim_{r \uparrow 1} M_p(r, f) \\ &= \begin{cases} \left(\dfrac{1}{2\pi} \displaystyle\int_0^{2\pi} |f(e^{j\theta})|^p d\theta \right)^{1/p} & \text{for } 0 < p < \infty\,, \\ \operatorname{ess\,sup}\limits_{\theta} |f(e^{j\theta})| & \text{for } p = \infty\,. \end{cases} \end{aligned} \tag{3.5}$$

For each $p > 0$, the collection of all functions $f(z)$ analytic in $|z| < 1$ and satisfying

$$\|f\|_{H^p} < \infty$$

is denoted by H^p, where the letter H is used to honor the celebrated British mathematician G. H. Hardy. Endowed with the norm defined in (3.4), each H^p, $1 \le p \le \infty$, is a Banach space which is also called a *Hardy space*. In particular, by using the inner product

$$\langle f, g \rangle = \lim_{r \uparrow 1} \frac{1}{2\pi} \int_0^{2\pi} f(re^{j\theta}) \overline{g(re^{j\theta})} d\theta\,, \tag{3.6}$$

H^2 is, in fact, a Hilbert space. It is also useful to know that $H^s \subset H^r$ for $0 < r < s \le \infty$ (Problem 3.2).

3.1.2 Inner and Outer Functions

A very important property of a Hardy space H^p is that each nontrivial function $f(z)$ in H^p has a *canonical factorization* into the product of an *inner function* and an *outer function.*

Let $f(z)$ be in H^p, $1 \le p \le \infty$, and suppose that $f(z)$ is not identically zero. To take care of the zeros of $f(z)$, we use the finite or infinite *Blaschke product*

$$B(z) = z^m \prod_i \frac{|z_i|}{z_i} \frac{z_i - z}{1 - \bar{z}_i z}, \tag{3.7}$$

where m (which may be zero) is the order of the zero of $f(z)$ at 0, and the z_i's are the other zeros of $f(z)$, listed according to multiplicities. For convenience, let $0 < |z_1| \le |z_2| \le \cdots < 1$. Note that there may be a finite number of z_i's. Then, it is well known that for $f(z)$ to be nontrivial, or equivalently, for the Blaschke product $B(z)$ to be convergent, it is necessary and sufficient that

$$\sum_i (1 - |z_i|) < \infty. \tag{3.8}$$

Now, suppose that (3.8) is satisfied. Then

$$|B(z)| < 1 \qquad \text{for} \quad |z| < 1,$$

and $|B(z)| = 1$ almost everywhere on the unit circle $|z| = 1$. Hence, $f(z)/B(z)$ has no zero in $|z| < 1$, and

$$|f(z)/B(z)| = |f(z)| \tag{3.9}$$

almost everywhere on $|z| = 1$ (Problem 3.3). From the properties that $|f(e^{j\theta})| \in L^p[0, 2\pi]$ and $\ln |f(e^{j\theta})| \in L^1[0, 2\pi]$, we see that the function $F(z)$ defined by

$$F(z) = \exp\left(\frac{1}{2\pi} \int_0^{2\pi} \frac{e^{jt} + z}{e^{jt} - z} \ln |f(e^{jt})| dt\right) \tag{3.10}$$

is a zero-free function in H^p. Hence, it follows that $f(z)/(B(z)F(z))$ is a zero-free analytic function in $|z| < 1$ and satisfies

$$|f(z)/(B(z)F(z))| = 1$$

almost everywhere on $|z| = 1$ (Problem 3.4). It can be shown that there is a finite positive Borel measure μ whose support is a set of Lebesgue measure zero (i.e., μ is singular with respect to the Lebesgue measure) such that

$$\frac{f(z)}{B(z)F(z)} = e^{j\alpha} \exp\left(-\frac{1}{2\pi} \int_0^{2\pi} \frac{e^{jt} + z}{e^{jt} - z} d\mu(t)\right) \tag{3.11}$$

where α is some real constant, see Rudin [1966, p.337]. Hence, we have the canonical factorization

$$f(z) = f_I(z) f_O(z) \tag{3.12}$$

where

$$f_O(z) = e^{j\alpha} F(z) = e^{j\alpha} \exp\left(\frac{1}{2\pi} \int_0^{2\pi} \frac{e^{jt} + z}{e^{jt} - z} \ln |f(e^{jt})| dt \right) \tag{3.13}$$

and

$$f_I(z) = B(z) S(z) \tag{3.14}$$

with

$$S(z) = \exp\left(-\frac{1}{2\pi} \int_0^{2\pi} \frac{e^{jt} + z}{e^{jt} - z} d\mu(t) \right). \tag{3.15}$$

$f_O(z)$ is called an *outer function* [the *outer factor* of $f(z)$], and $f_I(z)$ an *inner function* [the *inner factor* of $f(z)$]. Since the outer factor $f_O(z)$ does not vanish inside $|z| = 1$, all the zeros of $f(z)$ are zeros of the inner factor $f_I(z)$. The inner factor $f_I(z)$ is again factored into the product of a Blaschke product $B(z)$ given by (3.7) and a *singular inner function* $S(z)$ as shown in (3.15).

In general, $f_I(z)$ is said to be an *inner function* if it is the inner factor of some function in H^∞. Hence, any inner function is in H^∞ and has absolute value equal to 1 almost everywhere on $|z| = 1$.

Example 3.1. Consider the function $S_\delta(z)$ defined by (3.15) with the delta distribution $d\mu(t) = \delta(t)dt$, namely,

$$S_\delta(z) = \exp\left(-\frac{1}{2\pi} \int_0^{2\pi} \frac{e^{jt} + z}{e^{jt} - z} \delta(t) dt \right).$$

It can be easily verified that $S_\delta(z)$ is a singular inner function given by

$$S_\delta(z) = \exp\left(-\frac{1 + z}{1 - z} \right),$$

which is zero-free in $|z| < 1$ and satisfies

$$|S_\delta(e^{j\theta})| = \begin{cases} 1 & \text{if } e^{j\theta} \neq 1, \\ 0 & \text{if } e^{j\theta} = 1. \end{cases}$$

Recall that the radial limit (as $r \uparrow 1$) is used to define $S_\delta(e^{j\theta})$.

On the other hand, an outer function $f_O(z)$ is one which is defined by a positive measurable function $\varphi(e^{jt})$ such that $\ln \varphi(e^{jt})$ is in $L^1[0, 2\pi]$; namely,

$$f_O(z) = e^{j\alpha} \exp\left(\frac{1}{2\pi}\int_0^{2\pi} \frac{e^{jt}+z}{e^{jt}-z} \ln \varphi(e^{jt}) dt\right),$$

where α is a real constant. The outer function $f_O(z)$ is in H^p, $0 < p \leq \infty$, if and only if $\varphi(e^{jt})$ is in $L^p[0, 2\pi]$ for the same value of p. Furthermore, $f_O(z)$ is zero-free in $|z| < 1$, with

$$|f_O(e^{j\theta})| = \varphi(e^{j\theta})$$

almost everywhere (Problem 3.5).

Example 3.2. The function

$$f_a(z) = (1 - z)^{-a},$$

where $0 < a < \infty$ and the branch is chosen such that $f_a(0) = 1$, is an outer function with

$$\varphi_a(e^{jt}) = \frac{1}{2^a}\left|\sin\frac{t}{2}\right|^{-a}.$$

Hence, $f_a(z)$ is in H^p for $0 < p < 1/a$.

3.1.3 The Hausdorff-Young Inequalities

We will introduce in this section two useful inequalities. Since any function $f(z)$ analytic in $|z| < 1$ has a power series expansion

$$f(z) = \sum_{n=0}^{\infty} a_n z^n$$

for $|z| < 1$, it is important to obtain a relationship between a function norm such as the H^p norm of $f(z)$ and a sequence norm such as some ℓ^q norm of its corresponding sequence of coefficients $\{a_n\}$. For the Hilbert space setting, we have the relationship (Problem 3.6)

$$\|f\|_{H^2} = \|\{a_n\}\|_{l^2}. \tag{3.16}$$

However, for H^p and l^q where $p,\ q \neq 2$, the situation is not so simple. If p and q are conjugates of each other in the sense that $1 \leq p \leq 2$ and $1/p + 1/q = 1$, then for a function $f(z)$ in H^p, we have

$$\|\{a_n\}\|_{l^q} \leq \|f\|_{H^p}, \tag{3.17}$$

and for a sequence $\{a_n\}$ in l^p, we have

$$\|f\|_{H^q} \leq \|\{a_n\}\|_{l^p}. \tag{3.18}$$

These results are usually called the *Hausdorff-Young inequalities*, see, for example, Zygmund [1968, Vol.2, p.101].

Example 3.3. It is well known that

$$\sum_{n=1}^{\infty} \frac{1}{n^{\alpha}} < \infty$$

if and only if $\alpha > 1$. Consequently, the function

$$f(z) = \sum_{n=1}^{\infty} \frac{1}{n} e^{j\theta_n} z^n ,$$

where the θ_n's are real numbers, is in H^p for any p with $2 \leq p < \infty$ (Problem 3.7).

3.2 Least-Squares Approximation

In this section we will discuss approximation in the Hilbert space H^2 norm. Approximation in this norm is usually called *least-squares approximation.* We will begin with a general theorem due to Beurling which includes both "direct" and "inverse" approximations by polynomials. As an application, we introduce an all-pole filter design procedure that yields an optimal pole selection criterion for an optimal pole-zero design method. The procedure of double least-squares inverses will also be discussed. This procedure can be modified to yield an efficient stabilization algorithm.

3.2.1 Beurling's Approximation Theorem

One of the most basic results in the subject of approximation in the H^p norm is the so-called *Beurling's Approximation Theorem.*

Theorem 3.1. (Beurling)

Let $f(z)$ and $g(z)$ be nontrivial functions in H^p with inner factors $f_I(z)$ and $g_I(z)$, respectively, where $1 \leq p < \infty$. Then

$$\inf_{p_n \in \mathcal{P}_n} \|g - f p_n\|_{H^p} \longrightarrow 0 \tag{3.19}$$

as $n \to \infty$ if and only if $g_I(z)/f_I(z)$ is an inner function, where $\mathcal{P}_n$ is the family of polynomials of degree no greater than n as defined in (3.1).

Note that the minimization in (3.19) is, in some sense, a special case of Research Problem 5 posed in Sect. 2.5.3.

We now discuss the proof of Beurling's approximation theorem. The necessity of the theorem is at least intuitively clear. Indeed, setting $f(z) = f_O(z)f_I(z)$ and $g(z) = g_O(z)g_I(z)$, and assuming that (3.19) holds, then since

$$\|g - fp_n\|_{H^p} = \|g_O h - f_O p_n\|_{H^p},$$

where $h(z) = g_I(z)/f_I(z)$, and since convergence in H^p implies uniform convergence on every compact subset of $|z| < 1$, we conclude that $g_O(z)h(z)$ is analytic in $|z| < 1$, so that it is in H^p, where $1 \le p < \infty$ as indicated in the statement of the theorem. Also, since $g_O(z)$ is zero-free in $|z| < 1$, $h(z)$ must also be analytic in $|z| < 1$. Hence, from the fact that the radial limit of $|h(z)|$ as $|z| \uparrow 1$ is 1 almost everywhere on $|z| = 1$ and that $g_O(z)h(z) \in H^p$, it is not difficult to show that $h(z) = g_I(z)/f_I(z)$ is an inner function (Problem 3.8).

To verify the sufficiency of the theorem, we simply apply the Hahn-Banach theorem. Indeed, if (3.19) does not hold but $h(z) = g_I(z)/f_I(z)$ is an inner function, then from

$$\inf_{p_n \in P_n} \|g_O h - f_O p_n\|_{H^p} \ge \varepsilon > 0$$

for all n, we can "separate"

$$z^n f_O(z), \quad n = 0, 1, \cdots$$

from $g_O(z)h(z)$ by a nontrivial bounded linear functional on H^p, in the sense that there exists a function $k(z) \in L^q(|z| = 1)$, where $1/p + 1/q = 1$, such that

$$\text{(i)} \quad \int_{|z|=1} z^n f_O(z)k(z)\frac{dz}{z} = 0, \quad n = 0, 1, \cdots,$$

but

$$\text{(ii)} \quad \int_{|z|=1} g_O(z)h(z)k(z)\frac{dz}{z} \ne 0.$$

From (i), it is clear that $f_O(z)k(z)/z$ is analytic in $|z| < 1$, so that $k(z)/z$ is also analytic in $|z| < 1$. However, from (ii), it follows from Cauchy's theorem that $k(z)/z$ cannot be analytic in $|z| < 1$. This contradiction implies that (3.19) must hold whenever $h(z) = g_I(z)/f_I(z)$ is an inner function, completing the proof of the theorem.

There are two important consequences:

(1) If $f(z) \equiv 1$, then since $g_I(z)/f_I(z) = g_I(z)$ is already an inner function, we always have

$$\inf_{p_n \in P_n} \|g - p_n\|_{H^p} \longrightarrow 0 . \tag{3.20}$$

In other words, polynomials are "dense" in H^p.

(2) For $g(z) \equiv 1$, since $1/f_I(z)$ is an inner function if and only if $f_I(z) \equiv 1$ (Problem 3.8), we may conclude that a necessary and sufficient condition for

$$\inf_{p_n \in P_n} \|1 - f p_n\|_{H^p} \longrightarrow 0 \tag{3.21}$$

is that $f(z)$ is an outer function.

The approximation property in (3.20) may be called "direct approximation," while the approximation property in (3.21) will be called "inverse approximation." Note that if (3.21) holds, then

$$\frac{1}{p_n(z)} \longrightarrow f(z)$$

uniformly on $|z| \leq r_0$ for every $r_0 < 1$. Hence, (3.21) suggests a method for auto-regressive (or AR) filter design. The most efficient inverse approximation procedure is achieved, of course, by choosing $p = 2$.

3.2.2 An All-Pole Filter Design Method

Recall from Section 1.5 that if an ideal amplitude filter characteristic is given, one can raise the stopband by a small $\varepsilon > 0$, so that its transfer function [see (1.22)] becomes an outer function in H^∞. In fact, the reciprocal of this function is also an H^∞ outer function. We have the following more general AR filter design procedure, see Chui and Chan [1982].

Theorem 3.2. Let

$$f(z) = \sum_{n=0}^{\infty} a_n z^n$$

be an outer function in H^2. Then there is a unique polynomial $\hat{p}_n(z) = c_0 + \cdots + c_n z^n$ in P_n, such that

$$\|1 - f\hat{p}_n\|_{H^2} = \inf_{p_n \in P_n} \|1 - f p_n\|_{H^2} . \tag{3.22}$$

Furthermore, all the zeros of $\hat{p}_n(z)$ lie in $|z| > 1$, and

$$\frac{1}{\hat{p}_n(z)} \longrightarrow f(z)$$

uniformly on $|z| \leq r_0$ for every $r_0 < 1$. Moreover, $\hat{p}_n(z)$ can be computed by using the formula

$$\begin{bmatrix} d_0 & \bar{d}_1 & \cdots & \bar{d}_n \\ d_1 & \ddots & \ddots & \vdots \\ \vdots & \ddots & \ddots & \bar{d}_1 \\ d_n & \cdots & d_1 & d_0 \end{bmatrix} \begin{bmatrix} c_0 \\ c_1 \\ \vdots \\ c_n \end{bmatrix} = \begin{bmatrix} \bar{a}_0 \\ 0 \\ \vdots \\ 0 \end{bmatrix}, \tag{3.23}$$

where

$$d_k = \frac{1}{2\pi}\int_0^{2\pi} |f(e^{j\theta})|^2 e^{-jk\theta} d\theta = \sum_{i=0}^{\infty} a_{k+i}\bar{a}_i \tag{3.24}$$

for $k = 0, 1, \cdots, n$.

To prove this theorem, we introduce a positive measure

$$d\mu(t) = \frac{1}{2\pi}|f(e^{jt})|^2 dt,$$

so that the quantity $||1 - fp_n||_{H^2}$ may be reformulated as

$$||1 - fp_n||_{H^2} = \left\| \frac{1}{f} - p_n \right\|_{H^2(d\mu)},$$

where $H^2(d\mu)$ denotes the Hilbert space of functions $g(z)$ and $h(z)$ analytic in $|z| < 1$ with the inner product defined by

$$\langle g, h\rangle_\mu = \lim_{r\uparrow 1}\int_0^{2\pi} g(re^{jt})\overline{h(re^{jt})}d\mu(t),$$

such that $|\langle g, h\rangle_\mu| \le ||g||_{H^2(d\mu)}||h||_{H^2(d\mu)} < \infty$. Then using a standard Hilbert space argument, we conclude that $\hat{p}_n(z)$ exists, is unique, and satisfies

$$\left\langle \frac{1}{f} - \hat{p}_n, z^i \right\rangle_\mu = 0, \qquad i = 0, 1, \cdots, n.$$

It is clear that this condition is equivalent to (Problem 3.9)

$$\int_0^{2\pi} \hat{p}_n(e^{jt})e^{-jit}d\mu(t) = \begin{cases} 0 & \text{for } i = 1, \cdots, n, \\ \overline{f(0)} & \text{for } i = 0. \end{cases} \tag{3.25}$$

Next, consider the reciprocal polynomial $\hat{q}_n(z)$ of $\hat{p}_n(t)$ defined by

$$\hat{q}_n(z) = z^n \overline{\hat{p}_n\left(\frac{1}{\bar{z}}\right)} = \bar{c}_n + \cdots + \bar{c}_0 z^n.$$

Then $\hat{q}_n(z)$ satisfies (Problem 3.10)

$$\langle \hat{q}_n, z^i\rangle_\mu = 0, \qquad i = 0, 1, \cdots, n-1,$$

so that $\hat{q}_n(z)$ is an nth degree orthogonal polynomial with respect to the measure $d\mu(t)$. By using a classical result on orthogonal polynomials, see Szegö [1967], we may conclude that all the zeros of $\hat{q}_n(z)$ lie in $|z| < 1$, or equivalently, all the zeros of $\hat{p}_n(z)$ lie in $|z| > 1$. Of course, Theorem 3.1 implies that $\|1 - f\hat{p}_n\|_{H^2} \to 0$, so that $1/\hat{p}_n(z) \to f(z)$ uniformly on $|z| \le r_0$ for any $r_0 < 1$. Finally, it can be verified that (3.23) is equivalent to (3.25) with d_i's defined by (3.24) (Problem 3.11). This completes the proof of the theorem.

Note that the first Taylor coefficient a_0 of $f(z)$ can never be zero because $f(z)$, being an outer function, is zero-free in $|z| < 1$. Hence, (3.23) always has a (unique) nontrivial solution. In most applications, the Taylor coefficients a_k's of $f(z)$ are not known. Hence, the constants d_k's in (3.24) must be calculated by using the FFT and information on the square of the "magnitude spectrum" $|f(e^{j\theta})|$. Since the coefficient matrix in (3.23) is Toeplitz, the best inverse polynomial approximant $\hat{p}_n(t)$ can be calculated quite efficiently.

3.2.3 A Pole-Zero Filter Design Method

In digital filter design, $f(z)$ may be chosen as the Herglotz transform of a "raised" magnitude spectrum $|H_\epsilon(e^{-j\omega})|$, see Chui and Chan [1982]. Hence, the least-squares inverse scheme described in (3.23) and (3.24) yields a stable recursive digital filter with transfer function given by

$$H(z) = \frac{1}{\hat{p}(z^{-1})} = \frac{1}{c_0 + \cdots + c_n z^{-n}}.$$

The main objection to this efficient design procedure is, perhaps, that only an AR realization is obtained. However, by using any (standard or modified) zero-search algorithm, the (approximate) location of the n (distinct) zeros of $\hat{p}_n(z)$ can be used as poles of a stable ARMA model, which can be achieved by using the following result:

Let $z_1, \cdots, z_n$ be n distinct points in $|z| > 1$, and define

$$R_n(z_1, \cdots, z_n) = \{f(z) \in R_n \ : \ f(z) \text{ has poles at } z_1, \cdots, z_n\}. \tag{3.26}$$

Recall that the inversion $z \to z^{-1}$ takes all poles of $f(z)$ into the unit disk $|z| < 1$. Then we have the following, see Walsh [1960]:

Theorem 3.3. For every $f(z)$ in H^2 there exists a unique $\hat{r}_n(z)$ in $R_n(z_1, \cdots, z_n)$ such that

$$\|\hat{r}_n - f\|_{H^2} = \inf_{r_n \in R_n(z_1, \cdots, z_n)} \|r_n - f\|_{H^2}.$$

Furthermore, $\hat{r}_n(z)$ is uniquely determined by the interpolation conditions

$$\hat{r}_n(0) = f(0), \quad \hat{r}_n\left(\frac{1}{\bar{z}_i}\right) = f\left(\frac{1}{\bar{z}_i}\right), \quad i = 1, \cdots, n; \tag{3.27}$$

or equivalently, $\hat{r}_n(z)$ can be written as

$$\hat{r}_n(z) = b_0 + \sum_{i=1}^{n} \frac{b_i z_i}{z_i - z}$$

where the coefficients $b_0, \cdots, b_n$ can be computed by solving the following system of equations:

$$\begin{cases} b_0 + \cdots + b_n = f(0)\,, \\ b_0 + \displaystyle\sum_{i=1}^{n} \frac{z_i \bar{z}_k}{z_i \bar{z}_k - 1} b_i = f\Big(\frac{1}{\bar{z}_k}\Big), \quad k = 1, \cdots, n\,. \end{cases} \tag{3.28}$$

We first note that $R_n(z_1, \cdots, z_n)$ is a linear space with basis

$$\left\{1, \frac{1}{z_1 - z}, \cdots, \frac{1}{z_n - z}\right\}. \tag{3.29}$$

Hence, the best approximant $\hat{r}_n(z)$ of $f(z)$ in H^2 from $R_n(z_1, \cdots, z_n)$ exists and is unique. This theorem says that the best approximant $\hat{r}_n(z)$ is also an interpolant of $f(z)$ at the reflections $1/\bar{z}_k$ of z_k across the unit circle $|z| = 1$. The linear equations (3.28) give an efficient computational scheme for $\hat{r}_n(z)$ assuming that the values of $f(1/\bar{z}_k)$, $k = 1, \cdots, n$, are available. Of course, Taylor's formula

$$f\Big(\frac{1}{\bar{z}_k}\Big) = \sum_{i=0}^{\infty} a_i (\bar{z}_k)^{-i}$$

may be used to give approximate values of $f(1/\bar{z}_k)$, where the coefficients

$$a_i = \frac{1}{2\pi} \int_0^{2\pi} f(e^{jt}) e^{-jit} dt\,, \qquad i = 0, 1, \ldots\,,$$

can be estimated by using the FFT.

To prove the theorem, the usual Hilbert space argument can be used. Indeed, $\hat{r}_n(z)$ is uniquely determined by the orthogonality conditions

$$\frac{1}{2\pi j} \int_{|z|=1} (\hat{r}_n(z) - f(z)) \overline{g(z)} \frac{dz}{z} = 0\,, \tag{3.30}$$

where $g(z) = 1, (z_1 - z)^{-1}, \cdots, (z_n - z)^{-1}$. By applying Cauchy's formula, these orthogonality conditions can be translated into the interpolation conditions (3.27) (Problem 3.15). It should be clear that the interpolation conditions in (3.27) are equivalent to the system of linear equations (3.28) (Problem 3.15). This completes the proof of Theorem 3.3.

Example 3.4. Consider the transfer function $H(z) = f(z^{-1})$, where $f(z) = 5/(z-5)$, with $z_1 = 2$ and $z_2 = 3$. Design an ARMA filter $R(z) = \hat{r}(z^{-1})$ where $\hat{r}(z)$ is the best approximant to $f(z)$ in the H^2 norm from $R_2(z_1, z_2)$.

To derive $\hat{r}(z)$, we simply write

$$\hat{r}(z) = b_0 + \frac{b_1 z_1}{z_1 - z} + \frac{b_2 z_2}{z_2 - z} = b_0 + \frac{2b_1}{2-z} + \frac{3b_2}{3-z}\,.$$

and apply (3.28) to yield

$$\begin{cases} b_0 + b_1 + b_2 = -1\,, \\ b_0 + \dfrac{4}{3}b_1 + \dfrac{6}{5}b_2 = -\dfrac{10}{9}\,, \\ b_0 + \dfrac{6}{5}b_1 + \dfrac{9}{8}b_2 = -\dfrac{15}{14}\,, \end{cases}$$

or

$$\begin{cases} b_0 = -1.0100602\,, \\ b_1 = +0.6673392\,, \\ b_2 = -0.6572790\,. \end{cases}$$

Hence,

$$R(z) = \hat{r}(z^{-1}) = -1.010 + \frac{1.335}{2 - z^{-1}} - \frac{1.972}{3 - z^{-1}}\,.$$

3.2.4 A Stabilization Procedure

Let us return to the discussion of inverse approximation in Sect. 3.2.2. In (3.22), the best inverse approximant $\hat{p}_n(z)$ of $f(z)$ in the H^2 norm is usually called the *least-squares inverse* (LSI) of $f(z)$ from $\mathcal{P}_n$. With $f(z) = f_m(z) = a_0 + \cdots + a_m z^m$ itself a polynomial (of degree m with m larger than n), the LSI $\hat{p}_n(z) = c_0 + \cdots + c_n z^n$ of $f_m(z)$ from $\mathcal{P}_n$ was first introduced by Robinson to obtain a minimal-delay finite-length "wavelet" $(c_0, \cdots, c_n)$ whose convolution with a given "wavelet" $(a_0, \cdots, a_m)$ produces a best approximation of the "unit spike" $(1, 0, \cdots, 0)$ in the l^2 norm, see Robinson [1967, pp.167-174] and Problem 3.17. Later Shanks [1967] suggested taking the LSI of $\hat{p}_n(z)$ from $\mathcal{P}_m$ to produce $\hat{f}_m(z)$, namely, the LSI of the given $f_m(z)$, in order to "stabilize" $f_m(z)$. Note that we have $\hat{f}_m(z) \neq 0$ for $|z| \leq 1$ even if the original $f_m(z)$ may have zeros in $|z| < 1$. He observed that $|\hat{f}_m(e^{j\theta})|$ and $|f_m(e^{j\theta})|$ are approximately the same and called $\hat{f}_m(z)$ a *double least-squares inverse* (DLSI) of $f_m(z)$. Hence, if Shanks' observation were valid, then the DLSI method, which is a *linear* procedure, would definitely provide a useful scheme for stabilizing (the denominator of) an

ARMA model without changing its magnitude spectrum. In the following, we will see that Shanks' suggestion requires some minor modification.

Let $f_m(z)$ be in $\mathcal{P}_m$ with $f_m(0) \neq 0$ and $\hat{p}_n(z)$ be the LSI of $f_m(z)$ from $\mathcal{P}_n$. Also, let $\hat{f}_{m,n}(z) \in \mathcal{P}_m$ be the LSI of $\hat{p}_n(z)$ from $\mathcal{P}_m$. We say that "$\hat{f}_{m,n}(z)$ is the DLSI of $f_m(z)$ through $\mathcal{P}_n$". We first state the following result obtained in Chui [1980]:

Theorem 3.4. Let $f_m(z) \in \mathcal{P}_m$ with $f_m(0) \neq 0$, and let $\hat{f}_{m,n}(z) \in \mathcal{P}_m$ be the DLSI of $f_m(z)$ through $\mathcal{P}_n$. Then $\hat{f}_{m,n}(z) \to f_m(z)$ as $n \to \infty$ if and only if $f_m(z)$ does not vanish in $|z| < 1$.

In other words, if $\hat{f}_m(z)$ is already "stable," then the DLSI procedure does not change $f_m(z)$, which may be the denominator of an ARMA model, provided that n is chosen sufficiently large. However, if the given polynomial $f_m(z)$ has at least one zero in $|z| < 1$, then Shanks' observation is not correct. Here, since $\mathcal{P}_m$ is a finite-dimensional space, coefficient-wise convergence is equivalent to convergence in any H^p norm, and hence, it is not necessary to specify the type of convergence.

To prove the theorem, we first assume that $\hat{f}_{m,n}(z) \to f_m(z)$ uniformly on $|z| \leq r < 1$ as $n \to \infty$. Then, since $\hat{f}_{m,n}(z) \neq 0$ for $|z| < 1$ and $f_m(0) \neq 0$, we must have $f_m(z) \neq 0$ for $|z| < 1$. Conversely, suppose that $f_m(z)$ does not vanish anywhere in $|z| < 1$. Then $\ln |f_m|$ is a harmonic function there, so that

$$|f_m(0)| = \exp\left(\frac{1}{2\pi}\int_{-\pi}^{\pi} \ln |f_m(e^{j\theta})| d\theta\right).$$

Let $\{\phi_k\}$ be the orthonormal polynomials on the unit circle $|z| = 1$ with respect to the measure $d\mu(\theta) = \frac{1}{2\pi}|f_m(e^{j\theta})|^2 d\theta$. Then it is well known, see Szegö [1967, p.303], that

$$\frac{1}{|f_m(0)|^2} = \sum_{i=0}^{\infty} |\phi_i(0)|^2 . \tag{3.31}$$

Since $1/f_m(z)$ is analytic in $|z| < 1$ and is in $H^2(d\mu)$, it is the limit in $H^2(d\mu)$ of its Fourier series

$$\sum_{i=0}^{\infty} \left\langle \frac{1}{f_m}, \phi_i \right\rangle_{\mu} \phi_i .$$

Now, $\hat{p}_n$ is the best approximant of $1/f_m$ from $\mathcal{P}_n$ in $H^2(d\mu)$, see (3.22), and hence we have

$$\hat{p}_n = \sum_{i=0}^{n} \left\langle \frac{1}{f_m}, \phi_i \right\rangle_{\mu} \phi_i .$$

Therefore, if follows from (3.31) that

$$\begin{aligned}\|1 - f_m\hat{p}_n\|_{H^2}^2 &= \left\|\frac{1}{f_m} - \hat{p}_n\right\|_\mu^2 \\ &= \sum_{i=n+1}^{\infty} \left|\left\langle \frac{1}{f_m}, \phi_i\right\rangle_\mu\right|^2 \\ &= \sum_{i=n+1}^{\infty} \left|\frac{1}{2\pi}\int_{-\pi}^{\pi} f_m(e^{j\theta})\overline{\phi_i(e^{j\theta})}d\theta\right|^2 \\ &= \sum_{i=n+1}^{\infty} \left|\frac{1}{2\pi}\int_{-\pi}^{\pi} (f_m\phi_i)(e^{j\theta})d\theta\right|^2 \\ &= |f_m(0)|^2 \sum_{i=n+1}^{\infty} |\phi_i(0)|^2 \to 0 \end{aligned} \tag{3.32}$$

as $n \to \infty$. Next, since $\hat{f}_{m,n}$ is the LSI of $\hat{p}_n$ in $\mathcal{P}_m$ and $f_m \in \mathcal{P}_m$, we have

$$\begin{aligned}\|f_m - \hat{f}_{m,n}\|_{H^2} &\le \|(1 - \hat{f}_{m,n}\hat{p}_n)f_m\|_{H^2} + \|(1 - \hat{p}_n f_m)\hat{f}_{m,n}\|_{H^2} \\ &\le \|1 - \hat{f}_{m,n}\hat{p}_n\|_{H^2}\|f_m\|_{H^\infty} + \|1 - \hat{p}_n f_m\|_{H^2}\|\hat{f}_{m,n}\|_{H^\infty} \\ &\le \|1 - f_m\hat{p}_n\|_{H^2}(\|f_m\|_{H^\infty} + \|\hat{f}_{m,n}\|_{H^\infty}) . \end{aligned} \tag{3.33}$$

In addition, since all H^p norms for $\mathcal{P}_m$ are equivalent, there exists a constant c such that $\|\hat{f}_{m,n}\|_{H^\infty} \le c\|\hat{f}_{m,n}\|_{H^2}$. It then follows from (3.33) that

$$\begin{aligned}\|\hat{f}_{m,n}\|_{H^\infty} &\le c\|\hat{f}_{m,n}\|_{H^2} \\ &\le c\|f_m\|_{H^2} + c\|f_m - \hat{f}_{m,n}\|_{H^2} \\ &\le c\|f_m\|_{H^2} + c\|1 - f_m\hat{p}_n\|_{H^2}(\|f_m\|_{H^\infty} + \|\hat{f}_{m,n}\|_{H^\infty}) ,\end{aligned}$$

which implies that

$$(1 - c\|1 - \hat{p}_n f_m\|_{H^2})\|\hat{f}_{m,n}\|_{H^\infty} \le c\|f_m\|_{H^2} + c\|1 - \hat{p}_n f_m\|_{H^2}\|f_m\|_{H^\infty} ,$$

so that by using (3.32), we obtain

$$\overline{\lim_{n\to\infty}} \|\hat{f}_{m,n}\|_{H^\infty} \le c\|f_m\|_{H^2} .$$

This, together with (3.33), shows that there exists a constant $\alpha > 0$ such that

$$\|f_m - \hat{f}_{m,n}\|_{H^2} \le \alpha\|1 - \hat{p}_n f_m\|_{H^2} . \tag{3.34}$$

It follows now from (3.32) that $\hat{f}_{m,n} \to f_m$ as $n \to \infty$, completing the proof of the theorem.

It should be remarked that if the Beurling approximation theorem is applied, then the proof of the above theorem can be shortened by starting from (3.33). The reason for using the identity (3.31) is that the exact error of inverse approximation to the outer function $f_m(z)$ has an explicit formulation as given in (3.32).

In order to apply the DLSI procedure for stabilizing an ARMA filter, Shanks' suggestion must be modified. So, let us now consider the important situation when $f_m(z)$ has at least one zero in $|z| < 1$. In this case, we write $f_m(z) = (z_1 - z) \cdots (z_k - z) g_{m-k}(z)$, where $0 < |z_i| < 1$, $i = 1, \cdots, k$, and $g_{m-k}(z)$ is in $\mathcal{P}_{m-k}$ with $g_{m-k}(z) \neq 0$ for all z with $|z| < 1$. Then set

$$\tilde{f}_m(z) = \left(\frac{1}{\bar{z}_1} - z\right) \cdots \left(\frac{1}{\bar{z}_k} - z\right) g_{m-k}(z).$$

That is, let $\tilde{f}_m(z)$ be obtained from $f_m(z)$ by replacing the zeros of $f_m(z)$ in $|z| < 1$ with their reflections across the unit circle $|z| = 1$. As before, let $\hat{f}_{m,n}(z) \in \mathcal{P}_m$ be the DLSI of $f_m(z)$ through $\mathcal{P}_n$. Then we obtain the following result, see Chui [1980]:

Theorem 3.5. $\hat{f}_{m,n}(z) \to \tilde{f}_m(z)$ as $n \to \infty$.

This implies that, instead of reproducing $f_m(z)$, the DLSI $\hat{f}_{m,n}(z)$ of $f_m(z)$ "eventually" stabilizes $f_m(z)$ by replacing its zeros in $|z| < 1$ with their corresponding reflections across the unit circle $|z| = 1$. Here, again, since $\mathcal{P}_m$ is a finite-dimensional space, it is not necessary to specify the type of convergence.

So, what happens to the magnitude spectrum? To answer this question, we may investigate the ratio $f_m(z)/\tilde{f}_m(z)$. Let

$$B_k(z) = \prod_{i=1}^{k} \frac{|z_i|}{z_i} \frac{z_i - z}{1 - \bar{z}_i z}$$

be the Blaschke product of the zeros $z_1, \cdots, z_k$ of $f_m(z)$ that lie in $|z| < 1$, and set $c = |z_1 \cdots z_k|$. Then, since $0 < |z_i| < 1$, $i = 1, \cdots, k$, we have $0 < c < 1$. It is clear that

$$\frac{f_m(z)}{\tilde{f}_m(z)} = c B_k(z).$$

Since $|B_k(z)| = 1$ when $|z| = 1$, an immediate consequence of the above theorem is the following (Problem 3.18).

Corollary 3.1. $|\hat{f}_{m,n}(z)| \to \frac{1}{c}|f_m(z)|$, where $|z| = 1$, as $n \to \infty$.

This implies that the *shape* of the magnitude spectrum of a polynomial $f_m(z)$ with $f_m(0) \neq 0$ is "eventually" preserved by taking its DLSI, but if $f_m(z)$ has at least one zero in $|z| < 1$, then the DLSI must be scaled down by multiplying by a constant c to yield the desired result. Hence, Shanks' suggestion of using the linear computation scheme of DLSI provides a very efficient method for stabilizing (the denominator of) an ARMA realization, provided that a suitable multiplicative constant c, $0 < c < 1$, can be determined. In digital filter design, this constant can usually be obtained easily. For instance, if the filter characteristic is known at a certain frequency, then the value of the amplitude characteristic produced by DLSI should be adjusted accordingly.

Example 3.5. Consider a given (unstable) transfer function

$$H(z) = \frac{0.082 + 0.078z^3}{1 - 2.9z + 2.9z^2 - 1.16z^3},$$

which represents a low-pass filter with cutoff frequency at 0.25π.

For this transfer function, we may apply the DLSI to the denominator $q_3(z) = 1 - 2.9z + 2.9z^2 - 1.16z^3$ through $\mathcal{P}_n$, that is, we must solve (3.28) twice. With $n = 10$, we arrive at the denominator

$$\hat{q}_3(z) = 1 - 2.13z + 1.78z^2 - 0.54z^3,$$

yielding the transfer function

$$\tilde{H}(z) = \frac{0.082 + 0.078z^3}{1 - 2.13z + 1.78z^2 - 0.54z^3}.$$

Next, since we want to pass the zero frequency, we require the transfer function to have absolute value 1 at $z = 1$. Hence, from $|\tilde{H}(1)| = 0.16/0.11$, we obtain the transfer function of the desired stable filter

$$\hat{H}(z) = \frac{0.11}{0.16}\tilde{H}(z) = \frac{0.0563 + 0.0598z^3}{1 - 2.13z + 1.78z^2 - 0.54z^3}.$$

3.3 Minimum-Norm Interpolation

A problem which may be considered as a generalization of Beurling's approximation theorem to rational approximation is posed in this section. It turns out that the problem of minimum-norm interpolation is also a special case of this problem.

3.3.1 Statement of the Problem

Let R^a denote the collection of all rational functions which are bounded at ∞ and have all their poles in $|z| > 1$, that is,

$$R^a = \left\{ f(z) = \frac{p_0 z^n + p_1 z^{n-1} + \cdots + p_n}{z^n + q_1 z^{n-1} + \cdots + q_n} : \right.$$
$$\left. f(z) \text{ analytic on } |z| \le 1, n = 0, 1, \cdots \right\}. \tag{3.35}$$

It is clear that R^a is a linear space (with infinite dimension), and in fact, R^a is an algebra. The following problem is an SISO version of Research Problem 5 discussed in Sect. 2.5.3.

Research Problem 6. Let $r(z)$ and $g(z)$ be functions in R^a. Determine an $\hat{f}(z)$ in R^a such that

$$\|g - r\hat{f}\|_{H^p} = \inf_{f \in R^a} \|g - rf\|_{H^p},$$

where $1 \le p \le \infty$.

In the special case where $r(z) \neq 0$ for $z = \infty$ or $|z| = 1$, we may write

$$r(z) = \frac{(z - z_1)^{n_1} \cdots (z - z_m)^{n_m}}{z^n + q_1 z^{n-1} + \cdots + q_n} h(z),$$

where $n_1, \cdots, n_m$ are positive integers with $n_1 + \cdots + n_m \le n$, $z_1, \cdots, z_m$ are distinct points in $|z| < 1$, and $h(z)$ is a polynomial of degree exactly equal to $n - n_1 - \cdots - n_m$ such that $h(z) \neq 0$ for $|z| \le 1$. Then by setting

$$w_k^i = D^i g(z_k), \quad i = 0, \cdots, n_k - 1, \quad k = 1, \cdots, m, \tag{3.36}$$

where D^i denotes the ith order differentiation operator, it follows that Research Problem 6 reduces to the following "Minimum-norm interpolation" problem (Problem 3.22):

Research Problem 7. Let R_I^a be the collection of all functions $f(z)$ in R^a that satisfy the interpolation conditions

$$D^i f(z_k) = w_k^{(i)}, \quad i = 0, \cdots, n_k - 1, \quad k = 1, \cdots, m. \tag{3.37}$$

Determine an $\hat{f}(z)$ in R_I^a such that

$$\|\hat{f}\|_{H^p} = \inf_{f \in R_I^a} \|f\|_{H^p}, \tag{3.38}$$

where $1 \le p \le \infty$.

We remark that (3.37) is usually called a *Hermite interpolation condition*. The two important special cases are

(i) $n_1 = \cdots = n_m = 1$, or equivalently, with $w_k = w_k^{(0)}$,

$$f(z_k) = w_k, \quad k = 1, \cdots, m; \tag{3.39}$$

(ii) $m = 1$, or equivalently,

$$f^{(i)}(0) = i! b_i, \quad i = 0, \cdots, n-1\,, \tag{3.40}$$

where we have set $z_1 = 0$, $b_i = \frac{1}{i!} w_1^{(i)}$, and $n_1 = n$.

The interpolation condition (3.39) is called *Lagrange interpolation*, while (3.40) is simply matching the first n Taylor coefficients of $f(z)$ at the origin with $b_0, \cdots, b_{n-1}$.

3.3.2 Extremal Kernels and Generalized Extremal Functions

We will study Research Problem 7 via the following more general extremal problem studied by Macintyre and Rogosinski [1950]. Let $k(z)$ be a rational function which is pole-free on the unit circle $|z| = 1$ but has at least one pole in $|z| < 1$, and define a linear functional λ on H^q where $1 \le q \le \infty$, by

$$\lambda(f) = \frac{1}{2\pi \mathrm{j}} \int_{|z|=1} f(z) k(z) dz, \quad f \in H^q\,. \tag{3.41}$$

Then λ is nontrivial, and since

$$\int_{|z|=1} f(z)(k(z) - g(z)) dz = \int_{|z|=1} f(z) k(z) dz \tag{3.42}$$

for all $g(z)$ in H^p, where $p^{-1} + q^{-1} = 1$ and $1 \le p, q \le \infty$ (Problem 3.23a), a duality argument shows that (Problem 3.23b)

$$\sup_{\|f\|_{H^q} \le 1} |\lambda(f)| = \inf_{g \in H^p} \|k - g\|_{L^p(|z|=1)}\,. \tag{3.43}$$

Since the quantity on the left-hand side of (3.43) is the norm of the linear functional λ, we have

$$\|\lambda\| = \inf_{g \in H^p} \|k - g\|_{L^p(|z|=1)}\,. \tag{3.44}$$

A function $F(z)$ in H^q with $\|F\|_{H^q} = 1$ and $\lambda(F) = \|\lambda\|$ will be called a *normalized extremal function*. Similarly, for the rational function $k(z)$ defined above, a function $K(z)$ of the form $K(z) = k(z) - g(z)$, for some $g \in H^p$, will be called an *extremal kernel* if it satisfies

$$\|K\|_{L^p(|z|=1)} = \inf_{g \in H^p} \|k - g\|_{L^p(|z|=1)}\,. \tag{3.45}$$

The following result can be found in Duren [1970, p.138].

Theorem 3.6. Let $1 \le p, q \le \infty$ with $p^{-1} + q^{-1} = 1$ and $k(z)$ be a rational function with no poles on the unit circle $|z| = 1$. Suppose that $\beta_1, \cdots, \beta_n$ are the poles of $k(z)$ that lie in $|z| < 1$, where $1 \le n < \infty$ and each pole is repeated according to its multiplicity. Consider the functions

$$F(z) = a \prod_{i=s+1}^{\sigma} \frac{z - \alpha_i}{1 - \bar{\alpha}_i z} \prod_{i=1}^{n-1} (1 - \bar{\alpha}_i z)^{2/q} \prod_{i=1}^{n} (1 - \bar{\beta}_i z)^{-2/q} \tag{3.46}$$

and

$$K(z) = b \prod_{i=1}^{s} \frac{z - \alpha_i}{1 - \bar{\alpha}_i z} \prod_{i=1}^{n-1} (1 - \bar{\alpha}_i z)^{2/p} \prod_{i=1}^{n} \left(\frac{(1 - \bar{\beta}_i z)}{z - \beta_i} \right)^{1-2/p} \tag{3.47}$$

where a, b are complex numbers, $|a_i| < 1$ for $i = 1, \cdots, \sigma$, and $|\alpha_i| = 1$ for $i = \sigma + 1, \cdots, n-1$, and $1 \le s \le \sigma \le n-1$. If $\alpha_1, \cdots, \alpha_{n-1}$ and b are chosen such that

$$(K(z) - k(z)) \in H^p,$$

then $K(z)$ is an extremal kernel, and with an appropriate choice of the constant a, so that $\|F\|_{H^q} = 1$, $F(z)$ is a normalized extremal function. Furthermore, $K(z)$ is the unique extremal kernel for each $p, 1 \le p \le \infty$, and $F(z)$ is the unique normalized extremal function for $1 \le p < \infty$.

Note that if it turns out that $s = n - 1$, so that (3.46) becomes

$$F(z) = a \prod_{i=1}^{n-1} (1 - \bar{\alpha}_i z)^{2/q} \prod_{i=1}^{n} (1 - \bar{\beta}_i z)^{-2/q},$$

then even in the case $p = \infty$ (or $q = 1$), the normalized extremal function $F(z)$ is unique.

3.3.3 An Application to Minimum-Norm Interpolation

We now apply Theorem 3.6 to the minimum-norm interpolation problem. For simplicity, we only consider the Lagrange interpolation (3.39).

Let $z_1, \cdots, z_m$ be distinct points in $|z| < 1$ and $w_1, \cdots, w_m$ be arbitrary complex numbers. We are interested in finding a function $\hat{f}(z)$ in H^p such that $\hat{f}(z_i) = w_i$, $i = 1, \cdots, m$, and that $\| \hat{f} \|_{H^p}$ is minimum among all such functions. As we will see below, by using the extremal kernel (3.47), $\hat{f}(z)$ will actually turn out to be a rational function in R^a, see (3.35), if and only if $p = 1, 2, \infty$. To solve this problem, we set

$$B(z) = \prod_{i=1}^{m} \frac{z - z_i}{1 - \bar{z}_i z} \quad \text{and} \quad B_k(z) = \prod_{\substack{i=1 \\ (i \ne k)}}^{m} \frac{z - z_i}{1 - \bar{z}_i z}.$$

Then any function $f(z)$ in H^p that satisfies the interpolation condition $f(z_i) = w_i$, $i = 1, \cdots, m$, takes on the form

$$f(z) = \sum_{k=1}^{m} w_k \frac{B_k(z)}{B_k(z_k)} - B(z)g(z)\,, \tag{3.48}$$

where $g(z)$ is some function in H^p (Problem 3.24). Since $|B(z)| = 1$ for $|z| = 1$, we have

$$\inf_{\substack{f \in H^p \\ f(z_i)=w_i, i=1,\cdots,m}} \|f\|_{H^p} = \inf_{g \in H^p} \|k - g\|_{L^p(|z|=1)}\,, \tag{3.49}$$

where

$$k(z) = \frac{1}{B(z)} \sum_{k=1}^{m} w_k \frac{B_k(z)}{B_k(z_k)} = \sum_{k=1}^{m} \frac{w_k}{B_k(z_k)} \frac{1 - \bar{z}_k z}{z - z_k}\,. \tag{3.50}$$

Hence, by Theorem 3.6, $\hat{f}(z)$ is unique and is given by $\hat{f}(z) = B(z)K(z)$ where $K(z)$ is the extremal kernel. In our situation, the poles of $k(z)$ in $|z| < 1$ are precisely those z_k's with the corresponding nonzero w_k's. Hence, we drop each z_k if $w_k = 0$ and set the remaining ones to be β_i in Theorem 3.6. By (3.47) it is clear that $\hat{f}(z)$ is in R^a if and only if $p = 1, 2, \infty$.

Let us consider the special case where each $w_k \neq 0$ and $p = \infty$. Then we have

$$\hat{f}(z) = b \prod_{i=1}^{s} \frac{z - \alpha_i}{1 - \bar{\alpha}_i z}$$

for some integer $s \leq m-1$ and complex numbers $b, \alpha_1, \cdots, \alpha_s$ with $|\alpha_i| < 1$, $i = 1, \cdots, s$. This (unique) choice is determined by the condition that

$$K(z) - k(z) = \frac{1}{B(z)} \left(b \prod_{i=1}^{s} \frac{z - \alpha_i}{1 - \bar{\alpha}_i z} - \sum_{i=1}^{m} w_i \frac{B_i(z)}{B_i(z_i)} \right)$$

is in H^p, or equivalently, by the interpolation condition

$$b \prod_{i=1}^{s} \frac{z_k - \alpha_i}{1 - \bar{\alpha}_i z_k} = w_k, \quad k = 1, \cdots, m. \tag{3.51}$$

Although it is obvious that we must have $s = m - 1$, finding the solution of the values $\alpha_1, \cdots, \alpha_{m-1}$ and b with the constraints $|\alpha_i| < 1$, $i = 1, \cdots, m-1$, is not an easy task. For this reason, it is sometimes necessary to sacrifice the minimum-norm property for possible computational efficiency.

3.3.4 Suggestions for Computation of Solutions

In the engineering literature, a standard approach to solving Research Problem 7 posed in Sect. 3.3.1 is to apply an algorithm of Nevanlinna-Pick type, which we will discuss in the next section. Our suggestion is to scale down the data $\{w_k\}$ so that the Pick condition (to be derived in the next section) is satisfied. The scaling factor has to depend on the value of the norm of the minimum-norm interpolant, namely,

$$\|\hat{f}\|_{H^p} = \|K\|_{L^p(|z|=1)} = \lambda(F) = \|\lambda\|, \quad 1 \le p \le \infty,$$

where $F(z)$ and $K(z)$ are the corresponding normalized extremal function and extremal kernel defined in (3.46) and (3.47), respectively. Suppose that an upper bound M of this norm has been determined. For example, since

$$\begin{aligned}
\|\lambda\| &= \inf_{g \in H^p} \|k - g\|_{L^p(|z|=1)} \\
&\le \|k\|_{L^p(|z|=1)} \\
&\le \|k\|_{L^\infty(|z|=1)} \\
&\le \sum_{k=1}^{m} \frac{|w_k|}{|B_k(z_k)|} \left(\sup_{|z|=1} \left| \frac{1 - \bar{z}_k z}{z - z_k} \right| \right) \\
&\le \sum_{k=1}^{m} \frac{|w_k|}{|B_k(z_k)|},
\end{aligned}$$

see (3.50), we may choose $M = \sum_{k=1}^{m} |w_k|/|B_k(z_k)|$. Then,

$$\|\hat{f}\|_{H^p} \le M < \infty.$$

Consequently, the function $h(z) = \hat{f}(z)/M$ is an H^p function with $|h(z)| \le 1$ in $|z| < 1$ that satisfies the interpolation conditions

$$D^i h(z_k) = w_k^i/M, \quad i = 0, \cdots, n_k - 1, \quad k = 1, \cdots, m,$$

where, again, D^i is the ith order differentiation operator. Hence, by a generalization of the theorem of Pick (to be discussed in the next section), the data pairs $(z_k, w_k^i/M)$ satisfy the generalized Nevanlinna-Pick condition (GNP condition), so that a Pick algorithm can be applied to compute $h(z)$. Now, if we set $f(z) = Mh(z)$, then $f(z)$ satisfies the original interpolation conditions (3.37) and $\|f\|_{H^p} \le M$. Of course, if the actual value of the minimum norm is known, then by setting M to be this value, $f(z)$ not only satisfies the interpolation conditions (3.39) but actually has minimum norm. It is a well-known result due to D. Sarason [1967], that the minimum interpolant is unique, and hence, $f(z) \equiv \hat{f}(z)$.

To evaluate or estimate the value of the minimum norm, we recall that this value is the same as $\|\lambda\|$, where the linear functional λ on H^q, $1 \le q \le \infty$ and $1/p + 1/q = 1$, is defined by (3.41), namely,

$$\lambda(f) = \frac{1}{2\pi \mathrm{j}} \int_{|z|=1} f(z)k(z)dz, \quad f \in H^q.$$

Hence, the expression (3.46) for a normalized extremal function $F(z)$ may sometimes be useful for estimating $\|\lambda\|$. In the special case where only Lagrange interpolation is considered, the function $k(z)$ is given by (3.50), and by the Cauchy theorem we have

$$\lambda(f) = \sum_{k=1}^{m} \frac{w_k}{B_k(z_k)}(1 - |z_k|^2)f(z_k),$$

so that

$$\|\lambda\| = \max_{\|f\|_q \le 1} \left| \sum_{k=1}^{m} \frac{w_k}{B_k(z_k)}(1 - |z_k|^2)f(z_k) \right|. \tag{3.52}$$

Of course, this maximum is attained by choosing $f(z)$ to be a normalized extremal function $F(z)$ given by (3.46) with $s = m - 1$, so that the first product in (3.46) should be deleted. An algorithm for finding the value of minimum norm for H^∞ is given in Chui, Li and Zhong [1989].

3.4 Nevanlinna-Pick Interpolation

The problem of interpolation by a function (usually a rational function) analytic in the unit disk $|z| < 1$ with restricted range, such as the unit disk $|w| \le 1$ or the right half plane Re $w > 0$, has numerous important applications in electrical engineering problems. We have seen the importance of minimum-norm interpolation in the previous section and its relation with the rational approximation problem (Research Problem 6) encountered in systems theory, particularly in sensitivity consideration of feedback systems. It also has applications to broad-band matching, network modeling, cascade synthesis, etc. If a good upper bound of the value of minimum norm can be found, then we have also indicated that a "near" minimum-norm interpolant can be determined by deriving an algorithm which may be considered as an extension of the so-called *Pick algorithm* to be discussed in this section.

3.4.1 An Interpolation Theorem

We approach the interpolation problem by quoting a more general result which can be found in Rosenblum and Rovnyak [1985]. To facilitate our discussion, we use the following notation:

H_1^∞ will denote the collection of all functions $f(z)$ in H^∞ with $\|f\|_{H^\infty} \le 1$. Let X be a linear space with dual X^*, and $\mathcal{L}(X)$ the space of all linear operators from X into itself. For each A in $\mathcal{L}(X)$, A^* will denote its adjoint, that is, $(y, Ax) = (A^*y, x)$ for all x in X and y in X^*, where $(f, g) := f(g)$. As usual, a subspace Y^* of X^* is called an *invariant subspace under* A^*, if

$A^*Y^* \subseteq Y^*$. Now, in terms of the above notation we can state the important result given in Rosenblum and Rovnyak [1985] as follows:

Theorem 3.7. Let $A \in \mathcal{L}(X)$ with adjoint A^*, and $x_0 \in X$. Also, let Y^* be an invariant subspace of X^* under A^* such that

$$\sum_{i=0}^{\infty} |(y, A^i x_0)|^2 < \infty \tag{3.53}$$

for all y in Y^*. Then an element $w \in X$ can be represented by a symbol

$$f(z) = \sum_{i=0}^{\infty} a_i z^i \in H_1^\infty$$

with the formula

$$(y, w) = \sum_{i=0}^{\infty} a_i (y, A^i x_0) \tag{3.54}$$

for all $y \in Y^*$ if and only if the inequality

$$\sum_{i=0}^{\infty} |(y, A^i w)|^2 \leq \sum_{i=0}^{\infty} |(y, A^i x_0)|^2 \tag{3.55}$$

is satisfied for all $y \in Y^*$.

In the following we give two examples to demonstrate the generality of this theorem.

Corollary 3.2. (Carathéodory-Fejér's Theorem)

Let $b_0, \cdots, b_{n-1}$ be complex numbers. Then there exists a function

$$f(z) = \sum_{i=0}^{\infty} a_i z^i \in H_1^\infty$$

that satisfies $a_i = b_i$ for each $i = 0, \cdots, n-1$, if and only if the Toeplitz matrix

$$T(b_0, \cdots, b_{n-1}) = \begin{bmatrix} b_0 & 0 & \cdots & 0 \\ b_1 & \ddots & \ddots & \vdots \\ \vdots & \ddots & \ddots & 0 \\ b_{n-1} & \cdots & b_1 & b_0 \end{bmatrix} \tag{3.56}$$

satisfies $\|T(b_0, \cdots, b_{n-1})\| \leq 1$.

In other words, if we set

$$w_1^{(i)} = i!b_i, \quad i = 0, \cdots, n-1$$

as suggested by (3.40), this result characterizes all data values $w_1^{(i)}$, $i = 0, \cdots, n-1$, such that there exists an $f(z)$ in H_1^∞ that interpolates this data set at $z_1 = 0$, namely,

$$f^{(i)}(0) = w_1^{(i)}, \quad i = 0, \cdots, n-1.$$

To verify the theorem of Carathéodory-Féjer, we set $X = X^* = Y^* = \mathcal{C}^n$ and

$$A = \begin{bmatrix} 0 & 0 & \cdots & \cdots & 0 \\ 1 & \ddots & \ddots & & \vdots \\ 0 & \ddots & \ddots & \ddots & \vdots \\ \vdots & \ddots & \ddots & \ddots & 0 \\ 0 & \cdots & 0 & 1 & 0 \end{bmatrix}, \quad x_0 = \begin{bmatrix} 1 \\ 0 \\ \vdots \\ 0 \end{bmatrix}.$$

Then clearly, for every $y = [y_0 \cdots y_{n-1}]^\top$, condition (3.53) is satisfied, since $A^i = 0$ for all $i > n-1$; hence, Theorem 3.7 applies. Now since

$$A^i x_0 = [0 \quad \cdots \quad 0 \quad 1 \quad 0 \quad \cdots \quad 0]^\top, \tag{3.57}$$

where the value 1 is at the $(i+1)$st entry, the representation formula (3.54) for the data $w = [b_0 \cdots b_{n-1}]^\top$ in this example is

$$\sum_{i=0}^{n-1} a_i \bar{y}_i = \sum_{i=0}^{n-1} b_i \bar{y}_i$$

for all $y = [y_0 \cdots y_{n-1}]^\top$ in $\mathcal{C}^n$; or in other words,

$$a_i = b_i, \quad i = 0, \cdots, n-1,$$

as claimed. On the other hand, a necessary and sufficient condition for this assertion to hold is that (3.55) is satisfied. In this example, by using (3.57) again, the inequality (3.55) becomes

$$\|T(b_0, \cdots, b_{n-1})\| \leq 1$$

where $T(b_0, \cdots, b_{n-1})$ is the Toeplitz matrix given in (3.56) (Problem 3.25).

In the above example, we have considered the special case where all the sample points $z_1, \cdots, z_n$ for interpolation coalesce with the origin. In the following, we will consider the other extreme case where $z_1, \cdots, z_n$ are distinct.

Corollary 3.3. (Pick's Theorem)

Let $z_1, \cdots, z_n$ be distinct points in the unit disk $|z| < 1$, and $w_1, \cdots, w_n$ be complex numbers. Then there exists a function $f(z)$ in H_1^∞ that satisfies the interpolation condition:

$$f(z_i) = w_i, \quad i = 1, \cdots, n,$$

if and only if the $n \times n$ matrix

$$\left[\frac{1 - w_p \bar{w}_q}{1 - z_p \bar{z}_q}\right]_{1 \le p,q \le n} \tag{3.58}$$

is non-negative definite.

To verify this result, we again set $X = X^* = Y^* = \mathcal{C}^n$, and let

$$A = \begin{bmatrix} z_1 & 0 & \cdots & 0 \\ 0 & \ddots & \ddots & \vdots \\ \vdots & \ddots & \ddots & 0 \\ 0 & \cdots & 0 & z_n \end{bmatrix}, \quad x_0 = \begin{bmatrix} 1 \\ \vdots \\ 1 \end{bmatrix}, \quad w = \begin{bmatrix} w_1 \\ \vdots \\ w_n \end{bmatrix},$$

so that

$$A^i x_0 = [\, z_1^i \ \cdots \ z_n^i \,]^\top .$$

Then for any $y = [y_1 \cdots y_n]^\top$ in $\mathcal{C}^n$, we have

$$\begin{aligned}
\sum_{i=0}^{\infty} |(y, A^i x_0)|^2 &= \sum_{i=0}^{\infty} \left| \sum_{k=1}^{n} \bar{y}_k z_k^i \right|^2 \\
&\le \sum_{i=0}^{\infty} \left(\sum_{k=1}^{n} |\bar{y}_k|^2 \right) \left(\sum_{k=1}^{n} |z_k|^{2i} \right) \\
&= \|y\|_{l^2}^2 \sum_{k=1}^{n} \sum_{i=0}^{\infty} |z_k|^{2i} \\
&= \|y\|_{l^2}^2 \sum_{k=1}^{n} \frac{1}{1 - |z_k|^2} < \infty
\end{aligned}$$

so that (3.53) is satisfied and Theorem 3.7 applies. Now, the representation formula (3.54) in this example becomes

$$\begin{aligned}
\sum_{k=1}^{n} \bar{y}_k w_k = (y, w) &= \sum_{i=0}^{\infty} a_i \left(\sum_{k=1}^{n} \bar{y}_k z_k^i \right) \\
&= \sum_{k=1}^{n} \bar{y}_k \sum_{i=0}^{\infty} a_i z_k^i = \sum_{k=1}^{n} \bar{y}_k f(z_k)
\end{aligned}$$

for all $y = [y_1 \cdots y_n]^\top$ in $\mathcal{C}^n$, where $f(z) = \sum_{i=0}^{\infty} a_i z^i$ is in H_1^∞. In other words, there exists an $f(z)$ in H_1^∞ that satisfies $f(z_k) = w_k$, $k = 1, \cdots, n$.

On the other hand, the inequality in (3.55), which is a necessary and sufficient condition, becomes

$$\begin{aligned}
0 &\leq \sum_{i=0}^{\infty} \left[|(y, A^i x_0)|^2 - |(y, A^i w)|^2 \right] \\
&= \sum_{i=0}^{\infty} \left[\left| \sum_{k=1}^{n} \bar{y}_k z_k^i \right|^2 - \left| \sum_{k=1}^{n} \bar{y}_k w_k z_k^i \right|^2 \right] \\
&= \sum_{i=0}^{\infty} \left[\sum_{k=1}^{n} \sum_{\ell=1}^{n} \bar{y}_k y_\ell z_k^i \bar{z}_\ell^i (1 - w_k \bar{w}_\ell) \right] \\
&= \sum_{k=1}^{n} \sum_{\ell=1}^{n} \bar{y}_k y_\ell (1 - w_k \bar{w}_\ell) \sum_{i=0}^{\infty} z_k^i \bar{z}_\ell^i \\
&= \sum_{k=1}^{n} \sum_{\ell=1}^{n} \bar{y}_k \frac{1 - w_k \bar{w}_\ell}{1 - z_k \bar{z}_\ell} y_\ell
\end{aligned}$$

for all $[y_1 \cdots y_n]^\top$ in $\mathcal{C}^n$, or equivalently, the matrix in (3.58) is non-negative definite. This completes the proof of Pick's theorem.

We now turn to the study of the useful Pick's algorithm.

3.4.2 Nevanlinna-Pick's Theorem and Pick's Algorithm

The matrix in (3.58) is usually called *Pick's matrix* and the non-negative definiteness of this matrix is called *Pick's condition*. Pick's theorem was later extended by Nevanlinna to the case when the number of distinct sample points $\{z_1, z_2, \cdots\}$ in $|z| < 1$ is infinite. For this reason, the problem of interpolation by functions from H_1^∞ is usually called the problem of *Nevanlinna-Pick interpolation*. By choosing $X = X' = Y' = l^2$, an analogous matrix A and vector x_0, and using arbitrary finitely supported sequences y, it is not difficult to show that the following result is also a consequence of Theorem 3.7 (Problem 3.26).

Theorem 3.8. (Nevanlinna-Pick)

Let $\{z_1, z_2, \cdots\}$ be distinct points in $|z| < 1$ and $\{w_1, w_2, \cdots\}$ be complex numbers. Then there exists a function $f(z)$ in H_1^∞ such that $f(z_i) = w_1$, $i = 1, 2, \cdots$, if and only if the matrix

$$\left[\frac{1 - w_p \bar{w}_q}{1 - z_p \bar{z}_q} \right]_{1 \leq p,q < \infty} \tag{3.59}$$

is non-negative definite in the sense that for any finite sequence $\{a_1, \cdots, a_N\}$ of complex numbers, the inequality

$$\sum_{p=1}^{N}\sum_{q=1}^{N} a_p \frac{1 - w_p \bar{w}_q}{1 - z_p \bar{z}_q} \bar{a}_q \geq 0$$

is satisfied.

Of course, in applications, the sample points $\{z_i\}$ are not necessarily distinct and we must consider interpolation of derivative values, see (3.37). In this situation, the matrix A and vector x_0 in the proofs of the above corollaries must be changed to "block" matrices as follows: For instance, we may use the notation introduced in (3.37) and set

$$A = \begin{bmatrix} A_1 & & \\ & A_2 & \\ & & \ddots \end{bmatrix} \quad \text{and} \quad x_0 = \begin{bmatrix} x^1 \\ x^2 \\ \vdots \end{bmatrix},$$

where

$$A_k = \begin{bmatrix} z_1 & & & \\ 1 & z_k & & \\ & \ddots & & \\ & & (n_k - 1) & z_k \end{bmatrix}_{n_k \times n_k} \quad \text{with} \quad x^k = \begin{bmatrix} 1 \\ 0 \\ \vdots \\ 0 \end{bmatrix}.$$

The "values" $w_k^{(i)}$ may even be matrix values. We now pose the following problem:

Research Problem 8. Formulate a generalized matrix-valued Nevanlinna-Pick condition for arbitrary points $\{z_i\}$ in $|z| < 1$ which would yield the Carathéodory-Fejér condition and the Pick condition as special cases. Prove the corresponding interpolation theorem and derive a computational algorithm.

If $\{z_1, \cdots, z_n\}$ is a finite set of distinct complex numbers, then the following algorithm due to Pick is very efficient in computing an interpolation function $f(z)$. Of course, if $f(z)$ does not have a minimum H^p norm for some $1 \leq p \leq \infty$, it is not unique, as can be seen in the algorithm, where $f_0(z)$ does not have to be defined by (3.62) below.

Pick's algorithm. Let $z_1, \cdots, z_n$ be distinct points in $|z| < 1$ and $w_1, \cdots, w_n$ be complex numbers such that Pick's matrix in (3.58) is non-negative definite.

Step 1. Compute the complex numbers $w_{k,i}$, $i = k, \cdots, n$ and $k = 2, \cdots, n$, defined by

$$w_{k,i} = \frac{(1 - \bar{z}_{k-1} z_i)(w_{k-1,i} - w_{k-1,k-1})}{(z_i - z_{k-1})(1 - \bar{w}_{k-1,k-1} w_{k-1,i})} \tag{3.60}$$

by using the initial values

$$w_{1,i} = w_i, \qquad i = 1, \cdots, n.$$

Step 2. Compute

$$f_k(z) = \frac{w_{n-k,n-k}(1 - \bar{z}_{n-k}z) + (z - z_{n-k})f_{k-1}(z)}{(1 - \bar{z}_{n-k}z) + \bar{w}_{n-k,n-k}(z - z_{n-k})f_{k-1}(z)} \tag{3.61}$$

for $k = 1, \cdots, n-1$, by using the initial function

$$f_0(z) \equiv w_{n,n} \,. \tag{3.62}$$

Conclusion: Set $f(z) = f_{n-1}(z)$. Then $f(z)$ is a required function in H_1^∞ that satisfies $f(z_i) = w_i$, $i = 1, \cdots, n$.

3.4.3 Verification of Pick's Algorithm

Pick's algorithm will be verified in this section. First, let us remark that by using any other function $f_0(z)$ in H_1^∞ that satisfies the condition $f_0(z_n) = w_{n,n}$, we obtain another interpolation function $f(z) = f_{n-1}(z)$ in H_1^∞ (Problem 3.27). However, the choice of the constant function $f_0(z)$ in (3.62) guarantees that a rational function $f(z) = f_{n-1}(z)$ of minimum degree is obtained. The following problem is important since it may give rise to a possible solution to some special cases of Research Problem 7.

Research Problem 9. Is there an initial function $f_0(z)$ that can be used in Pick's algorithm so that the interpolation function $f(z) = f_{n-1}(z)$ obtained by using (3.61) has minimum H^p norm among all functions in H_1^p that satisfy the same interpolation condition? If so, determine the initial function for each p, $1 \le p \le \infty$. Is $f_0(z) \equiv w_{n,n}$ defined in (3.62) such an initial function for $p = 1, 2$, or ∞?

To verify the validity of Pick's algorithm, we first note that (3.61) is equivalent to

$$f_{k-1}(z) = \frac{1 - \bar{z}_{n-k}z}{z - z_{n-k}} \frac{f_k(z) - w_{n-k,n-k}}{1 - \bar{w}_{n-k,n-k}f_k(z)}, \tag{3.63}$$

$k = 1, \cdots, n-1$. By Pick's theorem (stated in Corollary 3.3), there exists a function in H_1^∞, which we denote by $\tilde{f}_{n-1}(z)$, that satisfies the interpolation condition

$$\tilde{f}_{n-1}(z_i) = w_i \,, \quad i = 1, \cdots, n \,.$$

Define

$$\tilde{f}_{n-2}(z) = \frac{1 - \bar{z}_1 z}{z - z_1} \frac{\tilde{f}_{n-1}(z) - w_1}{1 - \bar{w}_1 \tilde{f}_{n-1}(z)} \,.$$

Then by using a property of Möbius transformations and the maximum modulus theorem, it can be shown that $\tilde{f}_{n-2}(z)$ is again in H_1^∞ and satisfies the interpolation condition

$$\tilde{f}_{n-2}(z_i) = w_{2,i}, \quad i = 2, \cdots, n,$$

for the data $\{w_{2,i}\}$ defined by (3.60) (Problem 3.28). In particular, by the maximum modulus theorem again, we have $|w_{2,i}| < 1$ for $i = 2, \cdots, n$ if $n > 2$ (and $|w_{2,2}| \leq 1$ if $n = 2$). For $n > 2$, we proceed and define

$$\tilde{f}_{n-3}(z) = \frac{1 - \bar{z}_2 z}{z - z_2} \frac{\tilde{f}_{n-2}(z) - w_{2,2}}{1 - \bar{w}_{2,2}\tilde{f}_{n-2}(z)}$$

which, by the same argument, is in H_1^∞ and satisfies the interpolation condition

$$\tilde{f}_{n-3}(z_i) = w_{3,i}, \quad i = 3, \cdots, n,$$

etc. Finally, we define

$$\tilde{f}_0(z) = \frac{1 - \bar{z}_{n-1} z}{z - \bar{z}_{n-1}} \frac{\tilde{f}_1 - w_{n-1,n-1}}{1 - \bar{w}_{n-1,n-1}\tilde{f}_1(z)}$$

and note that again $\tilde{f}_0(z)$ is in H_1^∞ and satisfies the interpolation condition

$$\tilde{f}_0(z_n) = w_{n,n},$$

where $|w_{n,n}| \leq 1$. If $|w_{n,n}| = 1$, then by the maximum modulus theorem, $\tilde{f}_0(z)$ must be the constant function $\tilde{f}_0(z) = f_0(z) = w_{n,n}$, so that $f_{n-1}(z) \equiv \tilde{f}_{n-1}(z)$. That is, if $|w_{n,n}| = 1$, then with the initial function $f_0(z) \equiv w_{n,n}$ in (3.62), Pick's algorithm reproduces the original interpolation function $\tilde{f}_{n-1}(z)$ introduced in the proof. Suppose now that $|w_{n,n}| < 1$. Then it is clear from the above argument that

$$|w_{k,k}| < 1, \quad k = 1, \cdots, n. \tag{3.64}$$

Let us introduce the auxiliary functions

$$g_k(z) = -\frac{z - z_{n-k}}{1 - \bar{z}_{n-k} z} f_{k-1}(z), \quad k = 1, \cdots, n, \tag{3.65}$$

where $f_0(z) \equiv w_{n,n}$, and $f_1(z), \cdots, f_{n-1}(z)$ are defined recursively by (3.61). By (3.61) and (3.65), we have

$$f_k(z) = -\frac{g_k(z) - w_{n-k,n-k}}{1 - \bar{w}_{n-k,n-k} g_k(z)}. \tag{3.66}$$

Clearly, $g_1(z)$ is in H_1^∞, and by (3.64) and (3.66), we may also conclude that $f_1(z)$ is in H_1^∞. Using this fact in (3.65), we now have $g_2(z)$ in H_1^∞ which, in turn, implies that $f_2(z)$ is in H_1^∞ by (3.64) and (3.66). By repeated applications of the same argument, we may conclude that $f_i(z)$, $i = 3, \cdots, n-1$, are in H_1^∞. A careful investigation of the definition of $w_{k,i}$, $i = k, \cdots, n$ and $k = 1, \cdots, n$ in (3.60) and that of $f_k(z)$, $k = 0, \cdots, n-1$ in (3.61) and (3.62), allows us to conclude that $f_{n-1}(z_i) = w_i$, for $i = 1, \cdots, n$ (Problem 3.29). This verifies that Pick's algorithm is always valid as long as Pick's condition on the non-negative definiteness of the $n \times n$ matrix

$$\left[\frac{1 - w_p \bar{w}_q}{1 - z_p \bar{z}_q}\right]_{1 \le p,q \le n}$$

is satisfied.

Problems

Problem 3.1. For $p = 2$ and ∞, show that the function $M_p(r, f)$ defined in (3.3) is nondecreasing in r.

Problem 3.2. By using Hölder's inequality, verify that $H^s \subset H^r$ for $0 < r < s \le \infty$.

Problem 3.3. Consider the Blaschke product $B(z)$ defined by (3.7), namely,

$$B(z) = z^m \prod_i \frac{|z_i|}{z_i} \frac{z_i - z}{1 - \bar{z}_i z}$$

with $0 < |z_i| < 1$ and $0 \le m < \infty$.

(a) Let $|z| \le r < 1$ and $0 < |\zeta| < 1$. Show that

$$\left|\frac{\zeta + |\zeta| z}{(1 - \bar{\zeta} z)\zeta}\right| \le \frac{1+r}{1-r}.$$

(b) Let $\{z_i\}$ be such that $\sum_i (1 - |z_i|) < \infty$. Show that the Blaschke product $B(z)$ converges.

(c) Show that $|B(z)| < 1$ for $|z| < 1$ and $|B(z)| = 1$ almost everywhere on $|z| = 1$.

Problem 3.4. Suppose that $f(z) \in H^p, 1 \le p \le \infty$, is not identically zero, with an associated Blaschke product $B(z)$ defined by (3.7), and assume that $F(z)$ is the analytic function given by (3.10). Show that

$$|f(z)/(B(z)F(z))| = 1$$

almost everywhere on $|z| = 1$. [Hint: Consider the Dirichlet problem on the unit circle.]

Problem 3.5. Let $f_O(z)$ be an outer function defined by a positive measurable function $\varphi(e^{jt})$ such that $\ln\varphi(e^{jt}) \in L^1[0, 2\pi]$, namely,

$$f_O(z) = e^{j\alpha} \exp\left(\frac{1}{2\pi}\int_0^{2\pi} \frac{e^{jt}+z}{e^{jt}-z} \ln\, \varphi(e^{jt})dt\right),$$

where α is a real constant. Show that

$$\left|f_O(e^{j\theta})\right| = \varphi(e^{j\theta})$$

almost everywhere. [Hint: This problem is similar to Problem 3.4.]

Problem 3.6. Let $f(z)$ be analytic in $|z| < 1$ with the power series expansion

$$f(z) = \sum_{n=0}^{\infty} a_n z^n\,, \qquad |z| < 1\,.$$

By using the orthogonality of $\{z^n\}$ on $|z| = 1$, verify that $\|f\|_{H^2} = \|\{a_n\}\|_{l^2}$.

Problem 3.7. Verify that the function

$$f(z) = \sum_{n=1}^{\infty} \frac{1}{n} e^{j\theta_n} z^n\,,$$

where θ_n are real numbers, is in H^p with $2 \le p < \infty$. Is it possible that $f(z)$ is in H^p for some $p < 2$?

Problem 3.8. Let $f_I(z)$ be an inner function. Show that $1/f_I(z)$ is also an inner function if and only if $f_I(z) \equiv 1$.

Problem 3.9. Let $f(z) = \sum_{n=0}^{\infty} a_n z^n$. Verify (3.25) by showing that

$$\int_0^{2\pi} \frac{1}{f(e^{jt})} e^{-jit} d\mu(t) = \begin{cases} 0 & \text{for } i = 1, 2, \cdots\,, \\ \overline{f(0)} & \text{for } i = 0\,. \end{cases}$$

Problem 3.10. By using the orthogonal property of $\hat{p}_n(z)$ with $z, \cdots, z^n$ in (3.25), show that its reciprocal polynomial $\hat{q}_n(z)$ is orthogonal to $1, \cdots, z^{n-1}$ under the same inner product $\langle\ ,\ \rangle_\mu$.

Problem 3.11. Let $f(z) = \sum_{k=0}^{\infty} a_k z^k$, $\hat{p}_n(z) = \sum_{k=0}^{n} c_k z^k$, and $d_k = \sum_{i=0}^{\infty} a_{k+i}\bar{a}_i$. Verify the identity

$$\frac{1}{2\pi}\int_0^{2\pi} \hat{p}_n(e^{jt}) e^{-jit} |f(e^{jt})|^2 dt = \sum_{k=0}^{n} \left\{ \sum_{\substack{l,m \ge 0 \\ l+k=m+i}} a_l \bar{a}_m \right\} c_k\,.$$

Hence, conclude that (3.23) and (3.25) are equivalent statements.

Problem 3.12. Determine the best H^2 approximant $\hat{r}_2(z)$ of $f(z) = z$ from $R_2(2, -2)$. [See Sect. 3.2.3 for notation.]

Problem 3.13. Determine the best H^2 approximant of $f(z) = \ln(1 - z)$ from $R_3(2, 3, 4)$.

Problem 3.14. Let $|z_k| > 1$, $k = 1, \cdots, n$, and $f(z)$ be in H^2. Verify that

$$\frac{1}{2\pi \mathrm{j}} \int_{|z|=1} f(z) \overline{\left(\frac{1}{z_k - z}\right)} \frac{dz}{z} = \frac{1}{z_k} f\left(\frac{1}{\bar{z}_k}\right), \qquad k = 0, 1, \cdots, n.$$

[Hint: Consider $f_\rho(z) = f(\rho z)$, $0 < \rho < 1$, and apply Cauchy's formula to $f_\rho(z)$.]

Problem 3.15. By replacing $f(z)$ in problem 3.14 with $\hat{r}(z) - f(z)$, verify that (3.30) is equivalent to the interpolation conditions

$$\hat{r}(0) = f(0), \quad \hat{r}\left(\frac{1}{\bar{z}_k}\right) = f\left(\frac{1}{\bar{z}_k}\right), \quad k = 1, \cdots, n.$$

Also, verify that these interpolation conditions can be replaced by the following system of linear equations:

$$\begin{cases} b_0 + \cdots + b_n = f(0), \\ b_0 + \displaystyle\sum_{i=1}^{n} \frac{z_i \bar{z}_k}{z_i \bar{z}_k - 1} b_i = f\left(\frac{1}{\bar{z}_k}\right), \quad k = 1, \cdots, n. \end{cases}$$

Problem 3.16. Given a transfer function $H(z) = f(\frac{1}{z})$ where $f(z) = 1/[(z-2)(z-3)]$ and $z_1 = 4$ and $z_2 = 5$. Design an ARMA filter $R(z) = \hat{r}(\frac{1}{z})$ where $\hat{r}(z)$ is the best approximant to $f(z)$ in the H^2 norm over $R_2(z_1, z_2)$, see (3.26). Note that the poles of $R(z)$ are located at 1/4 and 1/5.

Problem 3.17. Show that $\hat{p}_n(z) = c_0 + \cdots + c_n z^n$ is the LSI of $f_m(z) = a_0 + \cdots + a_m z^m$ from $\mathcal{P}_n$ if and only if $\{c_i\}$ satisfies

$$||\{\delta_i\} - \{c_i\} * \{a_i\}||_{l^2} = \min \{||\{\delta_i\} - \{d_i\} * \{a_i\}||_{l^2} : d_0, \cdots, d_n\},$$

when $\{\delta_n\}$ is the "unit spike" or the convolution identity [See Sect. 1.2.1.]

Problem 3.18. Apply Theorem 3.5 to derive Corollary 3.2.

Problem 3.19. Verify that $\hat{q}_3(z)$ in Example 3.5 is the DLSI of $q_3(z) = 1 - 2.9z + 2.9z^2 - 1.16z^3$ through $\mathcal{P}_{10}$. Also verify that

$$\left|\frac{\hat{q}_3(e^{j\omega})}{q_3(e^{j\omega})}\right| \approx \text{constant, different from one}.$$

Problem 3.20. Given a transfer function

$$H(z) = \frac{0.8 + z}{1 - 2.5z + z^2}.$$

Use the DLSI procedure to design a stable transfer function $\tilde{H}(z)$ such that $|\tilde{H}(z)| \approx |H(z)|$.

Problem 3.21. Let R^a be defined as in (3.35) and let $g(z) \in R^a$. Show that

$$\inf_{f \in R^a} ||g(z) - z^2 f(z)||_{H^p}$$
$$= \inf\{||f||_{H^p} : f \in R^a, f(0) = g(0), f'(0) = g'(0)\}.$$

Problem 3.22. Verify that Reserch Problems 6 and 7 posed in Sect. 3.3.1 are equivalent for the special case when $r(z) \neq 0$ for $z = \infty$ or $|z| = 1$. [This generalizes Problem 3.21.]

Problem 3.23. Let $k(z)$ be a rational function which is pole-free on $|z| = 1$ but has at least one pole in $|z| < 1$.

(a) Show that the linear functional λ defined in (3.41) is bounded and nontrivial.

(b) Let $f(z) \in H^q$ and $g(z) \in H^p$ where $p^{-1}+q^{-1} = 1$ and $1 \le p, q \le \infty$. Show that

$$\int_{|z|=1} f(z)(k(z) - g(z))dz = \int_{|z|=1} f(z)k(z)dz.$$

(c) Apply the duality argument to the result obtained in part (b) to show that

$$\sup_{\|f\|_{H^q} \le 1} \left|\frac{1}{2\pi j}\int_{|z|=1} f(z)k(z)dz\right| = \inf_{g \in H^p} \|k - g\|_{L^p(|z|=1)}.$$

Problem 3.24. Let $z_1, \cdots, z_n$ be distinct points in $|z| < 1$ and $w_1, \cdots, w_n$ be arbitrary complex numbers. Set

$$B(z) = \prod_{i=1}^{m} \frac{z - z_i}{1 - \bar{z}_i z} \quad \text{and} \quad B_k(z) = \prod_{\substack{i=1 \\ (i \neq k)}}^{m} \frac{z - z_i}{1 - \bar{z}_i z}.$$

Verify that any function $f(z)$ in H^p that satisfies the interpolation conditions $f(z_i) = w_i, i = 1, \cdots, m$, takes on the form

$$f(z) = \sum_{k=1}^{m} w_k \frac{B_k(z)}{B_k(z_k)} - B(z)g(z)$$

for some function $g(z) \in H^p$.

Problem 3.25. Complete the proof of Carathéodory-Fejér's theorem studied in Sect. 3.4.1 by verifying that the condition (3.56) follows from the inequality (3.55).

Problem 3.26. Prove the Nevanlinna-Pick theorem by following the same procedure as that in the proof of Pick's theorem in Sect. 3.4.1.

Problem 3.27. Show that any function $f_0(z)$ in H_1^∞ that satisfies the condition $f_0(z_n) = w_{n,n}$ in Pick's algorithm will give an interpolation function $f(z) = f_{n-1}(z)$ in H_1^∞.

Problem 3.28. Let $z_1, \cdots, z_n$ be distinct points in $|z| < 1$ and $w_1, \cdots, w_n$ be complex numbers such that the Pick matrix (3.58) is non-negative definite. Suppose that $f(z)$ is in H_1^∞ and satisfies $f(z_i) = w_i, i = 1, \cdots, n$. Show that the function $g(z)$ defined by

$$g(z) = \frac{1 - \bar{z}_1 z}{z - z_1} \frac{f(z) - w_1}{1 - \bar{w}_1 f(z)}$$

is also in H_1^∞ and satisfies

$$g(z_i) = \frac{(1 - \bar{z}_1 z_i)(w_i - w_1)}{(z_i - \bar{z}_1)(1 - \bar{w}_1 w_i)}, \qquad i = 2, \cdots, n.$$

[Hint: Apply the Möbius transform and the maximum modulus theorem.]

Problem 3.29. Convince yourself that Pick's algorithm produces a function $f_{n-1}(z)$ that satisfies the required interpolation conditions

$$f_{n-1}(z_i) = w_i, \qquad i = 1, \cdots, n.$$

Problem 3.30. Find the unique interpolant $f(z) \in H^\infty$ with minimum norm such that

$$f(0) = \frac{1}{2} \quad \text{and} \quad f\left(\frac{1}{2}\right) = 1.$$

4. Optimal Hankel-Norm Approximation and H^∞-Minimization

The notion of the Hankel norm was briefly introduced in Sects. 1.5.3 and 2.3.3. By a fundamental result of Nehari (to be discussed later), it will be shown in this chapter that the Hankel norm provides a meaningful measurement for functions which are essentially bounded on the unit circle $|z| = 1$. However, since the Hankel norm of any H^∞ function is zero, in contrast to the L^p norms on $|z| = 1$, it is only a "semi-norm" in the sense that any two functions in L^∞ ($|z| = 1$) with the same singular part have identical Hankel norms. In other words, in applying the Hankel norm, one has to take into account that an additive H^∞ function must be determined by using a different method. This is not a serious draw-back in general. For instance, with the exception of an additive constant h_0, the transfer function of a causal SISO linear system has zero analytic part. Since h_0 can be easily determined and is really not very important, the Hankel norm is a very useful measurement for the study of causal linear systems. Indeed, as we have seen in Chap. 2, if the system is realizable in the sense that its transfer function is rational, then it is stable if and only if the Hankel norm of the transfer function has finite value. Moreover, the importance of this norm in the study of systems theory is apparent due to the fact that the best Hankel-norm (strictly proper) rational approximant with prescribed degree to the transfer function of a stable linear system along with the exact error of approximation can be described analytically. Important applications of best Hankel-norm approximation include system reduction, digital filter design, etc. This elegant analytical description is a fundamental result of the previously cited work (see Sect. 2.3.3) of Adamjan, Arov, and Krein [1971, 1978], usually known as the AAK approach. We will give a detailed constructive proof of the AAK theorem for finite-rank and real Hankel matrices in this chapter. A proof of the AAK's general theorem is much more complicated and its discussion will be delayed to the next chapter. Two important applications, namely, system reduction and H^∞ minimization for SISO systems, will be discussed in this chapter. The study of the matrix-valued setting of the AAK theorem for MIMO systems will be delayed to Chap. 6.

4.1 The Nehari Theorem and Related Results

We first recall the notion of Hankel norms which has been introduced briefly in Chaps. 1 and 2. In this section, we will introduce a very important result of Nehari on best rational approximation. Since the AAK theorem is a generalization of the Nehari theorem, no proof for the Nehari theorem will be given. Instead, the AAK theorem for approximation of finite-rank real Hankel matrices will be proved in Sect. 4.3 below.

4.1.1 Nehari's Theorem

We first recall that with any Fourier series

$$f(z) = \sum_{n=-\infty}^{\infty} f_n z^{-n},$$

where $z = e^{jw}$, we may associate an infinite Hankel matrix $\Gamma_f = [f_{i+\ell-1}]$, $i, \ell \geq 1$, or more explicitly,

$$\Gamma_f = \begin{bmatrix} f_1 & f_2 & f_3 & \cdots \\ f_2 & f_3 & \cdots & \cdots \\ f_3 & \cdots & \cdots & \cdots \\ \cdots & \cdots & \cdots & \cdots \end{bmatrix},$$

and consider Γ_f as an operator on l^2. Then the Hankel norm of $f(z)$, denoted by $\|f\|_\Gamma$, is defined to be the (spectral) norm of the operator Γ_f, namely,

$$\|f\|_\Gamma = \|\Gamma_f\|_s = \sup_{\|\mathbf{x}\|_{l^2}=1} \|\Gamma_f \mathbf{x}\|_{l^2}. \tag{4.1}$$

Hence, $\|f\|_\Gamma < \infty$ *if and only if* Γ_f *is a bounded linear operator on* l^2.

Example 4.1. Let $f(z) = z^{-3} + z^{-2} + \frac{1}{2}z^{-1} + f_0 + f_1 z + f_2 z^2 + \cdots$. Then the associated Hankel matrix is defined to be

$$\Gamma_f = \begin{bmatrix} 1/2 & 1 & 1 & 0 & \cdots \\ 1 & 1 & 0 & \cdots & \cdots \\ 1 & 0 & \cdots & \cdots & \cdots \\ 0 & \cdots & \cdots & \cdots & \cdots \\ \cdots & \cdots & \cdots & \cdots & \cdots \end{bmatrix},$$

and the Hankel norm of the function $f(z)$ is given by

$$\|f\|_\Gamma = \|\Gamma_f\|_s = \max\{\, |\lambda_1|, |\lambda_2|, |\lambda_3| \,\},$$

where $\lambda_1, \lambda_2, \lambda_3$ are the three nonzero eigenvalues of Γ_f. Since the eigenvalues of Γ_f are $\lambda_1 = 2, \lambda_2 = -1, \lambda_3 = 1/2$, and $\lambda_4 = \lambda_5 = \cdots = 0$, we have $\|f\|_\Gamma = 2$.

The following result was obtained by Nehari [1967]. For simplicity, we will use the abbreviation L^∞ for $L^\infty(|z|=1)$.

Theorem 4.1. (Nehari)

Let $f(z)$ be any function in L^∞. Then

$$\|f\|_\Gamma = \inf_{g\in H^\infty} \|f-g\|_{L^\infty}\,. \tag{4.2}$$

An immediate consequence of this important theorem is that every function in L^∞ has finite Hankel norm and a "singular" Fourier series

$$h_s(z) = \sum_{n=1}^{\infty} h_n z^{-n}\,, \qquad z = e^{j\omega}\,,$$

has finite Hankel norm if and only if there exists an analytic function

$$h_a(z) = \sum_{n=0}^{\infty} h_{-n} z^n$$

in $|z|<1$ such that $h_s(z)+h_a(z)$ is an L^∞ function.

Example 4.2. The following functions have finite Hankel norms:

$$\sin(z) \quad \text{and} \quad \sum_{k=-N}^{N} a_k z^k\,,$$

where a_k are constants. The following functions do not have finite Hankel norms:

$$e^{1/(z-j)} \quad \text{and} \quad \sum_{k=-\infty}^{\infty} k z^k\,.$$

In order to understand how good the Hankel norm is, it is interesting to compare the Hankel norm with the usual L^2 norm or L^∞ norm on $|z|=1$. We have the following results:

Theorem 4.2. Let

$$f(z) = \sum_{n=1}^{\infty} f_n z^{-n}$$

be a function in L^∞. Then

$$\|f\|_{L^2} \le \|f\|_\Gamma \le \|f\|_{L^\infty}\,. \tag{4.3}$$

The first inequality follows from the definition of $\|f\|_\Gamma$. Indeed, by setting

$$\mathbf{e}_1 = [1 \quad 0 \quad 0 \quad \cdots \;]^\top ,$$

we have

$$\begin{aligned}\|f\|_{L^2} &= \left(\sum_{n=1}^{\infty} |f_n|^2\right)^{1/2} = \|\Gamma_f \mathbf{e}_1\|_{l^2} \\ &\le \sup_{\|\mathbf{x}\|_{l^2}=1} \|\Gamma_f \mathbf{x}\|_{l^2} = \|\Gamma_f\|_s = \|f\|_\Gamma .\end{aligned}$$

The second inequality follows immediately from the Nehari theorem, namely, $\|f\|_\Gamma \le \|f - 0\|_{L^\infty}$ (see also Problem 4.3). This completes the proof of the theorem.

It is perhaps interesting to observe that if a function $f(z) \in L^\infty$ is used to define a linear functional λ_f on H^1 via the formula

$$\lambda_f(g) = \frac{1}{2\pi \mathrm{j}} \int_{|z|=1} g(z) f(z) dz, \quad g(z) \in H^1 , \tag{4.4}$$

then by (3.43) (where only rational functions were considered) and (4.2) we have (Problem 4.4)

$$\|\lambda_f\| = \|f\|_\Gamma .$$

Another interesting observation is that if $f(z)$ is the rational function

$$f(z) = \sum_{k=1}^{m} \left(w_k \prod_{i \ne k} \frac{1 - \bar{z}_i z_k}{z_k - z_i} \right) \frac{1 - \bar{z}_k z}{z - z_k} \tag{4.5}$$

like the function $k(z)$ defined in (3.43), then as we have seen in Sect. 3.3.3,

$$\begin{aligned}&\inf\{||h||_{H^\infty} : h \in H^\infty, h(z_i) = w_i, i = 1, \cdots, m\} \\ &= \inf\{||f - g||_{H^\infty} : g \in H^\infty\} ,\end{aligned}$$

where $z_1, \cdots, z_m$ are distinct points in $|z| < 1$ and $w_1, \cdots, w_m$ are arbitrary complex numbers. Combining this with (4.5) and Nehari's theorem, we have the following:

Theorem 4.3. Let $f(z)$ be a rational function as defined in (4.5) where $z_1, \cdots, z_m$ are distinct points in $|z| < 1$ and $w_1, \cdots, w_m$ are arbitrary complex numbers. Also, let λ_f be the corresponding linear functional on H^1 as defined in (4.4). Then

$$\begin{aligned}||f||_\Gamma = ||\lambda_f|| &= \inf\{||h||_{H^\infty} : h \in H^\infty, h(z_i) = w_i, i = 1, \cdots, m\} \\ &= \inf\{||f - g||_{H^\infty} : g \in H^\infty\}.\end{aligned}$$

Hence, approximation in the Hankel norm is also related to minimum-norm interpolation. Therefore, Theorem 3.6 and its consequences are useful in the study of the Hankel norms of rational functions whose poles lie in $|z| < 1$.

Another important observation is that the results of Kronecker (Theorem 2.8 in Sect. 2.3.1) and Nehari (Theorem 4.1) together form an important foundation of the AAK theory. Indeed, by using the notation $\mathcal{H}_m$ for the collection of all bounded infinite Hankel matrix operators (on l^2) with rank no greater than m as introduced in Sect. 2.4.1, and denoting by

$$\mathcal{R}_m^s = \left\{ r_n(z) = \frac{p_1 z^{n-1} + \cdots + p_n}{z^n + q_1 z^{n-1} + \cdots + q_n} : \right.$$

$$\left. \text{all poles of } r_n(z) \text{ lie in } |z| < 1,\ n \le m \right\}, \tag{4.6}$$

the collection of all stable and strictly proper rational functions of degree at most m, it is not difficult to see that the following result is a consequence of these two celebrated results, namely, Theorems 2.8 and 4.1 (Problem 4.5).

Corollary 4.1. A function

$$h(z) = \sum_{n=1}^{\infty} h_n z^{-n}$$

is in $\mathcal{R}_m^s$ if and only if its corresponding Hankel matrix Γ_h is in $\mathcal{H}_m$.

Example 4.3. Consider the function $f(z) = \sum_{n=1}^{\infty} z^{-n}$. It is easily seen that its associated Hankel matrix

$$\Gamma_f = \begin{bmatrix} 1 & 1 & 1 & \cdots \\ 1 & 1 & \cdots & \cdots \\ 1 & \cdots & \cdots & \cdots \\ \cdots & \cdots & \cdots & \cdots \end{bmatrix}$$

has rank 1. It is also clear that $f(z)$ has a pole at $z = 1$. Hence, $f(z)$ is not in $\mathcal{R}_m^s$, and by Corollary 4.1, $\Gamma_f \notin \mathcal{H}_m$ for any m.

Example 4.4. The function

$$f(z) = \sum_{n=1}^{\infty} (\rho^n - r^n) z^{-n}$$

is in $\mathcal{R}_2^s$ if and only if $|\rho| < 1$ and $|r| < 1$. It is in $\mathcal{R}_2^s$ but not in $\mathcal{R}_1^s$ if, in addition, $\rho \neq r$. Consequently, the associated Hankel matrix

$$\Gamma_f = \begin{bmatrix} \rho - r & \rho^2 - r^2 & \rho^3 - r^3 & \cdots \\ \rho^2 - r^2 & \rho^3 - r^3 & \cdots & \cdots \\ \rho^3 - r^3 & \cdots & \cdots & \cdots \\ \cdots & \cdots & \cdots & \cdots \end{bmatrix}$$

is in $\mathcal{H}_2$ if and only if $|\rho| < 1$ and $|r| < 1$; and Γ_f is not in $\mathcal{H}_1$, if, in addition, $\rho \neq r$.

Note, however, that if $\rho = r$, then $f(z) = 0$ and $\Gamma_f = 0$.

4.1.2 The AAK Theorem and Optimal Hankel-Norm Approximations

From the above discussion, it is reasonable to expect an intimate relationship between approximation of an L^∞ function by rational functions from $\mathcal{R}_m^s$ and approximation of its corresponding infinite Hankel matrix by finite-rank infinite Hankel matrices from $\mathcal{H}_m$. That this is indeed the case is a main result of AAK [1971]. To state this result more precisely, we need the following notation:

$$H_m^\infty = \{r(z) + g(z) : \ r(z) \in \mathcal{R}_m^s, \ g(z) \in H^\infty\}. \tag{4.7}$$

Note that $H_0^\infty \equiv H^\infty$, the Hardy space of bounded analytic functions in $|z| < 1$ (Chap. 3). Here, recall that we use the notation $L^\infty = L^\infty(|z| = 1)$ and for every function $f(z)$ in L^∞, we use Γ_f to denote the infinite Hankel matrix associated with (the singular part of) $f(z)$. The following is the main result of AAK [1971].

Theorem 4.4. Let $f(z)$ be a function in L^∞ and m a non-negative integer. Then

$$\inf_{r \in \mathcal{R}_m^s} \|f - r\|_\Gamma = \inf_{h \in H_m^\infty} \|f - h\|_{L^\infty}. \tag{4.8}$$

In the special case when $m = 0$, we have $\mathcal{R}_0^s = \{0\}$, and (4.8) becomes (4.2), so that the above result of AAK is indeed a generalization of Nehari's theorem. It must be emphasized that best approximation in the Hankel norm, namely, the extremal problem stated in the left-hand side of (4.8), is very important to our study of digital filter design and system reduction, since such best approximants are *stable* rational functions.

A very important contribution of the AAK approach is that not only the actual error of best approximation in (4.8) can be described and computed, but also the solution of the extremal problem on the right-hand side of (4.8) can be written in closed form. These will be further studied in the rest of this chapter.

Suppose that

$$\hat{h}(z) = \hat{r}_m(z) + \hat{g}(z), \tag{4.9}$$

where $\hat{r}_m(z) \in \mathcal{R}_m^s$ and $\hat{g}(z) \in H^\infty$, is a solution to the extremal problem stated in the right-hand side of (4.8), namely,

$$\|f - \hat{h}\|_{L^\infty} = \inf_{h \in H_m^\infty} \|f - h\|_{L^\infty}.$$

Then, by setting $\tilde{f}(z) = f(z) - \hat{r}_m(z)$, it is clear that $\|\tilde{f}\|_\Gamma = \|f - \hat{r}_m\|_\Gamma$ on one hand, and

$$\inf_{g \in H^\infty} \|\tilde{f} - g\|_{L^\infty} = \inf_{g \in H^\infty} \|f - (\hat{r}_m + g)\|_{L^\infty} = \inf_{h \in H_m^\infty} \|f - h\|_{L^\infty}$$

on the other. Hence, by Theorem 4.1 and Corollary 4.1, we may conclude that $\hat{r}_m(z)$ also solves the extremal problem stated in the left-hand side of (4.8), in the sense that

$$\|f - \hat{r}_m\|_\Gamma = \inf_{r \in \mathcal{R}_m^s} \|f - r\|_\Gamma = \inf_{h \in H_m^\infty} \|f - h\|_{L^\infty}. \tag{4.10}$$

In other words, to determine a best Hankel-norm approximant $\hat{r}_m(z)$ of $f(z) \in L^\infty$ from $\mathcal{R}_m^s$, we simply take the singular part of a best L^∞-norm approximant of $f(z)$ from H_m^∞.

In AAK [1971], it is proved that the actual error of best approximation in (4.8) is given by the $(m+1)$st *s-number* of Γ_f (arranged in nonincreasing order), and that $\hat{h}(z)$ [a solution to (4.8)] can be computed by using a corresponding *Schmidt pair*. Hence, to understand the AAK approach and construct computational schemes, we must understand s-numbers and Schmidt pairs, a topic to be discussed in detail in the next section. Here, the letter s may be considered as a letter to honor Schmidt who studied these numbers as a generalization of eigenvalues, see Gohberg and Krein [1969], but s-numbers are also called singular values of Γ. At this stage, we should remark that since all Hankel matrices are symmetric, if Γ_f is real and has finite rank, then the s-numbers of Γ_f are the same as the absolute values of its *eigenvalues*, with appropriate ordering. In addition, if $\xi \in l^2$ is an eigenvector corresponding to an eigenvalue λ of a Hankel matrix, then a Schmidt pair corresponding to the s-number (or singular value) $|\lambda|$ is given by

$$(\xi, \xi \operatorname{sgn} \lambda).$$

To demonstrate this remark and illustrate how the AAK approach works, let us first consider the following simple example in this section. More details on s-numbers and Schmidt pairs, the AAK theory, and computational techniques will be discussed later in Sects. 4.2 and 4.3.

Example 4.5. Let $f(z) = z^{-1} - z^{-3}$. Determine $\hat{r}_m(z) \in \mathcal{R}_m^s$ such that

$$\|f - \hat{r}_m\|_\Gamma = \inf_{r \in \mathcal{R}_m^s} \|f - r\|_\Gamma ,$$

$m = 0, 1, 2, 3, \cdots$.

We first note that Γ_f is given by

$$\Gamma_f = \begin{bmatrix} 1 & 0 & -1 & 0 & \cdots \\ 0 & -1 & 0 & \cdots & \cdots \\ -1 & 0 & \cdots & \cdots & \cdots \\ 0 & \cdots & \cdots & \cdots & \cdots \\ \cdots & \cdots & \cdots & \cdots & \cdots \end{bmatrix}$$

which has rank 3 and finite operator norm (on l^2). In fact, since the eigenvalues are

$$\lambda_1 = \frac{1+\sqrt{5}}{2}, \quad \lambda_2 = -1, \quad \lambda_3 = \frac{1-\sqrt{5}}{2},$$

and $\lambda_4 = \lambda_5 = \cdots = 0$, with $|\lambda_1| > |\lambda_2| > |\lambda_3| > |\lambda_4| = \cdots = 0$, we have

$$\|f\|_\Gamma = \|\Gamma_f\|_s = \frac{1+\sqrt{5}}{2} < \infty .$$

In addition, the corresponding eigenvectors may be chosen to be (Problem 4.6)

$$\xi_1 = \begin{bmatrix} 2 \\ 0 \\ 1-\sqrt{5} \\ 0 \\ \vdots \end{bmatrix}, \quad \xi_2 = \begin{bmatrix} 0 \\ 1 \\ 0 \\ 0 \\ \vdots \end{bmatrix}, \quad \xi_3 = \begin{bmatrix} -1+\sqrt{5} \\ 0 \\ 2 \\ 0 \\ \vdots \end{bmatrix}, \quad \xi_4 = \begin{bmatrix} 0 \\ 0 \\ 0 \\ 1 \\ 0 \\ \vdots \end{bmatrix}, \quad \cdots$$

Hence, we can now list the s-numbers (or singular values) of Γ_f (in decreasing order) and their corresponding Schmidt pairs as follows:

(i) *s-number*: $s_1 = |\lambda_1| = \frac{1+\sqrt{5}}{2}$, *Schmidt pair*: (ξ_1, ξ_1);
(ii) *s-number*: $s_2 = |\lambda_2| = 1$, *Schmidt pair*: $(\xi_2, -\xi_2)$;
(iii) *s-number*: $s_3 = |\lambda_3| = \frac{\sqrt{5}-1}{2}$, *Schmidt pair*: $(\xi_3, -\xi_3)$;
(iv) *s-numbers*: $s_4 = s_5 = \cdots = 0$, *Schmidt pairs*: $(\xi_4, \xi_4), (\xi_5, \xi_5), \cdots$.

To give the reader some feeling in advance for how the AAK theory works in solving the extremal problem (4.10) for this particular example, we further proceed as follows: According to AAK (see Sect. 4.3 for details), the errors of best Hankel-norm approximation for problem (4.10) are given by

$$s_{m+1} = \inf_{r \in \mathcal{R}_m^s} \|f - r\|_\Gamma, \quad m = 0, 1, \cdots . \tag{4.11}$$

In this example, the trivial cases are

$$\inf_{r \in \mathcal{R}_0^s} \|f - r\|_\Gamma = \|f\|_\Gamma = s_1 = \frac{1+\sqrt{5}}{2}$$

and

$$\inf_{r \in \mathcal{R}_m^s} \|f - r\|_\Gamma = 0, \quad m = 3, 4, \cdots ,$$

where the corresponding best Hankel-norm approximants must be

$$\hat{r}_0(z) = 0, \quad \hat{r}_m(z) = f(z) = \frac{z^2 - 1}{z^3}, \quad m = 3, 4, \cdots .$$

For each of the nontrivial cases $m = 1$ and 2, we will see from (4.55) in Sect. 4.3.3 that $\hat{r}_m(z)$ is actually given by the singular part of

$$\hat{h}_m(z) = \frac{[1 \;\; z \;\; z^2 \;\; \cdots \;]T_f \xi_{m+1}}{[1 \;\; z \;\; z^2 \;\; \cdots \;]\xi_{m+1}}, \tag{4.12}$$

where T_f is a Toeplitz matrix, defined by

$$T_f = \begin{bmatrix} 0 & f_1 & f_2 & f_3 & \cdots \\ 0 & 0 & f_1 & f_2 & \cdots \\ 0 & 0 & 0 & f_1 & \cdots \\ \cdots & \cdots & \cdots & \cdots & \cdots \end{bmatrix} \tag{4.13}$$

which is associated with the given L^∞ function

$$f(z) = \sum_{n=-\infty}^{\infty} f_n z^{-n} .$$

In this particular example, since

$$T_f = \begin{bmatrix} 0 & 1 & 0 & -1 & 0 & 0 & \cdots \\ 0 & 0 & 1 & 0 & -1 & 0 & \cdots \\ 0 & 0 & 0 & 1 & 0 & -1 & \cdots \\ \cdots & \cdots & \cdots & \cdots & \cdots & \cdots & \cdots \end{bmatrix},$$

it follows that

$$\hat{h}_1(z) = \frac{[1 \;\; z \;\; z^2 \;\; \cdots \;]T_f \begin{bmatrix} 0 \\ 1 \\ 0 \\ \vdots \end{bmatrix}}{[1 \;\; z \;\; z^2 \;\; \cdots] \begin{bmatrix} 0 \\ 1 \\ 0 \\ \vdots \end{bmatrix}} = \frac{1}{z}$$

and

$$\hat{h}_2(z) = \frac{[1 \quad z \quad z^2 \quad \cdots\ ]T_f \begin{bmatrix} -1+\sqrt{5} \\ 0 \\ 2 \\ 0 \\ \vdots \end{bmatrix}}{[1 \quad z \quad z^2 \quad \cdots\ ] \begin{bmatrix} -1+\sqrt{5} \\ 0 \\ 2 \\ 0 \\ \vdots \end{bmatrix}}$$

$$= \frac{z}{z^2 + (\sqrt{5}-1)/2} .$$

Since $\hat{h}_1(z)$ and $\hat{h}_2(z)$ have zero analytic parts, we conclude that

$$\hat{r}_1(z) = \hat{h}_1(z) = \frac{1}{z}$$

and

$$\hat{r}_2(z) = \hat{h}_2(z) = \frac{z}{z^2 + (\sqrt{5}-1)/2} ,$$

which are both stable, as expected.

4.2 s-Numbers and Schmidt Pairs

We have already seen that s-numbers and Schmidt pairs play a central role in the best Hankel-norm approximation. In this section, we give a detailed study of these two concepts. A discussion on best approximation of compact Hankel operators by finite-rank Hankel operators will also be given.

4.2.1 Adjoint and Normal Operators

An infinite matrix will always be considered as a linear operator on l^2, and it is bounded, if (considered as an operator) it is a bounded linear operator mapping l^2 to itself. As usual, the operator (or spectral) norm will be used for the norm of the infinite matrix. If A is an infinite matrix, then by using the same notation A as the corresponding linear operator on l^2, the spectrum of A is defined by

$$\sigma(A) = \{\lambda \in C : (A - \lambda I) \text{ is not invertible}\} . \tag{4.14}$$

It is well known in functional analysis that A *is bounded if and only if* $\sigma(A)$ *is a bounded set of complex numbers.*

Let $\mathcal{B}$ denote the space of all bounded linear operators (or infinite matrices) on l^2. Then for each $A \in \mathcal{B}$, its *adjoint*, denoted by A^*, is defined by:

$$\langle A^*\mathbf{x}, \mathbf{y}\rangle = \langle \mathbf{x}, A\mathbf{y}\rangle \tag{4.15}$$

for all $\mathbf{x}, \mathbf{y} \in l^2$, where $\langle\ ,\ \rangle$ is the usual inner product of two elements in l^2. Hence, as a matrix, we have $A^* = \bar{A}^\top$, where the complex conjugation in $\bar{A}$ is taken entry-wise. A is said to be *self-adjoint* (or *Hermitian*) if $A^* = A$.

Example 4.6. The infinite matrix $S : l^2 \to l^2$ defined by

$$S = \begin{bmatrix} 0 & 0 & 0 & \cdots \\ 1 & 0 & 0 & \cdots \\ 0 & 1 & 0 & \cdots \\ 0 & 0 & 1 & \cdots \\ \cdots & \cdots & \cdots & \cdots \end{bmatrix}$$

is called a *shift operator*. Clearly, for a vector $\mathbf{a} = [a_0, a_1, a_2, \cdots]^\top \in l^2$, we have

$$S: \ [a_0, a_1, a_2, \cdots]^\top \to [0, a_0, a_1, \cdots]^\top .$$

To find the adjoint operator S^* for S, we observe that for any vector $\mathbf{b} = [b_0, b_1, b_2, \cdots]^\top \in l^2$,

$$\begin{aligned} \langle \mathbf{b}, S\mathbf{a}\rangle &= \langle [b_0, b_1, b_2, \cdots]^\top, [0, a_0, a_1, \cdots]^\top\rangle \\ &= \sum_{k=0}^{\infty} b_{k+1}a_k \\ &= \langle [b_1, b_2, b_3, \cdots]^\top, [a_0, a_1, a_2, \cdots]^\top\rangle \\ &= \langle S^*\mathbf{b}, \mathbf{a}\rangle \end{aligned}$$

where

$$S^*: \ [b_0, b_1, b_2, \cdots]^\top \to [b_1, b_2, b_3, \cdots]^\top .$$

Hence,

$$S^* = \begin{bmatrix} 0 & 1 & 0 & 0 & \cdots \\ 0 & 0 & 1 & 0 & \cdots \\ 0 & 0 & 0 & 1 & \cdots \\ 0 & 0 & 0 & 0 & \cdots \\ \cdots & \cdots & \cdots & \cdots & \cdots \end{bmatrix}$$

is what we expected.

An operator (or infinite matrix) $A \in \mathcal{B}$ is said to be *normal* if

$$A^*A = AA^* . \tag{4.16}$$

Hence, any Hermitian operator A in $\mathcal{B}$ is normal, but the converse does not necessarily hold. For example, *if Γ is an infinite Hankel matrix in $\mathcal{B}$ then Γ is Hermitian if and only if it is real* (and by that we mean that all entries of the matrix are real, see Problem 4.10). However, there exist normal Hankel matrices in $\mathcal{B}$ which are not real and hence are not Hermitian.

Example 4.7. The Hankel matrix

$$\Gamma_1 = \begin{bmatrix} \mathrm{j} & \mathrm{j} & 0 & \cdots \\ \mathrm{j} & 0 & 0 & \cdots \\ 0 & 0 & 0 & \cdots \\ \cdots & \cdots & \cdots & \cdots \end{bmatrix},$$

$\mathrm{j} = \sqrt{-1}$, is normal. But since Γ_1 is not real, it is not Hermitian. Note that not all Hankel matrices are normal. In fact,

$$\Gamma_2 = \begin{bmatrix} 1 & \mathrm{j} & 0 & \cdots \\ \mathrm{j} & 0 & 0 & \cdots \\ 0 & 0 & 0 & \cdots \\ \cdots & \cdots & \cdots & \cdots \end{bmatrix}$$

is not normal (Problem 4.11).

It can be verified that *a Hankel matrix $\Gamma = [h_{i+\ell-1}]$ is normal if and only if $h_i\bar{h}_\ell$ is real for all i and ℓ* (Problem 4.11).

4.2.2 Singular Values of Hankel Matrices

Let Γ be an (infinite) Hankel matrix in $\mathcal{B}$. Then its adjoint is $\bar{\Gamma}$, and it is clear that $\bar{\Gamma}\Gamma$ is non-negative definite. We will use the notation $|\Gamma| = (\bar{\Gamma}\Gamma)^{1/2}$ to denote the *positive square root* of $\bar{\Gamma}\Gamma$, that is, $|\Gamma|^2 = \bar{\Gamma}\Gamma$ and $|\Gamma|$ is non-negative definite. Note that $|\Gamma|$ is not necessarily Hankel (Problem 4.12). The *spectrum of condensation*, $c(|\Gamma|)$, of $|\Gamma|$ is the union of the set of all limit points of the spectrum $\sigma(|\Gamma|)$ of $|\Gamma|$ and the set of all eigenvalues of $|\Gamma|$ with infinite multiplicity. Clearly, both $\sigma(|\Gamma|)$ and $c(|\Gamma|)$ are subsets of $[0, \infty)$, and we set

$$s_\infty(\Gamma) = \sup\{x : \ x \in c(|\Gamma|)\} . \tag{4.17}$$

Hence, if we label all the eigenvalues of $|\Gamma|$ by $s_m := s_m(\Gamma)$ and arrange them in non-increasing order, with multiplicities being listed, then we have

$$s_1 \geq s_2 \geq \cdots \geq s_\infty \geq 0 . \tag{4.18}$$

These values are called the *s-numbers* (or *singular values*) of Γ.

As we mentioned in Sect. 4.1.2, see (4.11), the s-number $s_{m+1}(\Gamma_f)$ gives the exact error of best approximation in the Hankel norm of an L^∞ function $f(z)$ from $\mathcal{R}_m^s$. Hence, for the collection

$$\mathcal{R}^s = \bigcup_{m=0}^{\infty} \mathcal{R}_m^s$$

of all stable and strictly proper rational functions to be dense in a collection $\mathcal{F}$ of functions in L^∞ under the Hankel norm, it is necessary and sufficient that $s_m(\Gamma_f) \to 0$ as $m \to \infty$ for all $f(z)$ in $\mathcal{F}$. A necessary and sufficient condition for $s_m(\Gamma_f) \to 0$ is that Γ_f is a *compact* operator on l^2, or equivalently, $f \in H^\infty + C$, where C is the family of continuous functions, see Partington [1988].

Recall that an operator $A \in \mathcal{B}$ is called a *compact* (or *completely continuous*) operator on l^2 if the closure of the image of any bounded set in l^2 under A is a compact set in l^2. In other words, A is compact if it maps weakly convergent sequences to strongly convergent ones. An important property of a compact operator A is that the only limit point of the sequence of eigenvalues of A is zero. We will use the notation $\mathcal{B}_c$ for the space of all compact operators on l^2. Of course, $\mathcal{B}_c \subset \mathcal{B}$ and any finite-rank operator $\mathcal{B}$ is in $\mathcal{B}_c$ (Problem 4.13).

4.2.3 Schmidt Series Representation of Compact Operators

Let Γ be an infinite Hankel matrix in $\mathcal{B}_c$ and $|\Gamma|$ the positive square root of $\bar{\Gamma}\Gamma$. Then by the (classical) polar decomposition theorem, see Rudin [1973], there exists a (not necessarily unique) unitary operator U (an infinite matrix) such that

$$\Gamma = U|\Gamma|. \tag{4.19}$$

Now, since it is clear that $|\Gamma|$ is compact, Hermitian, and non-negative definite, we may list all the eigenvalues of $|\Gamma|$, which are the same as the s-numbers $s_m = s_m(\Gamma)$ of Γ, as follows:

$$s_1 \geq s_2 \geq \cdots \geq s_m \geq \cdots \geq 0$$

where $s_m \to 0$ as $m \to \infty$. Corresponding to each s_m, we may choose an eigenvector $\mathbf{x}_m \in l^2$ of $|\Gamma|$, such that

$$|\Gamma|\mathbf{x}_m = s_m\mathbf{x}_m \tag{4.20}$$

and $\{\mathbf{x}_1, \mathbf{x}_2, \cdots\}$ forms a *complete orthonormal set* in the range of $|\Gamma|$. Hence, by setting

$$V = [\mathbf{x}_1 \ \mathbf{x}_2 \ \cdots \ ], \tag{4.21}$$

where the mth column of V is the eigenvector $\mathbf{x}_m$, we have a unitary operator V on l^2 that satisfies

$$|\Gamma|V = V \begin{bmatrix} s_1 & & \\ & s_2 & \\ & & \ddots \end{bmatrix}$$

or

$$|\Gamma| = V \begin{bmatrix} s_1 & & \\ & s_2 & \\ & & \ddots \end{bmatrix} V^* = \sum_{m=1}^{\infty} s_m \mathbf{x}_m \mathbf{x}_m^* , \tag{4.22}$$

where $\mathbf{x}_m^* = \bar{\mathbf{x}}_m^\top$, and the convergence of the infinite series will be clear by using an argument analogous to the proof of Theorem 4.5 to be discussed later in this section (Problem 4.15). Here, note that, for each m, $\mathbf{x}_m \mathbf{x}_m^*$ is an infinite matrix and so can be considered as an operator as well. The convergence of a series of operators, of course, means convergence in the operator norm. It is also called *strong convergence*. Hence, we may interchange the operator U with the summation to obtain

$$\Gamma = U|\Gamma| = \sum_{m=1}^{\infty} s_m (U\mathbf{x}_m)\mathbf{x}_m^* .$$

Moreover, by setting

$$\begin{cases} \xi_m = \mathbf{x}_m , \\ \eta_m = U\mathbf{x}_m , \end{cases} \tag{4.23}$$

we have

$$\Gamma = \sum_{m=1}^{\infty} s_m \eta_m \xi_m^* . \tag{4.24}$$

Also, it follows from (4.23) that the pair (ξ_m, η_m) satisfies the equations (Problem 4.16)

$$\begin{cases} \Gamma \xi_m = s_m \eta_m , \\ \bar{\Gamma} \eta_m = s_m \xi_m . \end{cases} \tag{4.25}$$

Any pair of elements in l^2 that satisfies (4.25) is called a *Schmidt pair* of Γ corresponding to the s-number $s_m = s_m(\Gamma)$, and the infinite series of the operator (4.24) is called a *Schmidt series representation* of Γ.

Example 4.8. Consider the rank-2 infinite Hankel matrix:

$$\Gamma = \begin{bmatrix} 1 & \mathrm{j} & 0 & \cdots \\ \mathrm{j} & 0 & 0 & \cdots \\ 0 & 0 & 0 & \cdots \\ \cdots & \cdots & \cdots & \cdots \end{bmatrix},$$

where $\mathrm{j} = \sqrt{-1}$.

We have already seen from Example 4.7 that Γ is *not* normal. Since the s-numbers $\{s_m\}$ of Γ are defined to be the eigenvalues of $|\Gamma|$, we will first determine $|\Gamma|$. It can be easily verified that the eigenvalues and corresponding eigenvectors (with unit length) of the matrix

$$\bar{\Gamma}\Gamma = \begin{bmatrix} 2 & \mathrm{j} & 0 & \cdots \\ -\mathrm{j} & 1 & 0 & \cdots \\ 0 & 0 & 0 & \cdots \\ \cdots & \cdots & \cdots & \cdots \end{bmatrix}$$

are given by (Problem 4.18)

$$\text{(i)}\ s_1^2 = \tfrac{3+\sqrt{5}}{2}, \quad \mathbf{x}_1 = \frac{1}{\sqrt{10-2\sqrt{5}}} \begin{bmatrix} -2\mathrm{j} \\ 1-\sqrt{5} \\ 0 \\ \vdots \end{bmatrix};$$

$$\text{(ii)}\ s_2^2 = \tfrac{3-\sqrt{5}}{2}, \quad \mathbf{x}_2 = \frac{1}{\sqrt{10+2\sqrt{5}}} \begin{bmatrix} -2\mathrm{j} \\ 1+\sqrt{5} \\ 0 \\ \vdots \end{bmatrix};$$

$$\text{(iii)}\ s_m^2 = 0, \quad \mathbf{x}_m = [0 \ \cdots \ 0 \ 1 \ 0 \ \cdots]^\top;$$

where 1 appears at the mth component, and $m = 3, 4, \cdots$. Hence, by setting

$$V = [\mathbf{x}_1 \ \mathbf{x}_2 \ \cdots]$$

we arrive at, see (4.22), the following:

$$|\Gamma| = (\bar{\Gamma}\Gamma)^{1/2} = VDV^* = \frac{1}{\sqrt{5}} \begin{bmatrix} 3 & \mathrm{j} & 0 & \cdots \\ -\mathrm{j} & 2 & 0 & \cdots \\ 0 & 0 & 0 & \cdots \\ \cdots & \cdots & \cdots & \cdots \end{bmatrix},$$

where

$$D = \begin{bmatrix} s_1 & 0 & 0 & \cdots \\ 0 & s_2 & 0 & \cdots \\ 0 & 0 & s_3 & \cdots \\ \cdots & \cdots & \cdots & \cdots \end{bmatrix}$$

with

$$s_1 = \sqrt{\frac{3+\sqrt{5}}{2}} = \frac{\sqrt{5}+1}{2},$$

$$s_2 = \sqrt{\frac{3-\sqrt{5}}{2}} = \frac{\sqrt{5}-1}{2},$$

$$s_3 = s_4 = \cdots = 0\,.$$

That is, the s-numbers and corresponding Schmidt pairs of Γ are

$$\text{(i)}\ \ s_1(\Gamma) = s_1 = \tfrac{\sqrt{5}+1}{2}\,, \qquad \begin{cases} \xi_1 = \mathbf{x}_1\,, \\ \eta_1 = U\mathbf{x}_1 = \dfrac{1}{\sqrt{10-2\sqrt{5}}}\begin{bmatrix} -2\mathrm{j} \\ \sqrt{5}-1 \\ 0 \\ \vdots \end{bmatrix}, \end{cases}$$

$$\text{(ii)}\ \ s_2(\Gamma) = s_2 = \tfrac{\sqrt{5}-1}{2}\,, \qquad \begin{cases} \xi_2 = \mathbf{x}_2 \\ \eta_2 = U\mathbf{x}_2 = \dfrac{1}{\sqrt{10+2\sqrt{5}}}\begin{bmatrix} 2\mathrm{j} \\ \sqrt{5}+1 \\ 0 \\ \vdots \end{bmatrix}, \end{cases}$$

$$\text{(iii)}\ \ s_m(\Gamma) = 0\,, \qquad \begin{cases} \xi_m = \mathbf{x}_m\,, \\ \eta_m = U\mathbf{x}_m = \mathbf{x}_m\,, \end{cases} \qquad m = 3, 4, \cdots,$$

where we have chosen

$$U = \begin{bmatrix} 1/\sqrt{5} & \mathrm{j}2/\sqrt{5} & 0 & \cdots \\ \mathrm{j}2/\sqrt{5} & 1/\sqrt{5} & 0 & \cdots \\ 0 & 0 & 1 & \cdots \\ \cdots & \cdots & \cdots & \cdots \end{bmatrix}$$

in the polar decomposition $\Gamma = U|\Gamma|$. Note that U is not unique since $|\Gamma|$ is not invertible.

We first comment that the eigenvalues s_1, s_2 of $|\Gamma|$ are usually different from the absolute value of the eigenvalues of Γ. Indeed, in the above example, the two nonzero eigenvalues of Γ are

$$\frac{1}{2}(1 \pm \mathrm{j}\sqrt{3})\,,$$

which have absolute value 1 that lies strictly between s_1 and s_2. We have also experienced in this example that determining $|\Gamma|$ and consequently its eigenvalues and eigenvectors to yield the Schmidt pairs is not an easy task. Fortunately, if Γ is normal, that is, $\bar{\Gamma}\Gamma = \Gamma\bar{\Gamma}$, then the s-numbers $\{s_m\}_{m=1}^{\infty}$ of Γ are simply the absolute values of the corresponding eigenvalues $\{\lambda_m\}_{m=1}^{\infty}$ of Γ, and the corresponding Schmidt pairs of Γ are $\{(\mathbf{x}_m, (\mathrm{sgn}\ \lambda_m)\mathbf{x}_m)\}_{m=1}^{\infty}$, where $\mathbf{x}_m$ is an eigenvector of Γ relative to the eigenvalue λ_m (Problem 4.19). Hence, the procedure to determine the s-numbers and their corresponding

Schmidt pairs of a compact normal Hankel operator can be outlined as follows:

(i) *Determine the eigenvalue and eigenvector pairs* $\{(\lambda_m, \mathbf{x}_m)\}_{m=1}^{\infty}$ *of* Γ, *where*

$$|\lambda_1| \geq |\lambda_2| \geq \cdots ,$$

with $|\lambda_m| \to 0$.

(ii) *The s-numbers and corresponding Schmidt pairs of* Γ *are:*

$$s_m = |\lambda_m| \quad \text{and} \quad (\mathbf{x}_m, (\text{sgn } \lambda_m)\mathbf{x}_m),$$

where sgn $\lambda_m = \lambda_m/|\lambda_m|$ *if* $\lambda_m \neq 0$ *and* 0 *if* $\lambda_m = 0$.

To verify this procedure, we simply note that if Γ is compact and normal, and has polar decomposition $\Gamma = U|\Gamma|$, then

$$\text{sgn } \lambda_m = \mathbf{x}_m^* U \mathbf{x}_m .$$

This identity follows from the fact that for a normal Γ the eigenvalues of $|\Gamma|$ are precisely the absolute value of those of Γ with the same eigenvectors (Problem 4.21). Hence, if we set $\xi_m = \mathbf{x}_m$ as in (4.23), then

$$\eta_m = U\mathbf{x}_m = (\text{sgn } \lambda_m)\mathbf{x}_m .$$

4.2.4 Approximation of Compact Hankel Operators

In this section, we consider approximation of compact Hankel operators by operators with finite rank.

Let $\mathcal{B}_m$ denote the collection of all bounded linear operators A on l^2 with rank $A \leq m$. In the following, we only consider approximation of compact Hankel operators from $\mathcal{B}_m$. (Approximation of arbitrary bounded Hankel operators will be studied in Chap. 5).

Theorem 4.5. Let Γ be a compact Hankel operator with s-numbers

$$s_1 \geq s_2 \geq \cdots$$

and corresponding Schmidt pairs

$$(\xi_1, \eta_1), (\xi_2, \eta_2), \cdots .$$

Then

$$s_{m+1} = \inf_{A \in \mathcal{B}_m} \|\Gamma - A\|_s , \tag{4.26}$$

$m = 0, 1, \cdots$. Furthermore, the infimum is attained by the mth partial sum

$$S_m = \sum_{k=1}^{m} s_k \eta_k \xi_k^* \tag{4.27}$$

of the Schmidt series (4.24) of Γ, that is,

$$\|\Gamma - S_m\|_s = s_{m+1} \,. \tag{4.28}$$

Note that in general the operator S_m is not a Hankel operator (Problem 4.23).

To prove this theorem, we first observe that since each column vector η_k or ξ_k has rank 1, the operator $\eta_k \xi_k^*$ is in $\mathcal{B}_1$. Also, since the sum of m one-dimensional subspaces is a subspace of dimension no greater than m, the operator S_m in (4.27) is in $\mathcal{B}_m$. We next show that $\|\Gamma - S_m\|_s \leq s_{m+1}$.

Let $N > m$ and set

$$E_m^N = \sum_{k=m+1}^{N} s_k \eta_k \xi_k^* \,.$$

For each $\mathbf{y}$ in l^2 with $\|\mathbf{y}\|_{l^2} = 1$, we consider a linear functional $\mathbf{w_y}(\cdot)$ defined on l^2 by $\mathbf{w_y}(\mathbf{u}) := \langle E_m^N \mathbf{y}, \mathbf{u} \rangle$ for all $\mathbf{u} \in l^2$. Now, let $\mathbf{y} \in l^2$ with $\|\mathbf{y}\|_{l^2} = 1$ be fixed. Then we have

$$\|\mathbf{w_y}\| = \sup_{\|\mathbf{u}\|_{l^2}=1} |\langle E_m^N \mathbf{y}, \mathbf{u} \rangle| \,.$$

Consequently, for all $\mathbf{u} \in l^2$ with $\|\mathbf{u}\|_{l^2} = 1$, we have

$$\begin{aligned}
|\mathbf{w_y}(\mathbf{u})| &= |\langle E_m^N \mathbf{y}, \mathbf{u} \rangle| \\
&= \left| \sum_{k=m+1}^{N} s_k \langle \mathbf{y}, \xi_k \rangle \langle \eta_k, \mathbf{u} \rangle \right| \\
&\leq s_{m+1} \sum_{k=m+1}^{N} |\langle \mathbf{y}, \xi_k \rangle \langle \eta_k, \mathbf{u} \rangle| \\
&\leq s_{m+1} \left(\sum_{k=m+1}^{N} |\langle \mathbf{y}, \xi_k \rangle|^2 \right)^{1/2} \left(\sum_{k=m+1}^{N} |\langle \eta_k, \mathbf{u} \rangle|^2 \right)^{1/2} \\
&\leq s_{m+1} \|\mathbf{y}\|_{l^2} \|\mathbf{u}\|_{l^2} \\
&= s_{m+1} \,,
\end{aligned}$$

where the Schwarz and Bessel inequalities have been applied and the latter can be used due to the fact that both $\{\xi_k\}$ and $\{\eta_k\}$ are orthonormal sets. It then follows that for all $\mathbf{y} \in l^2$ with $\|\mathbf{y}\|_{l^2} = 1$,

$$\|E_m^N \mathbf{y}\|_{l^2} = \sup_{\|\mathbf{u}\|_{l^2}=1} |\mathbf{w_y}(\mathbf{u})| \leq s_{m+1} \,,$$

and

$$\|E_m^N\|_s = \sup_{\|\mathbf{y}\|_{l^2}=1} \|E_m^N \mathbf{y}\|_{l^2} \le s_{m+1} \,.$$

Since the upper bound is independent of N, we may let $N \to \infty$, yielding

$$\|\Gamma - S_m\|_s \le s_{m+1} \,. \tag{4.29}$$

To show that s_{m+1} is also a lower bound, we recall the so-called "minimax characterization" of eigenvalues for non-negative definite compact operators, see Wilkinson [1965]. Since s_{m+1} is the $(m+1)$st eigenvalue of $|\Gamma|$, we have

$$s_{m+1} = \inf_{\dim S^\perp = m} \sup_{\substack{\mathbf{x}\in S \\ \|\mathbf{x}\|_{l^2}=1}} \||\Gamma|\mathbf{x}\|_{l^2} \,,$$

where the infimum is taken over each subspace S of l^2 such that the subspace $S^\perp$ defined by

$$S^\perp = \{\mathbf{x} \in l^2 : \langle \mathbf{x}, \mathbf{y} \rangle = 0 \text{ for all } \mathbf{y} \in S\}$$

has dimension m. Now since Γ is compact, it has a polar decomposition $\Gamma = U|\Gamma|$, where U is unitary, so that

$$\|\Gamma \mathbf{x}\|_{l^2} = \||\Gamma|\mathbf{x}\|_{l^2} \,,$$

or

$$s_{m+1} = \inf_{\dim S^\perp = m} \sup_{\substack{\mathbf{x}\in S \\ \|\mathbf{x}\|_{l^2}=1}} \|\Gamma\mathbf{x}\|_{l^2} \,.$$

In addition, since $s_1 \ge s_2 \ge \cdots$, this identity can be written as

$$s_{m+1} = \inf_{\dim S^\perp \le m} \sup_{\substack{\mathbf{x}\in S \\ \|\mathbf{x}\|_{l^2}=1}} \|\Gamma\mathbf{x}\|_{l^2} \,. \tag{4.30}$$

Now let $A \in \mathcal{B}_m$. Then rank $A \le m$, so that $\dim \nu(A)^\perp \le m$ where $\nu(A)$ indicates the null space of A. Hence, by (4.30), we have

$$\begin{aligned} s_{m+1} &\le \sup_{\substack{\mathbf{x}\in \nu(A) \\ \|\mathbf{x}\|_{l^2}=1}} \|\Gamma\mathbf{x}\|_{l^2} \\ &= \sup_{\substack{\mathbf{x}\in \nu(A) \\ \|\mathbf{x}\|_{l^2}=1}} \|(\Gamma - A)\mathbf{x}\|_{l^2} \\ &\le \|\Gamma - A\|_s \,. \end{aligned}$$

Since this inequality holds for all A in $\mathcal{B}_m$, it follows that

$$s_{m+1} \le \inf_{A \in \mathcal{B}_m} \|\Gamma - A\|_s \,. \tag{4.31}$$

Combining (4.29) and (4.31), together with the evident inequality

$$\inf_{A\in\mathcal{B}_m} \|\Gamma - A\|_s \le \|\Gamma - S_m\|_s ,$$

we have (4.26) and (4.28). This completes the proof of the theorem.

We remark that since Γ is compact so that $s_{m+1}(\Gamma) \to 0$ as $m \to \infty$, it follows from (4.29) that the Schmidt series (4.24) of Γ converges strongly (i.e., in norm) to Γ. However, if Γ is only bounded, the above proof shows that the series converges weakly (Problem 4.15).

4.3 System Reduction

As mentioned before, a remarkable contribution of AAK's result is its significance in system engineering. One of its various important applications is system reduction. As is well known, finding a lower-dimensional linear system to approximate a given high-dimensional one in a certain optimal sense is an important problem in systems engineering. The AAK result provides a closed-form characterization of this problem in the sense that the Hankel norm of the error between the given high-order transfer function and the approximant is minimized over all transfer functions of the same (lower) order. The advantage of this approach over other methods such as the Padé approximation is that an optimal criterion is followed and the resultant reduced-dimensional linear system is automatically stable. In this section, we will first give a detailed proof of the AAK theorem for finite-rank Hankel matrices, and then show how to apply it to solve system reduction problems for SISO systems. The study of system reduction problems for MIMO systems will be delayed to Chap. 6.

4.3.1 Statement of AAK's Theorem

Let $f(z)$ be a function in $L^\infty = L^\infty(|z|=1)$. We have seen in Section 4.1.3 that a best Hankel-norm approximant $\hat{r}_m(z)$ of $f(z)$ from $\mathcal{R}^s_m$ [defined in (4.6)] is given by the singular part of a best L^∞-norm approximant $\hat{h}(z)$ of $f(z)$ from H^∞_m [defined in (4.7)]. This is equivalent to solving the following extremal problem:

$$\|f - \hat{r}_m\|_\Gamma = \inf_{r\in\mathcal{R}^s_m} \|f - r\|_\Gamma ,$$

or

$$\|\Gamma_f - \Gamma_{\hat{r}_m}\|_s = \inf_{\text{rank } \Gamma_r \le m} \|\Gamma_f - \Gamma_r\|_s . \tag{4.32}$$

It is also known (Theorem 4.5) that if Γ_f is a compact (or more generally, bounded) operator, then

$$\inf_{A \in \mathcal{B}_m} \|\Gamma_f - A\|_s = s_{m+1},$$

where s_{m+1} is the $(m+1)$st s-number of the Hankel matrix Γ_f. Moreover, it is clear that a solution $\hat{A}$ of the extremal problem (4.32) is given by

$$\hat{A} = \sum_{i=1}^{m} s_i \eta_i \xi_i^*,$$

which is the mth partial sum of $\sum_{i=1}^{\infty} s_i \eta_i \xi_i^*$, the Schmidt series of the Hankel matrix Γ_f. However, the operator $\hat{A}$ is not a Hankel matrix in general ($\hat{A}$ is not even symmetric, Problem 4.23), and hence, does not give a solution to the extremal problem (4.32). In addition, since $\hat{A}$ is not Hankel, it is not realizable, and consequently, not useful in signal processing and systems engineering as we have already seen in Sect. 2.2.

According to a result of the AAK approach, the extremal problem (4.32) is solvable and moreover a Hankel matrix can be found as a solution to problem (4.32). More precisely, we have the following result in which only compact operators are considered.

Theorem 4.6. (Adamjan, Arov, and Krein)

Let $f(z)$ be a given function in $L^\infty = L^\infty(|z|=1)$ such that Γ_f is a compact operator with s-numbers $s_1 \geq s_2 \geq \cdots \geq s_\infty = 0$ and let (ξ_m, η_m), $\xi_m = [u_1^{(m)} u_2^{(m)} \cdots]^\top$ and $\eta_m = [v_1^{(m)} v_2^{(m)} \cdots]^\top$, be the Schmidt pair corresponding to s_m. Then a solution to the extremal problem

$$\|f - \hat{r}_m\|_\Gamma = \inf_{r \in \mathcal{R}_m^s} \|f - r\|_\Gamma,$$

where $\mathcal{R}_m^s$ is defined in (4.6), is given by the singular part $\hat{r}_m(z)$ of $\hat{h}(z)$, that is,

$$\hat{r}_m(z) = [\hat{h}(z)]_s, \tag{4.33}$$

where $[\cdot]_s$ denotes the singular part of the argument, and

$$\hat{h}(z) = f(z) - s_{m+1} \frac{\eta_-(z)}{\xi_+(z)} \tag{4.34a}$$

with

$$\xi_+(z) = \sum_{i=1}^{\infty} u_i^{(m+1)} z^{i-1} \tag{4.34b}$$

and

$$\eta_-(z) = \sum_{i=1}^{\infty} v_i^{(m+1)} z^{-i}. \tag{4.34c}$$

Moreover,

$$\|f - \hat{r}_m\|_\Gamma = \inf_{r \in \mathcal{R}_m^s} \|f - r\|_\Gamma = s_{m+1} \,. \tag{4.35}$$

We remark that the assumption on compactness of the Hankel matrix Γ_f can be relaxed to boundedness according to AAK's original result (Chap. 5). However, in this case s_∞ may not be equal to zero, and the analogous statements should be phrased more carefully.

Another remark is that an illustration of AAK's proof of the above theorem for the special case where the Hankel matrix Γ_f is real and has finite rank, and an application of this result of AAK to the system reduction problems were given in Kung [1980]. In the following, we give a detailed proof for this special case by following Kung's procedure, which was derived by some appropriate modifications of AAK's original proof. This may help the reader to understand the AAK results more easily. A more thorough discussion of the general theory of AAK will be given in the next chapter.

4.3.2 Proof of the AAK Theorem for Finite-Rank Hankel Matrices

In this section, we will supply a detailed proof of the AAK theorem (namely, Theorem 4.6) for the special case where the given Hankel matrix is real and has rank M with $M < \infty$.

First, to facilitate our treatment, let $f(z) = f_a(z) + f_s(z)$ where $f_a(z)$ and $f_s(z)$ are the analytic and singular parts of $f(z)$, respectively. Since Γ_f has rank M, it follows that

$$f_s(z) = \frac{P_{M-1}(z)}{Q_M(z)} \,, \tag{4.36}$$

where $P_{M-1}(z)$ and $Q_M(z)$ are irreducible real-coefficient polynomials in z with degree P_M $<$ degree $Q_M = M$, and that Γ_f has M nonzero (real) eigenvalues, which we denote by $\lambda_1, \cdots, \lambda_M$. For convenience, set $|\lambda_1| \geq |\lambda_2| \geq \cdots \geq |\lambda_M| > \lambda_{M+1} = \lambda_{M+2} = \cdots = 0$, and let $\{\mathbf{x}_i\}_{i=1}^M$ be the corresponding eigenvectors. Recall from Problems 4.10 and 4.13 that Γ_f is compact and normal, so that from Sect. 4.2.3, the s-numbers and the corresponding Schmidt pairs $\{\eta_i, \xi_i\}_{i=1}^\infty$ of Γ_f are given by

$$s_i = |\lambda_i|, \qquad \xi_i = \mathbf{x}_i, \qquad \eta_i = (\text{sgn}\,\lambda_i)\mathbf{x}_i, \qquad i = 1, 2, \cdots \,. \tag{4.37}$$

It is important to note that if $m \geq M$, then since f_s is the singular part of the given function $f(z)$ whose corresponding Hankel matrix has rank M, we have $f_s = P_{M-1}/Q_M \in \mathcal{R}_m^s$, where $\mathcal{R}_m^s$ is defined in (4.6). Hence, by letting $\hat{r}_m = f_s$, we have

$$\|f - \hat{r}_m\|_\Gamma = \|f - f_s\|_\Gamma = \|f_a\|_\Gamma = 0 \,,$$

which solves the extremal problem. In other words, the proof of the theorem in this case is trivial. For this reason, it is sufficient to study the problem for $m < M$. In doing so, let $m < M$ be an arbitrary but fixed positive integer. Then, $s_i > 0$ for $i = 1, \cdots, m, m+1, \cdots, M$. In the rest of this chapter, since m will not be changed, it is deleted in the following notation:

$$\xi := \xi_{m+1} = [u_1^{(m+1)} u_2^{(m+1)} \cdots]^{\top} := [\, u_1 \; u_2 \; \cdots \,]^{\top},$$

$$\eta := \eta_{m+1} = [v_1^{(m+1)} v_2^{(m+1)} \cdots]^{\top} := [\, v_1 \; v_2 \; \cdots \,]^{\top}.$$

As in (4.34b,c), let

$$\xi_+(z) = \sum_{i=1}^{\infty} u_i z^{i-1} \tag{4.38}$$

and

$$\eta_-(z) = \sum_{i=1}^{\infty} v_i z^{-i}. \tag{4.39}$$

Then the AAK theorem can be established by verifying the following sequence of lemmas.

Lemma 4.1. Let $f_s(z)$, $\xi_+(z)$, and $\eta_-(z)$ be given by (4.36), (4.38), and (4.39), respectively, and let the singular part of a given function $g(z)$ be denoted by $[g(z)]_s$. Then

$$[f_s(z)\xi_+(z) - s_{m+1}\eta_-(z)]_s = 0. \tag{4.40}$$

To prove this result, we first observe the following three facts: First, by (4.37-39) we have

$$s_{m+1}\eta = |\lambda_{m+1}|(\mathrm{sgn}\lambda_{m+1})\xi = \lambda_{m+1}\xi,$$

so that

$$s_{m+1}\eta_-(z) = \lambda_{m+1} \sum_{i=1}^{\infty} u_i z^{-i}.$$

Second, by a direct calculation it can be verified that

$$\sum_{i=1}^{\infty}\sum_{\ell=1}^{\infty} f_i u_\ell z^{\ell-i-1} = \sum_{i=1}^{\infty}\sum_{\ell=1}^{\infty} f_{i+\ell-1} u_\ell z^{-i} + \sum_{i=1}^{\infty}\sum_{\ell=1}^{\infty} f_{\ell-i+1} u_{\ell+1} z^{i-1},$$

in which $f_p = 0$ for all $p \leq 0$ (Problem 4.24). Third,

$$\begin{bmatrix} f_1 & f_2 & \cdots \\ f_2 & \cdots & \cdots \\ \cdots & \cdots & \cdots \end{bmatrix} \begin{bmatrix} u_1 \\ u_2 \\ \vdots \end{bmatrix} = \lambda_{m+1} \begin{bmatrix} u_1 \\ u_2 \\ \vdots \end{bmatrix}$$

implies that

$$\sum_{\ell=1}^{\infty} f_{i+\ell-1} u_\ell = \lambda_{m+1} u_i, \quad i = 1, 2, \cdots .$$

Hence, we finally have

$$\begin{aligned} & [f_s(z)\xi_+(z) - s_{m+1}\eta_-(z)]_s \\ &= \left[\sum_{i=1}^{\infty} \sum_{\ell=1}^{\infty} f_i u_\ell z^{\ell-i-1} - \sum_{i=1}^{\infty} \lambda_{m+1} u_i z^{-i} \right]_s \\ &= \left[\sum_{i=1}^{\infty} \sum_{\ell=1}^{\infty} f_{i+\ell-1} u_\ell z^{-i} + \sum_{i=1}^{\infty} \sum_{\ell=i}^{\infty} f_{\ell-i+1} u_{\ell+1} z^{i-1} - \sum_{i=1}^{\infty} \lambda_{m+1} u_i z^{-i} \right]_s \\ &= \left[\sum_{i=1}^{\infty} \sum_{\ell=i}^{\infty} f_{\ell-i+1} u_{\ell+1} z^{i-1} \right]_s \\ &= \left[\sum_{i=1}^{\infty} \sum_{\ell=i+1}^{\infty} f_i u_\ell z^{\ell-i-1} \right]_s \\ &= 0 . \end{aligned}$$

This establishes the lemma.

Lemma 4.2. Let $Q_M(z)$ and $\eta_-(z)$ be given by (4.36) and (4.39), respectively. Then the function

$$B(z) := (\text{sgn } \lambda_{m+1}) Q_M(z) \eta_-(z) \tag{4.41}$$

is in $\mathcal{P}_{M-1}$ with real coefficients.

Here, $\mathcal{P}_n$ is the family of polynomials in z with the highest degree not exceeding n.

We first show that

$$[B(z)]_s = (\text{sgn } \lambda_{m+1})[Q_M(z)\eta_-(z)]_s = 0 .$$

Indeed, since $P_{M-1}(z)\xi_+(z)$ has no negative powers of z, we have

$$[P_{M-1}(z)\xi_+(z)]_s = 0 ,$$

so that by applying (4.40),

$$\begin{aligned}
[B(z)]_s &= (\text{sgn } \lambda_{m+1})[Q_M(z)\eta_-(z)]_s \\
&= -s_{m+1}^{-1}(\text{sgn } \lambda_{m+1})[P_{M-1}(z)\xi_+(z) - s_{m+1}Q_M(z)\eta_-(z)]_s \\
&= -s_{m+1}^{-1}(\text{sgn } \lambda_{m+1})[Q_M(z)(f_s(z)\xi_+(z) - s_{m+1}\eta_-(z))]_s \\
&= 0\,.
\end{aligned}$$

Hence, $B(z) = (\text{sgn}\lambda_{m+1})Q_M(z)\eta_-(z)$ has only an analytic part. Since

$$\lim_{|z|\to\infty} \frac{B(z)}{Q_M(z)} = \lim_{|z|\to\infty} (\text{sgn } \lambda_{m+1})\eta_-(z) = 0\,,$$

the degree of $B(z)$ is at most $M-1$. This completes the proof of the lemma.

Next, for any polynomial $P(z) = \sum_{i=0}^n p_i z^i$ with possibly complex coefficients, let $\check{P}(z)$ be its reciprocal polynomial defined by $\check{P}(z) = \sum_{i=0}^n \bar{p}_i z^{n-i}$ (Sect. 3.2.2). Note that $\check{P}(z) = z^n \overline{P((\bar{z})^{-1})}$. Since all the polynomials in the rest of this chapter will have real coefficients, it is also clear that $\check{P}(z) = z^n P(\frac{1}{z})$. We then have the following:

Lemma 4.3. Let $P_{M-1}(z)$, $Q_M(z)$, and $B(z)$ be the polynomials defined as in Lemmas 4.1 and 4.2. Then the function

$$C(z) = \frac{P_{M-1}(z)\check{B}(z) - \lambda_{m+1}\check{Q}_M(z)B(z)}{Q_M(z)} \tag{4.42}$$

is in $\mathcal{P}_{M-1}$ with real coefficients.

To show this, we first note that

$$\begin{aligned}
\xi_+(z) &= \sum_{i=1}^{\infty} u_i z^{i-1} = \sum_{i=1}^{\infty} (\text{sgn } \lambda_{m+1}) v_i z^{i-1} \\
&= (\text{sgn } \lambda_{m+1}) z^{-1}\eta_-(z^{-1}) = \frac{z^{M-1}B(z^{-1})}{z^M Q_M(z^{-1})} \\
&= \frac{\check{B}(z)}{\check{Q}_M(z)}\,.
\end{aligned} \tag{4.43}$$

Substituting (4.41) and (4.43) into (4.40) and noting that $\lambda_{m+1} = s_{m+1}(\text{sgn } \lambda_{m+1})$, we have

$$\left[\frac{P_{M-1}\check{B}(z) - \lambda_{m+1}\check{Q}_M(z)B(z)}{Q_M(z)\check{Q}_M(z)}\right]_s = 0\,. \tag{4.44}$$

Since the polynomial in the numerator has a degree not greater than $2M-1$, we have by using the partial fraction method,

$$\frac{P_{M-1}(z)\check{B}(z) - \lambda_{m+1}\check{Q}_M(z)B(z)}{Q_M(z)\check{Q}_M(z)} = \frac{A(z)}{Q_M(z)} + \frac{C(z)}{\check{Q}_M(z)}$$

for some polynomials $A(z)$ and $C(z)$ of degree not greater than $M-1$. We will see that $A(z) \equiv 0$, so that it follows from (4.44) that the polynomial $C(z)$ satisfies (4.42). Indeed, since $\check{Q}_M(z)$ is analytic at 0 and $Q_M(z)$ has all its zeros in $|z|<1$ (recall that $f_s = P_{M-1}/Q_M \in \mathcal{R}_m^s$), we have

$$\frac{1}{\check{Q}_M(z)} = \sum_{i=0}^{\infty} q_i z^i$$

for some real constants $q_0, q_1, \cdots$, with $q_0 \neq 0$ (since $M > 0$), so that

$$\frac{1}{Q_M(z)} = \frac{1}{z^M \check{Q}_M(z^{-1})} = z^{-M} \sum_{i=0}^{\infty} q_i z^{-i}, \qquad (q_0 \neq 0).$$

Note that the degree of $A(z)$ does not exceed $M-1$, and so $A(z)/Q_M(z)$ has only a singular part. But, since $C(z)/\check{Q}_M(z) = C(z)\sum_{i=0}^{\infty} q_i z^i$ is analytic, it follows from (4.44) that $A(z) \equiv 0$. This completes the proof of the lemma.

Lemma 4.4. Let $B(z)$ and $C(z)$ be the polynomials with real coefficients as defined in Lemmas 4.2 and 4.3, respectively, and set

$$h(z) = \frac{C(z)}{\check{B}(z)}. \tag{4.45}$$

Then

$$h(z) = f(z) - s_{m+1}\frac{\eta_-(z)}{\xi_+(z)} \tag{4.46}$$

and

$$\|f_s - h_s\|_\Gamma = s_{m+1}. \tag{4.47}$$

Note that (4.46) follows immediately from (4.41-43). To prove (4.47), we first combine (4.42), (4.43), and (4.41) to yield

$$\begin{aligned} f_s(z) - h_s(z) &= \frac{P_{M-1}(z)}{Q_M(z)} - \left[\frac{C(z)}{\check{B}(z)}\right]_s \\ &= \left[\lambda_{m+1}\frac{B(z)}{Q_M(z)}\frac{\check{Q}_M(z)}{\check{B}(z)}\right]_s \\ &= s_{m+1}\left[\frac{\eta_-(z)}{\xi_+(z)}\right]_s. \end{aligned} \tag{4.48}$$

In addition, since $\xi_+(z) = \overline{z\eta_-(z)}$ for $|z|=1$, see (4.37), we have

$$\left|\frac{\eta_-(z)}{\xi_+(z)}\right| = 1, \qquad \text{for } |z| = 1. \tag{4.49}$$

Hence, it follows from (4.48), (4.3), and (4.49) that

$$\begin{aligned} \|f_s - h_s\|_\Gamma &= \left\| s_{m+1} \left[\frac{\eta_-}{\xi_+}\right]_s \right\|_\Gamma \\ &= \left\| s_{m+1} \frac{\eta_-}{\xi_+} \right\|_\Gamma \\ &\le \left\| s_{m+1} \frac{\eta_-}{\xi_+} \right\|_{L^\infty} \\ &= s_{m+1} . \end{aligned}$$

On the other hand, we can prove that s_{m+1} is also an s-number [which may not be the $(m+1)st$ one] of the Hankel matrix $\Gamma_{f_s - h_s} = \Gamma_{f_s} - \Gamma_{h_s}$, so that $\|f_s - h_s\|_\Gamma \ge s_{m+1}$, and hence (4.47) is obtained.

To show that s_{m+1} is also an s-number of $\Gamma_{f_s - h_s}$, we first apply (4.48) to obtain

$$\begin{aligned} \left[[f_s(z) - h_s(z)]\, \xi_+(z)\right]_s &= s_{m+1} \left[\left[\frac{\eta_-(z)}{\xi_+(z)}\right]_s \xi_+(z)\right]_s \\ &= s_{m+1} \left[\frac{\eta_-(z)}{\xi_+(z)} \xi_+(z)\right]_s \\ &= s_{m+1} \eta_-(z) , \end{aligned}$$

which is equivalent to the identity

$$\begin{bmatrix} f_1 - h_1 & f_2 - h_2 & \cdots \\ f_2 - h_2 & \cdots & \cdots \\ \cdots & \cdots & \cdots \end{bmatrix} \begin{bmatrix} u_1 \\ u_2 \\ \vdots \end{bmatrix} = s_{m+1} (\mathrm{sgn}\ \lambda_{m+1}) \begin{bmatrix} u_1 \\ u_2 \\ \vdots \end{bmatrix} ,$$

where $f_s(z) = \sum_{i=1}^\infty f_i z^{-i}, h_s(z) = \sum_{i=1}^\infty h_i z^{-i}$, and $\xi_+(z)$ and $\eta_-(z)$ are defined in (4.38-39). Since $\Gamma_{f_s} = [f_{i+\ell-1}]$ and $\Gamma_{h_s} = [h_{i+\ell-1}]$ are both real, it implies that s_{m+1} is an s-number of the Hankel matrix $\Gamma_{f_s - h_s}$. This completes the proof of the lemma.

Now, by comparing (4.34a) and (4.35) with (4.46) and (4.47), we see that in order to establish Theorem 4.6, what is left to prove is that the degree of the denominator of $h_s(z)$ is exactly equal to m, so that by the Kronecker theorem we have $\Gamma_{h_s} \in \mathcal{R}_m^s$, where $\mathcal{R}_m^s$ is defined in (4.6). In fact, we have the following:

Lemma 4.5. Let $h(z)$ be defined as in Lemma 4.4 and $h_s(z)$ be the singular part of $h(z)$. Then the degree of the denominator of $h_s(z)$ is exactly equal to m.

For simplicity, we only verify this result for the case where $s_{m+2} < s_{m+1} < s_m$, although the result in general holds when the multiplicity of s_{m+1} is larger than 1. The proof is based on the standard perturbation technique, but its verification becomes somewhat tedious (Problem 4.25).

Let n be the degree of the denominator of $h_s(z)$ and ℓ the number of zeros of $\check{B}(z)$ in $|z| < 1$. Since $h_s(z) = \left[C(z)/\check{B}(z)\right]_s$ may have pole-zero cancellation, we have $n \leq \ell$. In view of (4.26) and (4.47), we also have

$$s_{n+1} = \inf_{A \in \mathcal{B}_n} \|\Gamma_f - A\|_s \leq \|\Gamma_f - \Gamma_{h_s}\|_s = s_{m+1} \,.$$

Since s_{m+1} is simple and the s-numbers are arranged in non-increasing order, it is necessary that $m \leq n$. Hence, if we can establish $\ell \leq m$, then we have $n = m$, and the proof of the lemma is complete.

To show that $\ell \leq m$, let us first write the rational function $\xi_+(z) = \check{B}(z)/\check{Q}_M(z)$, see (4.43), in the form

$$\xi_+(z) = \xi_I(z)\xi_O(z)$$

where $\xi_I(z) = \prod_{i=1}^{\ell}(z - \alpha_i)/(1 - \bar{\alpha}_i z)$, with $|\alpha_i| < 1$, is the inner factor and $\xi_O(z)$ the outer factor of $\xi_+(z)$ with $\xi_O^{-1}(z) \in H^\infty$ (Chap. 3). Then we will show that

(i) for each $i = 1, 2, \cdots$,

$$s_i(\Gamma_f) \geq s_i(\Gamma_{f_s\xi_I}) \,,$$

where $s_i(\Gamma_f)$ and $s_i(\Gamma_{f_s\xi_I})$ are, respectively, the s-numbers of the Hankel matrices Γ_f and $\Gamma_{f_s\xi_I}$; and

(ii) $s_{m+1}(\Gamma_f)$ is the largest s-number of the Hankel matrix $\Gamma_{f_s\xi_I}$ with multiplicity $\ell + 1$, that is,

$$s_1(\Gamma_{f_s\xi_I}) = \cdots = s_{\ell+1}(\Gamma_{f_s\xi_I}) = s_{m+1}(\Gamma_f).$$

Once these two assertions are proved, we may then conclude that

$$s_{\ell+1}(\Gamma_f) \geq s_{\ell+1}(\Gamma_{f_s\xi_I}) = s_{m+1}(\Gamma_f) > s_{m+2}(\Gamma_f) \,,$$

so that $\ell + 1 < m + 2$, or $\ell \leq m$, as required.

The proof of the first assertion is simple. Since $|\xi_I(e^{j\theta})| = 1$, for any $w(z) \in H^2$, we have

$$\|[f_s\xi_I]_s w\|_{L^2} \leq \|f_s\xi_I w\|_{L^2} = \|f_s w\|_{L^2}$$

which is equivalent to

$$\|\Gamma_{f_s\xi_I}\mathbf{x}\|_{l^2} \leq \|\Gamma_f \mathbf{x}\|_{l^2}$$

for all $\mathbf{x} \in l^2$. Based on a result from linear algebra, see Wilkinson [1965], this implies (Problem 4.26)

$$s_i(\Gamma_{f_s\xi_I}) \le s_i(\Gamma_f), \quad i = 1,2,\cdots .$$

The proof of the second assertion is relatively more complicated. On one hand, it follows from (4.40) that

$$[f_s(z)\xi_I(z)\xi_O(z) - s_{m+1}\eta_-(z)]_s = 0\,, \tag{4.50}$$

or

$$[f_s(z)\xi_I(z)]_s = s_{m+1}[\eta_-(z)\xi_O^{-1}(z)]_s\,.$$

It then follows from Lemma 4.2, (4.49), and the fact $|\xi_I^{-1}(e^{j\theta})| = 1$ that

$$\begin{aligned}\|[f_s\xi_I]_s\|_\Gamma &= \|s_{m+1}[\eta_-\xi_O^{-1}]_s\|_\Gamma \\ &\le \|s_{m+1}\eta_-\xi_O^{-1}\xi_I^{-1}\|_{L^\infty} \\ &= s_{m+1}\left\|\frac{\eta_-}{\xi_+}\right\|_{L^\infty} \\ &= s_{m+1}\,. \end{aligned} \tag{4.51}$$

On the other hand, recall that since $\xi_+(z)$ has ℓ distinct zeros $\alpha_1,\cdots,\alpha_\ell$, we may decompose $\xi_I(z)$ into

$$\xi_I(z) = \prod_{i=1}^{\ell+1}\xi_i(z)$$

with $\xi_i(z) = (z-\alpha_i)/(1-\bar{\alpha}_i z), i = 1,\cdots,\ell$, and $\xi_{\ell+1}(z) \equiv 1$. For any $i = 1,\cdots,\ell+1$, define

$$x_i(z) = \frac{\xi_O(z)\bar{\xi}_i(z)}{\|\bar{\xi}_i\eta_-\|_{L^2}}$$

and

$$y_i(z) = \frac{\bar{\xi}_i(z)\eta_-(z)}{\|\bar{\xi}_i\eta_-\|_{L^2}}\,,$$

where $\bar{\xi}_i(z) := \overline{\xi_i(\bar{z})}$. Then, since $\eta_{m+1} = (\mathrm{sgn}\lambda_{m+1})\xi_{m+1}$, by (4.37) we again have $\|\xi_+\|_{L^2} = \|\eta_-\|_{L^2}$, and since $|\xi_i(e^{j\theta})| = 1$ for all $i = 1,\cdots,\ell+1$, we also have $\|x_i\|_{L^2} = \|y_i\|_{L^2} = 1$. From (4.50), it follows that

$$\begin{aligned}&[[f_s(z)\xi_I(z)]_s x_i(z)]_s \\ &= [[f_s(z)\xi_I(z)]_s\xi_O(z)\bar{\xi}_i(z)]_s \Big/ \|\bar{\xi}_i\eta_-\|_{L^2} \\ &= [[f_s(z)\xi_I(z)\xi_O(z)]_s\bar{\xi}_i(z)]_s \Big/ \|\bar{\xi}_i\eta_-\|_{L^2} \\ &= [[s_{m+1}\eta_-(z)]_s\bar{\xi}_i(z)]_s \Big/ \|\bar{\xi}_i\eta_-\|_{L^2} \\ &= s_{m+1}\eta_-(z)\bar{\xi}_i(z) \Big/ \|\bar{\xi}_i\eta_-\|_{L^2} \\ &= s_{m+1}y_i(z)\,. \end{aligned} \tag{4.52}$$

Similarly, we have (Problem 4.27)

$$[[f_s(z)\xi_I(z)]_s y_i(z)]_s = s_{m+1}x_i(z)\,. \tag{4.53}$$

This implies that s_{m+1} is an s-number of the (real) Hankel matrix $\Gamma_{f_s\xi_I}$ [and hence, (4.51) implies that s_{m+1} is the largest s-number of $\Gamma_{f_s\xi_I}$] with the corresponding Schmidt pair given by $x_i(z)$ and $y_i(z)$. Equations (4.52) and (4.53) together imply that $(x_i(z), y_i(z))$ is a Schmidt pair corresponding to the s-number s_{m+1}, and is determined by $\xi_i(z) = (z-\alpha_i)/(1-\bar{\alpha}_i z)$, $i = 1,\cdots,\ell+1$ with $\xi_{\ell+1}(z) \equiv 1$. Since $\{\xi_i(z)\}_{i=1}^{\ell+1}$ is a linearly independent set, $\xi_I(z)$ has $\ell+1$ linearly independent Schmidt pairs given by $(x_i(z), y_i(z)),\cdots,(x_{\ell+1}(z), y_{\ell+1}(z))$. Hence, s_{m+1} is the largest s-number of $\Gamma_{f_s\xi_i}$ with multiplicity $\ell+1$, namely,

$$s_1(\Gamma_{f_s\xi_i}) = \cdots = s_{\ell+1}(\Gamma_{f_s\xi_I}) = s_{m+1} = s_{m+1}(\Gamma_f)\,.$$

This completes the proof of the second assertion, and hence the proof of the lemma.

Now, combining Lemmas 4.4 and 4.5, we have actually established all results stated in Theorem 4.6.

4.3.3 Reformulation of AAK's Result

As pointed out in Sect. 4.1.2, the optimal solution $\hat{r}_m(z)$ in the AAK theorem can be reformulated as follows: For $\Gamma_f = [f_{i+\ell-1}]_{i,\ell=1}^{\infty}$, define the corresponding Toeplitz matrix

$$T_f = \begin{bmatrix} 0 & f_1 & f_2 & f_3 & \cdots \\ \cdots & 0 & f_1 & f_2 & \cdots \\ \cdots & \cdots & 0 & f_1 & \cdots \\ \cdots & \cdots & \cdots & 0 & \cdots \\ \cdots & \cdots & \cdots & \cdots & \cdots \end{bmatrix}, \tag{4.54}$$

set $\tilde{\mathbf{z}} = [\ 1\ z\ z^2\ \cdots\ ]^\top$, and let $\mathbf{u} = [\ u_1\ u_2\ \cdots\ ]^\top$ be the eigenvector corresponding to the $(m+1)$st eigenvalue λ_{m+1} of the Hankel matrix Γ_f. Then we have the following:

Theorem 4.7. The solution to the extremal problem

$$\|f-\hat{r}_m\|_\Gamma = \inf_{r\in\mathcal{R}_m^s} \|f-r\|_\Gamma\,,$$

where $\mathcal{R}_m^s$ is defined in (4.6), is given by

$$\hat{r}_m(z) = [\hat{h}(z)]_s = \left[\frac{\tilde{\mathbf{z}}^\top T_f \mathbf{u}}{\tilde{\mathbf{z}}^\top \mathbf{u}}\right]_s, \tag{4.55}$$

where $[\cdot]_s$ denotes, as usual, the singular part of the argument.

To verify this result, we first observe that (4.34a) yields

$$[\hat{h}(z)]_s = \left[\frac{f_s(z)\xi_+(z) - s_{m+1}\eta_-(z)}{\xi_+(z)}\right]_s . \tag{4.56}$$

From (4.37-39) and the relationship

$$\begin{bmatrix} f_1 & f_2 & f_3 & \cdots \\ f_2 & f_3 & \cdots & \cdots \\ f_3 & \cdots & \cdots & \cdots \\ \cdots & \cdots & \cdots & \cdots \end{bmatrix} \begin{bmatrix} u_1 \\ u_2 \\ \vdots \end{bmatrix} = \lambda_{m+1} \begin{bmatrix} u_1 \\ u_2 \\ \vdots \end{bmatrix},$$

we have

$$s_{m+1}\eta_-(z) = \lambda_{m+1} \sum_{i=1}^{\infty} u_i z^{-i} = \sum_{i=1}^{\infty} \left(\sum_{\ell=1}^{\infty} f_{i+\ell-1} u_\ell \right) z^{-i},$$

so that

$$\begin{aligned} & f_s(z)\xi_+(z) - s_{m+1}\eta_-(z) \\ =& \left(\sum_{i=1}^{\infty} f_i z^{-i}\right)\left(\sum_{\ell=1}^{\infty} u_\ell z^{\ell-1}\right) - \sum_{i=1}^{\infty}\left(\sum_{\ell=1}^{\infty} f_{i+\ell-1} u_\ell\right) z^{-i} \\ =& \sum_{i=1}^{\infty}\left(\sum_{\ell=i+1}^{\infty} f_{\ell-i} u_{\ell+1}\right) z^i \\ =& \tilde{\mathbf{z}}^\top T_f \mathbf{u} . \end{aligned}$$

Substituting this identity and the identity $\xi_+(z) = \tilde{\mathbf{z}}^\top \mathbf{u}$ into (4.56), we obtain (4.55) as claimed.

Now, we consider an important application of AAK's result to the system reduction problem for SISO systems, by illustrating with the following simple example.

Example 4.9. Let the transfer function $f(z)$ of a linear system be given by

$$f(z) = \frac{2z - a}{z(z-a)} = \frac{1}{z} + \frac{1}{z-a}, \quad 0 < a < 1 .$$

We wish to find best Hankel-norm approximants $\hat{r}_0(z)$ and $\hat{r}_1(z)$ of ranks 0 and 1, respectively, to the transfer function $f(z)$.

First, it can be easily verified that the eigenvalues and the corresponding eigenvectors of the Hankel matrix

$$\Gamma_f = \begin{bmatrix} 2 & a & a^2 & \cdots \\ a & a^2 & \cdots & \cdots \\ a^2 & \cdots & \cdots & \cdots \\ \cdots & \cdots & \cdots & \cdots \end{bmatrix}$$

are

$$\lambda_i = \frac{2 - a^2 \pm \sqrt{5a^4 - 8a^2 + 4}}{2(1 - a^2)} > 0\,,$$

and

$$\xi_i = \begin{bmatrix} -a/(1-a^2) \\ 2 - \lambda_i \\ (2-\lambda_i)a \\ (2-\lambda_i)a^2 \\ \vdots \end{bmatrix}, \qquad i = 1, 2\,,$$

(Problem 4.28a). By Formula (4.55), we have (Problem 4.28b)

$$\hat{r}_i(z) = [\hat{h}_i(z)]_s = \left[\frac{(2-\lambda_i)(1 + 1/(1-a^2))}{(2-\lambda_i)z - a(1-az)/(1-a^2)}\right]_s, \qquad i = 0, 1.$$

These two transfer functions, $\hat{r}_0(z)$ and $\hat{r}_1(z)$, are stable in the sense that all their poles are located in $|z| < 1$ (Problem 4.28c).

Finally, as mentioned before, the best rank-m Hankel approximants $\hat{r}_m(z)$ of $f(z)$ with $m \geq 2$ are all given by the singular part of the given function $f(z)$, namely,

$$\hat{r}_m(z) = f_s(z) = \frac{2z - a}{z(z-a)}\,.$$

4.4 H^∞-Minimization

We now turn to the study of a general H^∞-minimization problem which was formulated in Research Problem 5 in Sect. 2.4.3, where the norm of the objective functional has not been specified. In the following, we will restate this problem and specify the norm for further investigation.

4.4.1 Statement of the Problem

Let us first recall some notations. Let $\mathcal{M}$ be the collection of all finite rectangular matrices (of arbitrary dimensions) of rational functions such that for each element $M \in \mathcal{M}$ there exists a non-negative integer n and a polynomial $q_n(z) = z^n + q_1 z^{n-1} + \cdots + q_n$, so that

$$M = \frac{N_n(z)}{q_n(z)}\,,$$

where each entry of the matrix $N_n(z)$ is a polynomial of degree no greater than n. That is,

$$\mathcal{M} = \left\{ \frac{N_n(z)}{q_n(z)} \;:\; \begin{array}{l} q_n(z) = z^n + q_1 z^{n-1} + \cdots + q_n\,, \quad q_1, \cdots, q_n \in \mathcal{C} \\ N_n(z) \text{ a rectangular matrix of polynomials} \\ \text{of degree } \le n, \text{and } n = 0, 1, \cdots \end{array} \right\}. \tag{4.57}$$

Also, let

$$\mathcal{M}^s = \{\, M \in \mathcal{M} \;:\; \text{each entry of } M \text{ is analytic on } |z| \ge 1 \,\}. \tag{4.58}$$

For any given $H, P, Q \in \mathcal{M}^s$, consider

$$\mathcal{G} = \{\, G \in \mathcal{M}^s \;:\; H - PGQ \in \mathcal{M}^s \,\}. \tag{4.59}$$

Furthermore, recall that for a constant matrix $M = [m_{ij}]$, the l^2 norm of M is given by its length $|M| := \left(\sum_{i,j} |m_{ij}|^2\right)^{1/2}$, and that for a rational matrix $M(z)$ in $\mathcal{M}^s$, the H^∞ norm of $M(z)$ is defined to be

$$\|M(z)\|_{H^\infty} = \|\, |M(\mathrm{e}^{-\mathrm{j}\theta})| \,\|_{L^\infty[0,2\pi]}\,, \tag{4.60}$$

where $z = \mathrm{e}^{\mathrm{j}\theta}$, see (2.46).

Now, the general sensitivity minimization problem, or the so-called H^∞-optimization problem, can be stated as follows:

Research Problem 10. Let $P, H, Q \in \mathcal{M}^s$ with appropriate dimensions. Find a $\hat{G}$ in $\mathcal{G}$ such that

$$\|H - P\hat{G}Q\|_{H^\infty} = \inf_{G \in \mathcal{G}} \|H - PGQ\|_{H^\infty}\,. \tag{4.61}$$

In Sect. 4.4.3, a complete characterization of the solution to Research Problem 10 for certain SISO systems will be given. Note that for certain simple SISO systems, the two scalar-valued rational functions P and Q can be combined so that (4.61) becomes the approximation problem of finding a $\hat{G} \in \mathcal{G}$ such that

$$\|H - P\hat{G}\|_{H^\infty} = \inf_{G \in \mathcal{G}} \|H - PG\|_{H^\infty}\,, \tag{4.62}$$

where PQ is replaced by P. For the scalar-valued polynomial setting in H^p with $1 \le p < \infty$, this problem was already studied in Sect. 3.2.1 (Beurling's approximation theorem).

4.4.2 An Example of H^∞-Minimization

To motivate our study of the minimization problem stated in (4.61), let us discuss an example of "robust stabilization feedback design" of an SISO linear system. This consideration is somewhat different from the sensitivity minimization problem stated in Research Problem 5 in Sect. 2.4.3, but as we will see, it also leads to the same H^∞-optimization formulation.

Here, we recall that a rational matrix is said to be *proper* (or *strictly proper*) if the degree of the numerator of each entry is no greater (or strictly less) than that of its denominator. We also recall that a rational matrix is said to be *stable* if all its poles are located inside the open unit circle on the complex plane, or equivalently, if each entry is analytic on $|z| \geq 1$.

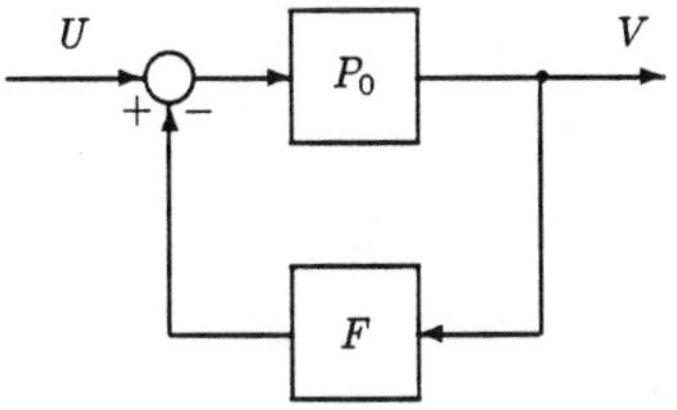

Fig. 4.1. A nominal feedback system

Consider the robust stabilization problem of the simple feedback system shown in Fig. 4.1, where $P_0 = P_0(z)$ is a given strictly proper rational matrix with real coefficients, which is not necessarily stable. P_0 is called the *nominal plant transfer function*. Suppose that the actual plant has a transfer function $P = P(z)$, with

$$P = P_0 + \Delta P\,,$$

where ΔP stands for an unknown error due to certain uncertainties such as modeling misfit, random perturbation, or parameter variation. Let $R = R(z)$ be a so-called *radius function*, which is a proper and stable rational function that dominates ΔP in the sense that

$$\|\Delta P\|_{H^\infty} < \|R\|_{H^\infty}\,. \tag{4.63}$$

The robust stabilization problem is to design a stable feedback function F for the purpose of maximizing the bounding radius $\|R\|_{H^\infty}$ of the error $\|\Delta P\|_{H^\infty}$ so that the system remains stable under the condition (4.63). Roughly speaking, we wish to design a stable feedback function F such that the overall closed-loop system is stable and that the allowable tolerance of the perturbation of the nominal plant transfer function P_0 is maximized.

Let us give a mathematical formulation of this robust stabilization problem. First, it can be easily verified (Problem 4.30) that the transfer function $H(z)$ for the ideal system (with plant error $\Delta P = 0$) is given by $V(z) = H(z)U(z)$ where

$$H = (I + P_0 F)^{-1} P_0 .$$

Next, it can also be shown that the system described in Fig. 4.2 is equivalent to the one described in Fig. 4.3, in the sense that the same input u yields the same output v, or that they have the same transfer function (Problem 4.31a,b), so that they have the same stability behavior. Moreover, it can be verified that the system shown in Fig. 4.3 is stable if both F and $(I+P_0F)^{-1}$ are stable and

$$\|\Delta P\|_{H^\infty} \|F(I + P_0 F)^{-1}\|_{H^\infty} < 1 , \tag{4.64}$$

(Problem 4.31c). This sufficient condition is actually a special case of the so-called *small gain theorem* (Desoer and Vidyasagar 1975).

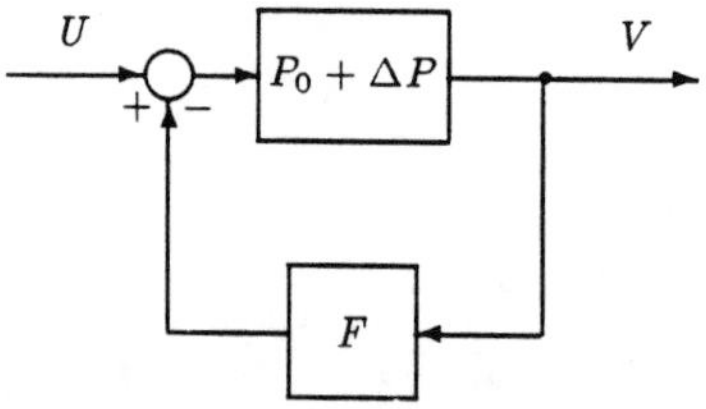

Fig. 4.2. A perturbed feedback system

In view of (4.63) and (4.64), we see that the radius function R should be chosen such that

$$\|R\|_{H^\infty} \le \frac{1}{\|F(I + P_0 F)^{-1}\|_{H^\infty}} .$$

It is also clear that in order to have the largest possible value of $\|R\|_{H^\infty}$, the quantity $\|F(I + P_0 F)^{-1}\|_{H^\infty}$ must be minimized over the set of all stable feedback functions $F \in \mathcal{M}^s$. In other words, we arrive at an H^∞-optimization problem

$$\min_{F \in \mathcal{F}} \|F(I + P_0 F)^{-1}\|_{H^\infty} , \tag{4.65}$$

where

$$\mathcal{F} = \{\ F \in \mathcal{M}^s \ : \quad (I + P_0 F)^{-1} \in \mathcal{M}^s \ \} .$$

Note that if both F and $(I + P_0 F)^{-1}$ are stable, then so is $F(I + P_0 F)^{-1}$. Under these stability conditions, the linear feedback system shown in Fig.

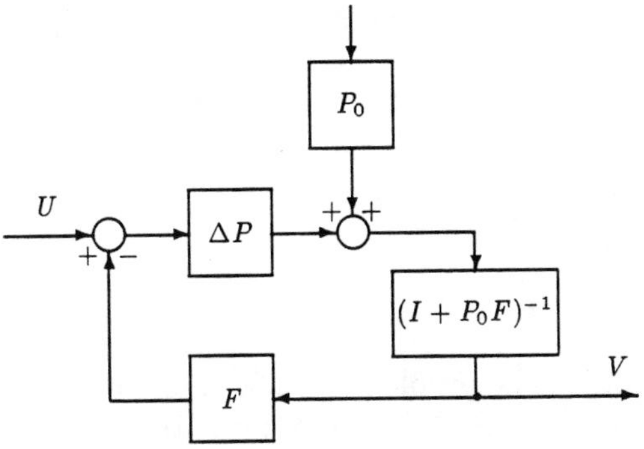

Fig. 4.3. An equivalent feedback system

4.1 is said to be *internally stable*, which means that the system is stable not only globally [from input to output through the system transfer function $F(I + P_0F)^{-1}$] but also locally [through F and $(I + P_0F)^{-1}$, respectively].

By following the idea shown at the end of Chap. 2, it is possible to consider a simpler optimization problem instead of the one described in (4.65). For this purpose, we first reformulate the problem of (4.65). Let

$$F = G(I - P_0G)^{-1} .$$

Then, since

$$I = (I - P_0G)(I - P_0G)^{-1} = (I - P_0G)^{-1} - P_0F ,$$

we have

$$(I + P_0F)^{-1} = I - P_0G .$$

Moreover, if we set

$$\mathcal{G} = \{ G \in \mathcal{M}^s : \quad I - P_0G \in \mathcal{M}^s \} ,$$

then (4.65) is reduced to

$$\min_{G \in \mathcal{G}} \|I - P_0G\|_{H^\infty} , \tag{4.66}$$

which is a special case of Research Problem 10 [see (4.62) and (2.52)]. A general form of such optimization problems may be stated as follows. For further information, the reader is referred to Vidyasagar [1985].

Research Problem 11. Given $P \in \mathcal{M}$ and $H, Q \in \mathcal{M}^s$ with appropriate dimensions, find a $\hat{G}$ in

$$\tilde{\mathcal{G}} = \{G \in \mathcal{M}^s : \; H - PGQ \in \mathcal{M}^s\}$$

such that

$$\|H - P\hat{G}Q\|_{H^\infty} = \inf_{G \in \tilde{\mathcal{G}}} \|H - PGQ\|_{H^\infty} . \tag{4.67}$$

Let us return to (4.66). Obviously, the problem of (4.66) is usually simpler than that of (4.65). If an optimal $\hat{G} \in \mathcal{G}$ is obtained, then we can determine the corresponding

$$\hat{F} = \hat{G}(I - P_0\hat{G})^{-1}$$

by the simple implementation shown in Fig. 2.8 at the end of Chap. 2. Consequently, a radius function R satisfying

$$\|R\|_{H^\infty} \le 1/\mu ,$$

where

$$\mu = \|\hat{F}(I + P_0\hat{F})^{-1}\|_{H^\infty} = \inf_{F \in \mathcal{F}} \|F(I + P_0F)^{-1}\|_{H^\infty} ,$$

gives rise to the robust stabilization. The above discussion shows how a typical engineering problem can be formulated as an H^∞-optimization problem. In the following, we only discuss Research Problem 10 and give a complete characterization to its solution.

4.4.3 Existence, Uniqueness, and Construction of Optimal Solutions

Let us return to the H^∞-optimization problem in (4.61). Before we discuss how to attack this problem, the question of solvability arises. Research Problem 10 is solvable with the infimum $\mu = 0$, if it happens that the matrix algebraic equation

$$H - PGQ = 0$$

has a solution $\hat{G} \in \mathcal{M}^s$, for any given $H, P, Q \in \mathcal{M}^s$. However, this trivial setting seldom occurs, especially for MIMO systems.

To study the solvability for the generic cases, let us first consider the following non-trivial example for SISO systems, where the factor Q has been combined with P for simplicity. This is possible, since in the scalar-valued setting functions G and Q commute.

Example 4.10. Let a plant function $P(z) = (z - \beta)/(z + \alpha)$, with $0 \le |\alpha|, \; |\beta| < 1$, be given and let $H \in \mathcal{M}^s$ with $|H(\beta)| < \infty$. Then, it is obvious that $P \in \mathcal{M}^s$ and $P(\beta) = 0$. For this example, a solution of the optimization problem

$$\min_{G\in\mathcal{G}} \|H - PG\|_{H^\infty}, \quad \text{where} \quad \mathcal{G} = \{\, G \in \mathcal{M}^s : \quad H - PG \in \mathcal{M}^s \,\}$$

is given by

$$\hat{G}(z) = P^{-1}(z)(H(z) - H(\beta)).$$

Indeed, it is clear that

$$\|H - PG\|_{H^\infty} \geq |H(\beta) - P(\beta)G(\beta)| = |H(\beta)|$$

for every $G \in \mathcal{G}$, so that

$$\mu = \inf_{G\in\mathcal{G}} \|H - PG\|_{H^\infty} \geq |H(\beta)|.$$

However, the function $\hat{G}$ defined above is clearly in $\mathcal{G}$ and satisfies

$$\|H - P\hat{G}\|_{H^\infty} = \|H(\beta)\|_{H^\infty} = |H(\beta)|.$$

Hence, $\hat{G}$ is an optimal solution of the problem.

In the above example, if it happens that

$$\|H\|_{H^\infty} = |H(\beta)|,$$

then $\hat{G} = 0$ is also an optimal solution. Hence, solutions to this problem are not unique.

Note also that the transfer function of the corresponding system with the optimal feedback $\hat{G}(z) = P^{-1}(z)(H(z) - H(\beta))$ is given by

$$H - P\hat{G} = H(\beta),$$

which is a constant (such a transfer function is said to be *all-pass*).

It is important to observe that the transfer function of the given plant P in the above example does not vanish on the unit circle $|z| = 1$. It will be shown in the next section that under this condition, Research Problem 10 always has a solution. However, the following example tells us that an optimal solution to Research Problem 10 (or 11) may not exist if the given plant P has at least one zero on $|z| = 1$.

Example 4.11. Let a plant $P(z) = (z-1)^2/4z^2$ be given and let $H(z) = (z-1)/2z$. Then, $P, H \in \mathcal{M}^s$. For any arbitrarily small $\epsilon > 0$, define

$$G_\epsilon(z) = \frac{2z}{\epsilon(z+1) + z - 1}.$$

Then, since the only pole for G_ϵ is $z = (1-\epsilon)/(1+\epsilon)$, we have $G_\epsilon \in \mathcal{M}^s$. Observe that

$$\begin{aligned}\|H - PG_\epsilon\|_{H^\infty} &= \left\| \frac{z-1}{2z} - \frac{(z-1)^2}{4z^2} \cdot \frac{2z}{\epsilon(z+1)+z-1} \right\|_{H^\infty} \\ &= \left\| \frac{\epsilon(z+1)(z-1)}{2z[\epsilon(z+1)+z-1]} \right\|_{H^\infty} \\ &\le \epsilon \left\| \frac{z+1}{2z} \right\|_{H^\infty} \left\| \frac{z-1}{\epsilon(z+1)+z-1} \right\|_{H^\infty} \le \epsilon\,,\end{aligned}$$

since for $|z| = 1$, the ratio $|z-1|/|z-(1-\epsilon)/(1+\epsilon)|$ is dominated by $1+\epsilon$. It follows that

$$\mu = \inf_{G \in \mathcal{G}} \|H - PG\|_{H^\infty} \le \inf_{\epsilon > 0} \|H - PG_\epsilon\|_{H^\infty} = 0\,,$$

where $\mathcal{G} = \{G \in \mathcal{M}^s : \; H - PG \in \mathcal{M}^s\}$ as defined before, so that the only solution of the minimization problem

$$\min_{G \in \mathcal{G}} \left\| \frac{z-1}{2} - \frac{(z-1)^2}{4z^2} G(z) \right\|_{H^\infty}$$

is

$$\hat{G}(z) = \lim_{\epsilon \to 0} G_\epsilon(z) = \frac{2z}{z-1}\,,$$

which is however not stable. In other words, this H^∞-optimization problem has no solution.

It follows from the above two examples that Research Problem 10 (or 11) may not have a solution, and even if it does, the solution may not be unique.

We will establish two theorems in the following on existence and uniqueness for Research Problem 10 for SISO systems. Let us first recall from Sect. 3.1.2 that a function $f \in H^\infty$ has the canonical factorization

$$f(z) = f_I(z) f_O(z)\,,$$

where $f_I(z)$ is an inner function which is the product of a Blaschke product and a singular inner function, and the Blaschke product contains all the zeros of $f(z)$ (that are inside the unit circle).

Example 4.12. The following functions in H^∞ have absolute value 1 for $|z| = 1$.

$$1\,, \quad \frac{4(z-1/2)^2}{(2-z)^2}\,, \quad \prod_{k=1}^{n} \frac{(k+1)z-1}{(k+1)-z}\,, \quad \frac{6z^2-z-1}{z^2+z-6}\,.$$

It is clear that they can be written as constant multiples of Blaschke products, and hence they are inner functions without singular factors. On the other hand, the function $S_\delta(z)$ in Example 3.1 is an inner function without a Blaschke product factor.

Let us now return to the H^∞-optimization problem for SISO systems and consider stable transfer rational functions $H, P \in \mathcal{M}^s$. The objective is to find a stable rational function $\hat{G} \in \mathcal{M}^s$ such that

$$\|H - P\hat{G}\|_{H^\infty} = \inf_{G \in \mathcal{G}} \|H - PG\|_{H^\infty}, \tag{4.68}$$

where

$$\mathcal{G} = \{ G \in \mathcal{M}^s : \quad H - PG \in \mathcal{M}^s \}.$$

Here, since we are considering only SISO systems, the function Q in the statement of Research Problem 10 has been combined with P.

It is important to point out that any rational function $M \in \mathcal{M}^s$ is stable in the sense that all its poles are located inside the unit circle, and hence, $M \notin H^\infty$. However, in the study of the H^∞-optimization problem for SISO systems, we are dealing with transfer functions of linear systems which are proper rational functions of finite degrees. Hence, although the functions $M \in \mathcal{M}^s$ do not have the canonical inner-outer factorizations, they can be written as

$$M(z) = cz^m \frac{(z - \tilde{z}_1)\cdots(z - \tilde{z}_\ell)(z - \hat{z}_1)\cdots(z - \hat{z}_p)}{(z - z_1)\cdots(z - z_n)}, \tag{4.69}$$

where $0 < |z_1|, \cdots, |z_n|, |\tilde{z}_1|, \cdots, |\tilde{z}_\ell| < 1$, $1 \leq |\hat{z}_1|, \cdots, |\hat{z}_p| < \infty$, $1 \leq \ell + m + p \leq n$, and c is a constant. Let

$$M_{I1}(z) = \prod_{i=1}^{n} \frac{|z_i|}{z_i} \frac{z - z_i}{1 - \bar{z}_i z}, \tag{4.70}$$

$$M_{I2}(z) = z^m \prod_{i=1}^{\ell} \frac{|\tilde{z}_i|}{\tilde{z}_i} \frac{z - \tilde{z}_i}{1 - \bar{\tilde{z}}_i z}, \tag{4.71}$$

and

$$M_O(z) = c \prod_{i=1}^{n} \frac{|z_i|}{z_i - |z_i|^2 z} \prod_{k=1}^{\ell} \frac{\tilde{z}_k}{|\tilde{z}_k|}(1 - \bar{\tilde{z}}_k z) \prod_{q=1}^{p} (z - \hat{z}_q). \tag{4.72}$$

Then, we have the factorization

$$M(z) = [M_{I1}(z)]^{-1} [M_{I2}(z)] M_O(z), \tag{4.73}$$

where $M_{I1}(z)$ and $M_{I2}(z)$ are Blaschke products (and hence inner functions), and $M_O(z)$ is an outer function.

Now, let $P, H \in \mathcal{M}^s$ be given as in (4.68) with $P = P_{I1}^{-1} P_{I2} P_O$, where $P_{I1}(z), P_{I2}(z)$ are Blaschke products and $P_O(z)$ is an outer function. Then, since P_{I1}^{-1} and P_{I2} are both in $\mathcal{M}^s$ with $|P_{I1}^{-1}(z)| = |P_{I2}(z)| = 1$ for all $|z| = 1$, we can reformulate the optimization problem in (4.68) as

$$\begin{aligned} \|H - PG\|_{H^\infty} &= \|P_{I1}^{-1} P_{I2}(P_{I1} P_{I2}^{-1} H - P_O G)\|_{H^\infty} \\ &= \|P_{I1} P_{I2}^{-1} H - P_O G\|_{H^\infty} \\ &= \|F - P_O G\|_{H^\infty} , \end{aligned} \tag{4.74}$$

where

$$F(z) = P_{I1}(z) P_{I2}^{-1}(z) H(z) . \tag{4.75}$$

Under the additional assumption that $P(z)$ does not vanish on $|z| = 1$, namely, $P_O(z)$ has no zero on $|z| = 1$, we then have $P_O(z)G(z) \in H^\infty$ if and only if $G(z) \in H^\infty$. Therefore, the problem of (4.68) reduces to finding a $\hat{G} \in H^\infty$ such that

$$\|F - \hat{G}\|_{H^\infty} = \inf_{G \in H^\infty} \|F - G\|_{H^\infty} . \tag{4.76}$$

Note that this is a scalar-valued (SISO) approximation problem. Hence, we are back to Nehari's theorem discussed in Sect. 4.1.2, and since the Hankel matrix Γ_F has finite rank, it can be solved completely by using the AAK techniques[see Theorem 4.6, or equivalently, (4.55)] with the s-number $s_1 = \|\Gamma_F\|$. More precisely, we can first decompose the function F in (4.76) as

$$F = F_a + F_s ,$$

where

$$F_a = \sum_{k=0}^{\infty} f_{-k} z^k$$

is the analytic part of F and

$$F_s = \sum_{k=1}^{\infty} f_k z^{-k}$$

is the singular part of F. Then, it follows from Nehari's theorem (Theorem 4.1) that

$$\inf_{G \in H^\infty} \|F - G\|_{L^\infty} = \|F_s\|_\Gamma = \|\Gamma_F\| , \tag{4.77}$$

where $\|F_s\|_\Gamma$ is the Hankel norm of F_s, or equivalently, the spectral norm of the corresponding Hankel matrix Γ_F. Since F_s is a strictly proper rational function of finite degree, say n, it follows from the Kronecker theorem in Sect. 2.3.1 that Γ_F has rank n, so that $\|F_s\|_\Gamma = s_1$, where s_1 is the largest singular value of the rank-n Hankel matrix Γ_F. Note also that if F_s has real coefficients so that Γ_F is real and hence normal (Sect. 4.2.1), then it follows from Sect. 4.2.3 that the s-number s_1 and its corresponding Schmidt pair $(\underline{\xi}, \underline{\eta})$ can be obtained by first solving the eigenvalue problem

$$\Gamma_F \mathbf{x} = \lambda \mathbf{x} \tag{4.78}$$

and then setting

$$s_1 = |\lambda_M|, \quad \underline{\xi} = \mathbf{x}, \quad \text{and} \quad \underline{\eta} = (\text{sgn}\ \lambda_M)\mathbf{x},$$

where λ_M is a solution of (4.78) with maximum absolute value. To this end, the AAK techniques (Theorem 4.6) can be applied to yield the solution $\hat{G}$ for problem (4.76) as follows:

$$\hat{G} = \left[F_s - s\frac{\eta_-(z)}{\xi_+(z)} \right]_s, \tag{4.79}$$

where $[\cdot]_s$ indicates the singular part of the argument,

$$\xi_+(z) = \sum_{i=1}^{\infty} \xi_i z^{i-1},$$

and

$$\eta_-(z) = \sum_{i=1}^{\infty} \eta_i z^{-i}$$

with $\underline{\xi} = [\xi_1\ \xi_2\ \cdots\ ]^\top$ and $\underline{\eta} = [\eta_1\ \eta_2\ \cdots\ ]^\top$. Finally, in view of (4.74), the solution to problem (4.68) is given by

$$P_O^{-1}(z)\hat{G}(z), \tag{4.80}$$

assuming, of course, that $P(z)$ is zero-free on $|z| = 1$.

In summary, we have established the following result:

Theorem 4.8. Let $H, P \in \mathcal{M}^s$ and assume that P has no zeros on the unit circle $|z| = 1$. Then, Research Problem 10, or equivalently the minimization problem (4.68), is solvable. Moreover, an optimal solution $\hat{G}$ to Research Problem 10 can be obtained by the following procedure:

(i) Apply formulas (4.69-73) to factorize $P(z)$ as

$$P(z) = P_{I1}^{-1}(z)P_{I2}(z)P_O(z),$$

and let

$$F(z) = P_{I2}^{-1}(z)P_{I1}(z)H(z) := F_a(z) + F_s(z) = \sum_{k=0}^{\infty} f_{-k}z^k + \sum_{k=1}^{\infty} f_k z^{-k}.$$

(ii) Solve the eigenvalue problem

$$\Gamma_F \mathbf{x} = \lambda \mathbf{x}$$

and identify the eigenvalue-eigenvector pair $(\lambda_M, \mathbf{x})$ where λ_M is an eigenvalue with maximum absolute value.

(iii) Let $s = |\lambda_M|$, $\underline{\xi} = \mathbf{x} := [\xi_1\ \xi_2\ \cdots]^\top$, and $\underline{\eta} = (\mathrm{sgn}\lambda_M)\mathbf{x} := [\eta_1\ \eta_2\ \cdots]^\top$. Set

$$\xi_+(z) = \sum_{i=1}^{\infty} \xi_i z^{i-1} \quad \text{and} \quad \eta_-(z) = \sum_{i=1}^{\infty} \eta_i z^{-i} .$$

(iv) Compute

$$\hat{G}(z) = \left[F_s(z) - s\frac{\eta_-(z)}{\xi_+(z)} \right]_s ,$$

where $[\cdot]_s$ denotes the singular part of the argument. In view of (4.55) in Theorem 4.7, $\hat{G}(z)$ has the following reformulation:

$$\hat{G}(z) = \left[\frac{\tilde{\mathbf{z}}^\top T_F \mathbf{x}}{\tilde{\mathbf{z}}^\top \mathbf{x}} \right]_s ,$$

where $\tilde{\mathbf{z}} = [1\ z\ z^2\ \cdots]^\top$ and

$$T_F = \begin{bmatrix} 0 & f_1 & f_2 & \cdots \\ \cdots & 0 & f_1 & \cdots \\ \cdots & \cdots & 0 & \cdots \\ \cdots & \cdots & \cdots & \cdots \end{bmatrix} .$$

(v) Finally, we obtain a stable rational solution $P_O^{-1}(z)\hat{G}(z)$.

We remark that it follows from (4.74) and (4.49) that

$$\left|F(e^{j\omega}) - \hat{G}(e^{j\omega})\right| = \left| s\frac{\eta_-(e^{j\omega})}{\xi_+(e^{j\omega})} \right| = s .$$

This implies that $F(z) - \hat{G}(z)$ is an all-pass transfer function with magnitude s.

We have already seen from Example 4.10 that optimal solutions to Research Problem 10, or the extremal problem in (4.68), may not be unique. Nevertheless, we have the following result concerning non-uniqueness.

Theorem 4.9. If $\hat{G}_1$ and $\hat{G}_2$ are two optimal solutions to Research Problem 10 or the extremal problem in (4.68), then any convex combination of $\hat{G}_1$ and $\hat{G}_2$ in the form of

$$\hat{G} = \alpha\hat{G}_1 + \beta\hat{G}_2\,, \qquad \alpha\,,\beta \geq 0 \quad \text{and} \quad \alpha + \beta = 1\,,$$

is also an optimal solution to the problem.

First, we see that $\hat{G} \in \mathcal{G}$. Moreover, we have

$$\begin{aligned}
\|H - P\hat{G}\|_{H^\infty} &= \|\alpha(H - P\hat{G}_1) + \beta(H - P\hat{G}_2)\|_{H^\infty} \\
&\leq \alpha\|H - P\hat{G}_1\|_{H^\infty} + \beta\|H - P\hat{G}_2\|_{H^\infty} \\
&= \alpha \inf_{G\in\mathcal{G}} \|H - PG\|_{H^\infty} + \beta \inf_{G\in\mathcal{G}} \|H - PG\|_{H^\infty} \\
&= \inf_{G\in\mathcal{G}} \|H - PG\|_{H^\infty}\,.
\end{aligned}$$

This implies that $\hat{G}$ is also an optimal solution of the problem, completing the proof of the theorem.

The above theorem illustrates that Research Problem 10 either has a unique solution or has infinitely many solutions when it is solvable. It is easily seen that this non-uniqueness theorem is also valid for MIMO systems to be studied later in Chap. 6.

Problems

Problem 4.1. Determine the Hankel norm $\|f\|_\Gamma$ of the functions $f(z) = e^z/z^3$ and $f(z) = 2z - e^{-1}/z(z - e^{-1})$.

Problem 4.2. Give examples to show that the rank of a Hankel matrix is very sensitive to perturbation of the values of its entries, in the sense that a Hankel matrix generated by a transfer function with noisy measurements on its parameters will be of infinite rank in general. This motivates the importance of best approximation to a transfer function whose corresponding Hankel matrix is of finite or infinite rank by a finite-rank Hankel matrix with specified finite rank.

Problem 4.3. Give a direct proof of the second inequality in (4.3) without using Nehari's theorem. [Hint:

$$\begin{aligned}
\|f\|_\Gamma &= \sup_{\|\mathbf{x}\|_{l^2}=1} \|\Gamma_f\mathbf{x}\|_{l^2} = \sup_{\|\mathbf{x}\|_{l^2}=1} \left(\sum_{i=0}^{\infty}\sum_{n=1}^{\infty} |f_{n+i}x_n|^2\right)^{1/2} \\
&= \sup_{\|X\|_{L^2}=1} \|fX\|_{L^2} \leq \|f\|_{L^\infty}\,,
\end{aligned}$$

where $X(z) = \sum_{n=1}^{\infty} x_n z^{-n}$ and fX is the product function of $f(z)$ and $X(z)$.]

Problem 4.4. Let $f \in L^\infty$ and λ_f be the linear functional defined by (4.4). Then, we have $\|\lambda_f\| = \|f\|_\Gamma$, see Duren [1970]. Construct a simple example to verify this identity.

Problem 4.5. Let $\mathcal{H}_m$ be the collection of all bounded infinite Hankel matrix operators on l^2 with rank no greater than m and $\mathcal{R}_m^s$ the collection of rational functions as defined in (4.6). Show that a function

$$h(z) = \sum_{n=1}^{\infty} h_n z^{-n}$$

is in $\mathcal{R}_m^s$ if and only if its corresponding Hankel matrix Γ_h is in $\mathcal{H}_m$. This supplies a proof of Corollary 4.1 stated in Sect. 4.1.1.

Problem 4.6. Verify that the eigenvalue and eigenvector pairs of the Hankel matrix

$$\Gamma = \begin{bmatrix} 1 & 0 & -1 & 0 & \cdots \\ 0 & -1 & 0 & \cdots & \cdots \\ -1 & 0 & \cdots & \cdots & \cdots \\ 0 & \cdots & \cdots & \cdots & \cdots \\ \cdots & \cdots & \cdots & \cdots & \cdots \end{bmatrix}$$

are $(\lambda_1, \xi_1), (\lambda_2, \xi_2), \cdots$, with

$$\lambda_1 = \frac{1+\sqrt{5}}{2}, \; \lambda_2 = -1, \; \lambda_3 = \frac{1-\sqrt{5}}{2}, \; \lambda_4 = \lambda_5 = \cdots = 0$$

and

$$\xi_1 = \begin{bmatrix} 2 \\ 0 \\ 1-\sqrt{5} \\ 0 \\ \vdots \end{bmatrix}, \; \xi_2 = \begin{bmatrix} 0 \\ 1 \\ 0 \\ 0 \\ \vdots \end{bmatrix}, \; \xi_3 = \begin{bmatrix} -1+\sqrt{5} \\ 0 \\ 2 \\ 0 \\ \vdots \end{bmatrix},$$

$$\xi_4 = \begin{bmatrix} 0 \\ 0 \\ 0 \\ 1 \\ 0 \\ \vdots \end{bmatrix}, \; \xi_5 = \begin{bmatrix} 0 \\ 0 \\ 0 \\ 0 \\ 1 \\ \vdots \end{bmatrix}, \; \cdots .$$

Problem 4.7. Determine the s-numbers and the corresponding Schmidt pairs for the Hankel matrix

$$\Gamma = \begin{bmatrix} -4.5 & 3j & 1 & 0 & \cdots \\ 3j & 1 & 0 & \cdots & \cdots \\ 1 & 0 & \cdots & \cdots & \cdots \\ 0 & \cdots & \cdots & \cdots & \cdots \\ \cdots & \cdots & \cdots & \cdots & \cdots \end{bmatrix}.$$

Problem 4.8. Let $f(z) = z^{-1} + z^{-2} + z^{-3}$. Determine $\hat{r}_m(z) \in \mathcal{R}_m^s$ such that

$$\|f - \hat{r}_m\|_\Gamma = \inf_{r \in \mathcal{R}_m^s} \|f - r\|_\Gamma ,$$

where $m = 0, 1, 2, \cdots$, and $\mathcal{R}_m^s$ is defined as in (4.6).

Problem 4.9. A linear operator $S : H^2 \to H^2$ defined by

$$S : \; f(z) \to z f(z)$$

is also called a *shift operator*. Show that the adjoint operator S^* of S is given by

$$S^* : \; f(z) \to z^{-1}(f(z) - f(0)) .$$

[Compare it with Example 4.6.]

Problem 4.10. Show that a bounded infinite Hankel matrix Γ is Hermitian if and only if Γ is real in the sense that all the entries of Γ are real numbers.

Problem 4.11. Verify that the Hankel matrix

$$\Gamma_1 = \begin{bmatrix} j & j & 0 & \cdots \\ j & 0 & 0 & \cdots \\ 0 & 0 & 0 & \cdots \\ \cdots & \cdots & \cdots & \cdots \end{bmatrix}$$

is normal, but the Hankel matrix

$$\Gamma_2 = \begin{bmatrix} 1 & j & 0 & \cdots \\ j & 0 & 0 & \cdots \\ 0 & 0 & 0 & \cdots \\ \cdots & \cdots & \cdots & \cdots \end{bmatrix}$$

is not. Prove that a Hankel matrix $\Gamma = [h_{i+\ell-1}]$ is normal if and only if $h_i \bar{h}_\ell$ is real for all i and ℓ.

Problem 4.12. Let $|\Gamma| = (\bar{\Gamma}\Gamma)^{1/2}$ be the positive square root of $\bar{\Gamma}\Gamma$ for the Hankel matrix Γ. Construct some examples to show that $|\Gamma|$ is not a Hankel matrix in general.

Problem 4.13. Let $A : l^2 \to l^2$ be bounded and have finite rank. Prove that A is a compact operator on l^2 in the sense that the closure of the image of any bounded set in l^2 under A is a compact set in l^2.

Problem 4.14. Let Γ be a bounded infinite Hankel matrix with $s_\infty(\Gamma) = 0$. Show that Γ, as an operator from l^2 to l^2, is compact.

Problem 4.15. Prove that the infinite series (4.22) (i.e., the spectral decomposition) of the positive operator $|\Gamma|$ converges strongly if Γ is compact, and weakly if Γ is bounded. [Hint: See the proof of Theorem 4.5 in Sect. 4.2.4.]

Problem 4.16. Use the Schmidt series representation of a compact Hankel matrix Γ in (4.24) with the Schmidt pairs (ξ_m, η_m) defined by (4.23) to verify the equalities in (4.25).

Problem 4.17. Let Γ be a compact normal Hankel matrix with s-numbers $s_1, s_2, \cdots$, and corresponding Schmidt pairs $(\xi_1, \eta_1), (\xi_2, \eta_2), \cdots$. Show that

$$\Gamma^k \xi_m = \begin{cases} s_m^k \xi_m & \text{if } k \text{ is even}, \\ s_m^k \eta_m & \text{if } k \text{ is odd}, \end{cases}$$

and

$$\Gamma^k \eta_m = \begin{cases} s_m^k \eta_m & \text{if } k \text{ is even}, \\ s_m^k \xi_m & \text{if } k \text{ is odd}. \end{cases}$$

Problem 4.18. Let

$$\Gamma = \begin{bmatrix} 1 & \mathrm{j} & 0 & \cdots \\ \mathrm{j} & 0 & 0 & \cdots \\ 0 & 0 & 0 & \cdots \\ \cdots & \cdots & \cdots & \cdots \end{bmatrix}.$$

Show that the eigenvalues and corresponding eigenvectors of $\bar{\Gamma}\Gamma$ are given by

$$(1)\ \lambda_1 = \frac{3+\sqrt{5}}{2}, \quad \mathbf{x}_1 = \frac{1}{\sqrt{10-2\sqrt{5}}} \begin{bmatrix} -2\mathrm{j} \\ 1-\sqrt{5} \\ 0 \\ \vdots \end{bmatrix},$$

$$(2)\ \lambda_2 = \frac{3-\sqrt{5}}{2}, \quad \mathbf{x}_2 = \frac{1}{\sqrt{10+2\sqrt{5}}} \begin{bmatrix} -2\mathrm{j} \\ 1+\sqrt{5} \\ 0 \\ \vdots \end{bmatrix},$$

$$(3)\ \lambda_m = 0, \quad \mathbf{x}_m = [\,0\ \cdots\ 0\ 1\ 0\ \cdots\,]^\top,$$

where 1 appears at the mth component, and $m = 3, 4, \cdots$.

Problem 4.19. Show that if Γ is a compact normal Hankel matrix, then the s-numbers $\{s_m\}_{m=1}^{\infty}$ of Γ are the absolute values of the corresponding eigenvalues $\{\lambda_m\}_{m=1}^{\infty}$ of Γ, and the corresponding Schmidt pairs of Γ are $(\mathbf{x}_m, (\text{sgn}\lambda_m)\,\mathbf{x}_m)$, where $\mathbf{x}_m$ is the eigenvector of Γ relative to the eigenvalue λ_m, $m = 1, 2, \cdots$.

Problem 4.20. Characterize all compact normal Hankel operators on l^2 by considering their s-numbers and the corresponding Schmidt pairs.

Problem 4.21. Let the eigenvalue-eigenvector pairs of a compact normal Hankel matrix Γ be given by $(\lambda_m, \mathbf{x}_m), m = 1, 2, \cdots$, and the eigenvalue-eigenvector pairs of $|\Gamma|$ be given by $(s_m, \mathbf{x}_m), m = 1, 2, \cdots$, with the same eigenvectors. Show that $s_m = |\lambda_m|$ for all m.

Problem 4.22. Let

$$\Gamma = \begin{bmatrix} \mathrm{j} & \mathrm{j} & 0 & \cdots \\ \mathrm{j} & 0 & 0 & \cdots \\ 0 & 0 & 0 & \cdots \\ \cdots & \cdots & \cdots & \cdots \end{bmatrix}.$$

Then, Γ is a compact normal operator on l^2. Find the s-numbers and corresponding Schmidt pairs for Γ.

Problem 4.23. Consider the Schmidt series $\sum_{i=1}^{\infty} s_i \eta_i \bar{\xi}_i$ of the Hankel matrix

$$\Gamma = \begin{bmatrix} 1 & \mathrm{j} & 0 & \cdots \\ \mathrm{j} & 0 & 0 & \cdots \\ 0 & 0 & 0 & \cdots \\ \cdots & \cdots & \cdots & \cdots \end{bmatrix}$$

(see Example 4.8). Verify that the partial sums $\hat{A}_1 = s_1 \eta_1 \bar{\xi}_1$ and $\hat{A}_2 = \sum_{i=1}^{2} s_i \eta_i \bar{\xi}_i$ are not Hankel matrices.

Problem 4.24. Verify, by a direct calculation, that

$$\sum_{i=1}^{\infty}\sum_{\ell=1}^{\infty} f_i u_\ell z^{\ell-i-1} = \sum_{i=1}^{\infty}\sum_{\ell=1}^{\infty} f_{\ell+i-1} u_\ell z^{-i} + \sum_{i=1}^{\infty}\sum_{\ell=1}^{\infty} f_{\ell-i+1} u_{\ell+1} z^{i-1},$$

where $f_p = 0$ for all $p \leq 0$.

Problem 4.25. Verify that Lemma 4.5 holds for the case where the multiplicity of s_{m+1} is larger than 1.

Problem 4.26. Let $\{s_i\}$ and $\{\tilde{s}_i\}$ be the s-numbers of the compact Hankel matrices Γ and $\tilde{\Gamma}$, respectively. Suppose that

$$\|\Gamma \mathbf{x}\|_{l^2} \leq \|\tilde{\Gamma}\mathbf{x}\|_{l^2}, \qquad \text{for all } \mathbf{x} \in l^2 .$$

Prove that

$$s_i(\Gamma) \leq \tilde{s}_i(\tilde{\Gamma}), \qquad i = 1, 2, \cdots .$$

(See Wilkinson [1965].)

Problem 4.27. By imitating the proof of (4.52), verify (4.53).

Problem 4.28. (a) Verify that the eigenvalues and corresponding eigenvectors of the Hankel matrix

$$\Gamma = \begin{bmatrix} 2 & a & a^2 & \cdots \\ a & a^2 & \cdots & \cdots \\ a^2 & \cdots & \cdots & \cdots \\ \cdots & \cdots & \cdots & \cdots \end{bmatrix}, \qquad 0 < a < 1,$$

are given by

$$\lambda_i = \frac{2 - a^2 \pm \sqrt{5a^4 - 8a^2 + 4}}{2(1 - a^2)} > 0$$

and

$$\xi_i = \begin{bmatrix} -a/(1-a^2) \\ 2 - \lambda_i \\ (2 - \lambda_i)a \\ \\ \vdots \end{bmatrix}, \qquad i = 1, 2.$$

(b) Apply (4.55) to show that the best rank-0 and rank-1 Hankel norm approximants $\hat{r}_i(z), i = 0, 1$, for the transfer function

$$f(z) = \frac{2z - a}{z(z - a)}$$

are given by

$$\hat{r}_i(z) = \left[\frac{(2 - \lambda_i)(1 + 1/(1 - a^2))}{(2 - \lambda_i)z - a(1 - az)/(1 - a^2)}\right]_s, \qquad i = 0, 1.$$

(c) Simplify $\hat{r}_0(z)$ and $\hat{r}_1(z)$ and identify their poles.

Problem 4.29. Consider a linear system with transfer function given by $f(z) = z^{-1} - z^{-3}$. Find the rank-0, rank-1, and rank-2 best Hankel-norm approximants $\hat{r}_i(z), i = 0, 1, 2$, to $f(z)$.

Problem 4.30. Consider the simple linear feedback system shown in Fig. 4.4.

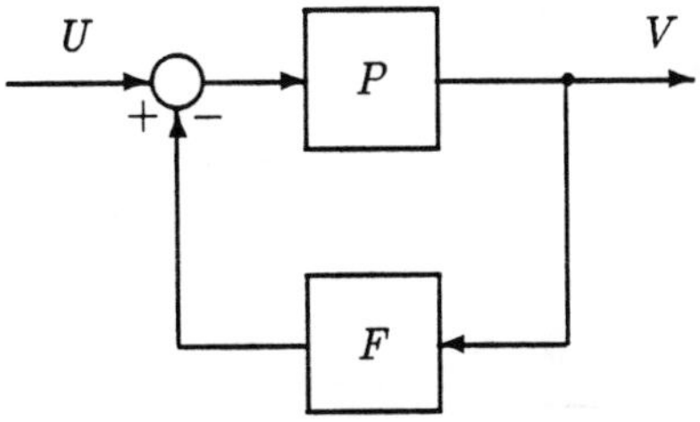

Fig. 4.4. A linear feedback system

Show that the transfer function $H(z)$ for this system is given by

$$H = (I + PF)^{-1}P\,.$$

Problem 4.31. Consider a simple linear feedback system with a nominal plant P_0 and a plant error ΔP as shown in Fig. 4.5. The system transfer function is given by

$$H = (I + (P_0 + \Delta P)F)^{-1}(P_0 + \Delta P)\,.$$

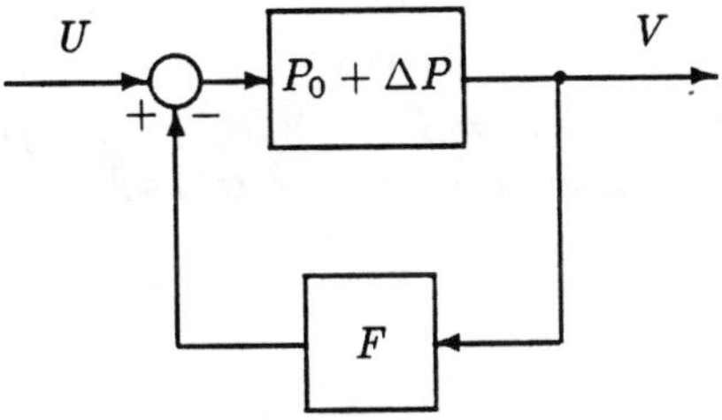

Fig. 4.5. A perturbed feedback system

[Hint: See Problem 4.30.]

(a) Show that this system is equivalent to the one shown in Fig. 4.6 in the sense that they have the same system transfer function, namely, the same input yields the same output.

[Hint: Verify that

$$\begin{cases} W = P_0U + \Delta P(U - X)\,, \\ X = FV\,, \\ V = (I + P_0F)^{-1}W\,, \end{cases}$$

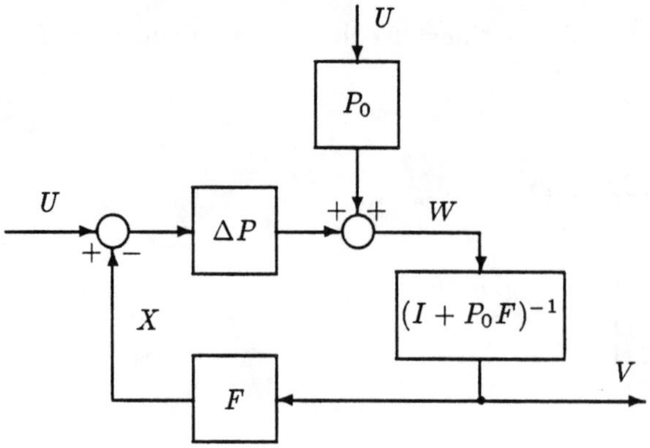

Fig. 4.6. An equivalent feedback system

and then express the output V in terms of the input U by eliminating the variable W.]

(b) Show that the linear system shown in Fig. 4.6 is stable if F and $(I + P_0F)^{-1}$ are both stable and

$$\|\Delta P\|_{H^\infty} \, \|F(I + P_0F)^{-1}\|_{H^\infty} < 1 .$$

[Hint: Imitate the proof of the small gain theorem in feedback control system theory, see Desoer and Vidyasagar [1975]: The entities u_1, u_2, H_1, H_2 therein on page 41 are respectively $u_1 = U, u_2 = P_0U, H_1 = \Delta P$, and $H_2 = F(I+P_0F)^{-1}$, when applied to Fig. 4.6 in this problem.]

Problem 4.32. Given a plant $P(z) = (z - \alpha)^m/(z + \alpha)^n$ with $0 < |\alpha| < 1$, and an $H \in \mathcal{M}^s$ with $|H(\alpha)| < \infty$. Show that $\hat{G}(z) = P^{-1}(z)(H(z) - H(\alpha))$ is an optimal solution of Research Problem 10, and that this optimal solution is not unique if $\|H\|_{H^\infty} = |H(\alpha)|$.

5. General Theory of Optimal Hankel-Norm Approximation

In the previous chapter we have studied in some detail the AAK theorem for the special case where the given Hankel matrix Γ has finite rank. An elementary proof of the theorem under this assumption was given, and an application of the result to the problem of system reduction has also been illustrated. In this chapter, we will study the general AAK theory, see AAK [1971], where the Hankel matrix Γ is not necessarily of finite rank and may not even be compact. We will supply a proof in detail of the general theorem. In order to give an elementary exposition that requires as little knowledge of linear operator theory as possible, our presentation is rather lengthy. As will be seen in the first section below, the solvability of the problem can be easily proved because it is easy to show that any given bounded Hankel matrix Γ has a best approximation from the set of Hankel operators with any specified finite rank. We remark, however, that one of the main contributions of the AAK theory is that an explicit closed-form solution to the problem is formulated. The derivation of this formulation is much more difficult and will contribute to the major portion of this chapter.

5.1 Existence and Preliminary Results

In this section, we include some basic results which are needed later in this chapter. We first show that for a given bounded Hankel matrix Γ, the general problem of best approximation of Γ by a Hankel matrix with specified finite rank is always solvable. We then characterize all bounded operators that commute with the shift operator, and apply this characterization to establish an important theorem due to Beurling concerning the l^2-closure of the shifts of a fixed l^2 sequence. Based on this result, we formulate the operator norm of any bounded Hankel matrix in terms of the outer factor of an analytic function. Finally, we will study various properties of this formulation.

5.1.1 Solvability of a Best Approximation Problem

In this section, we will show that the following best approximation problem is solvable. Let $G^{[k]}$ denote the family of bounded Hankel matrices with rank not exceeding k, namely,

$$G^{[k]} := \{\Gamma: \quad \Gamma \text{ a bounded Hankel matrix, } \operatorname{rank}(\Gamma) \leq k\}. \tag{5.1}$$

For any given bounded Hankel matrix Γ which may have infinite rank and may not even be compact, we consider the optimization problem

$$\inf_{\Lambda \in G^{[k]}} \|\Gamma - \Lambda\|, \tag{5.2}$$

where $\|\cdot\| = \|\cdot\|_s$ is the usual operator or spectral norm. Define

$$d_k(\Gamma) = \inf_{\Lambda \in G^{[k]}} \|\Gamma - \Lambda\|. \tag{5.3}$$

We will prove that the optimal error of approximation, $d_k(\Gamma)$, can always be attained, and so the best approximation problem in (5.2) is solvable. Recall that if the given Hankel matrix Γ is normal and has finite rank, then the solvability of this problem has been verified in Sect. 4.3.2 and a closed-form solution was also given in Sects. 4.3.2 and 4.3.3. For an arbitrary bounded Hankel matrix Γ, the following theorem gives an expected answer. However, the derivation of a closed-form solution to this general best approximation problem requires a sequence of preliminary results and will not be completed until the end of this chapter.

Theorem 5.1. Let Γ be a bounded Hankel matrix and $G^{[k]}$ be defined as in (5.1). Then for every non-negative integer k, there exists a matrix Λ_k^* in $G^{[k]}$ such that

$$d_k(\Gamma) = \|\Gamma - \Lambda_k^*\|. \tag{5.4}$$

We will describe a procedure to define Λ_k^*. Pick any approximating sequence $\{\Lambda_k^{(n)}\}_{n=1}^{\infty}$ in $G^{[k]}$, namely,

$$\lim_{n\to\infty} \|\Gamma - \Lambda_k^{(n)}\| = d_k(\Gamma),$$

and set

$$\Lambda_k^{(n)} = \begin{bmatrix} \gamma_1^{(n)} & \gamma_2^{(n)} & \gamma_3^{(n)} & \cdots \\ \gamma_2^{(n)} & \gamma_3^{(n)} & \cdots & \cdots \\ \gamma_3^{(n)} & \cdots & \cdots & \cdots \\ \cdots & \cdots & \cdots & \cdots \end{bmatrix}.$$

Then, for the unique vector $\mathbf{e}_1 = [\,1\ 0\ 0\ \cdots\,]^\top \in l^2$, we have

$$\Lambda_k^{(n)} \mathbf{e}_1 = \begin{bmatrix} \gamma_1^{(n)} \\ \gamma_2^{(n)} \\ \gamma_3^{(n)} \\ \vdots \end{bmatrix} .$$

It follows from the Banach-Alaoglu theorem, which can be found in Conway [1985], that there exists a subsequence $\{\Lambda_k^{(n_i)} \mathbf{e}_1\}_{i=1}^{\infty}$ of $\{\Lambda_k^{(n)} \mathbf{e}_1\}_{n=1}^{\infty}$ which converges weakly to an element $\underline{\xi} = [\ \xi_1\ \xi_2\ \xi_3\ \cdots\ ]^\top \in l^2$, that is,

$$\langle \Lambda_k^{(n_i)} \mathbf{e}_1, \underline{\eta} \rangle_{l^2} \to \langle \underline{\xi}, \underline{\eta} \rangle_{l^2} \quad \text{for all } \underline{\eta} \in l^2 \tag{5.5}$$

as $i \to \infty$. Hence, by choosing $\underline{\eta} = \mathbf{e}_1 = [\ 1\ 0\ 0\ \cdots\ ]^\top, \underline{\eta} = \mathbf{e}_2 = [\ 0\ 1\ 0\ \cdots\ ]^\top, \ldots$, consecutively, we have

$$\lim_{i \to \infty} \gamma_\ell^{(n_i)} = \xi_\ell, \qquad \text{for each } \ell = 1, 2, \ldots .$$

Now, define

$$\Lambda_k^* = \begin{bmatrix} \xi_1 & \xi_2 & \xi_3 & \cdots \\ \xi_2 & \xi_3 & \cdots & \cdots \\ \xi_3 & \cdots & \cdots & \cdots \\ \cdots & \cdots & \cdots & \cdots \end{bmatrix} .$$

Then, Λ_k^* is the desired infinite matrix in $G^{[k]}$ which satisfies (5.4). First, since each minor of Λ_k^* is the limit of the sequence of the corresponding minors of $\Lambda_k^{(n_i)}$, we have $\Lambda_k^* \in G^{[k]}$ (Problem 5.1). Secondly, (5.5) implies that the first column of $\Lambda_k^{(n_i)}$ converges weakly to the first column of Λ_k^*. In a similar manner, it is clear that the ℓth column of $\Lambda_k^{(n_i)}$ converges weakly to the ℓth column of Λ_k^*. Hence, it can be shown that

$$\langle (\Gamma - \Lambda_k^{(n_i)}) \underline{\xi}, \underline{\eta} \rangle_{l^2} \to \langle (\Gamma - \Lambda_k^*) \underline{\xi}, \underline{\eta} \rangle_{l^2}$$

as $i \to \infty$ for all $\underline{\xi}, \underline{\eta} \in l^2$ (Problem 5.2), so that

$$\|\Gamma - \Lambda_k^*\|_s \leq \overline{\lim_{i \to \infty}} \|\Gamma - \Lambda_k^{(n_i)}\|_s = d_k(\Gamma) .$$

Finally, since $\Lambda_k^* \in G^{[k]}$ and $d_k(\Gamma)$ is the infimum, we have

$$\|\Gamma - \Lambda_k^*\|_s = d_k(\Gamma) ,$$

completing the proof of the theorem.

5.1.2 Characterization of the Bounded Operators that Commute with the Shift Operator

Recall from Example 4.6 that the shift operator S defined on l^2 is given by

$$S = \begin{bmatrix} 0 & 0 & 0 & \cdots \\ 1 & 0 & 0 & \cdots \\ 0 & 1 & 0 & \cdots \\ 0 & 0 & 1 & \cdots \\ \cdots & \cdots & \cdots & \cdots \end{bmatrix} . \tag{5.6}$$

As before, let $\mathcal{B}$ denote the family of all bounded infinite matrices (or linear operators on l^2) and consider those that commute with the shift operators, namely,

$$\mathcal{B}_S = \{A \in \mathcal{B}: \quad AS = SA\} . \tag{5.7}$$

Then we will see that $\mathcal{B}_S$ is isometrically isomorphic to H^∞. Based on this isomorphism, we will give a characterization theorem to describe the class $\mathcal{B}_S$.

To start with, we observe that for $A = [a_{\ell m}]$ in $\mathcal{B}$, the condition $AS = SA$ implies that

$$\begin{aligned} A\mathbf{e}_1 &= [\, a_{11} \; a_{21} \; a_{31} \; a_{41} \; \cdots \,]^\top , \\ A\mathbf{e}_2 &= A(S\mathbf{e}_1) = S(A\mathbf{e}_1) = [\, 0 \; a_{11} \; a_{21} \; a_{31} \; \cdots \,]^\top , \end{aligned}$$

and in general,

$$A\mathbf{e}_\ell = [\, \underbrace{0 \cdots 0}_{\ell-1} \; a_{11} \; a_{21} \; a_{31} \; \cdots \,]^\top ,$$

so that A must be a lower triangular Toeplitz matrix as follows (Problem 5.3):

$$A = \begin{bmatrix} a_{11} & 0 & 0 & \cdots \\ a_{21} & a_{11} & 0 & \cdots \\ a_{31} & a_{21} & a_{11} & \cdots \\ \cdots & \cdots & \cdots & \cdots \end{bmatrix} = \sum_{i=1}^{\infty} a_{i1} S^{i-1} . \tag{5.8}$$

It is easy to show that such a matrix A must satisfy the identity

$$(A\underline{\xi})_+(z) = \phi(z)\xi_+(z) \qquad \text{for all } \underline{\xi} \in l^2 , \tag{5.9}$$

where

$$\phi(z) = \sum_{i=1}^{\infty} a_{i1} z^{i-1} \tag{5.10}$$

is called the *symbol* of the Toeplitz matrix A, and the "plus function" has been defined in (4.38), namely, for any element $f = \{f_1, f_2, \ldots\}$ in l^2,

$$f_+(z) := \sum_{i=1}^{\infty} f_i z^{i-1} . \tag{5.11}$$

To verify (5.9), we simply observe that for any $\underline{\xi} = [\ \xi_1\ \xi_2\ \xi_3\ \cdots\ ]^\top \in l^2$, we have

$$A\underline{\xi} = [\ a_{11}\xi_1,\ a_{21}\xi_1 + a_{11}\xi_2,\ \cdots,\ \sum_{i=1}^{\ell} a_{i1}\xi_{\ell-i+1},\ \cdots\ ]^\top ,$$

so that

$$\begin{aligned} (A\underline{\xi})_+(z) &= \sum_{\ell=1}^{\infty} \left(\sum_{i=1}^{\ell} a_{i1}\xi_{\ell-i+1} \right) z^{\ell-1} \\ &= \left(\sum_{i=1}^{\infty} a_{i1} z^{i-1} \right) \left(\sum_{i=1}^{\infty} \xi_i z^{i-1} \right) \\ &= \phi(z)\xi_+(z) . \end{aligned}$$

Now, we can state and prove the following theorem which shows that the class $\mathcal{B}_S$ of bounded linear operators that commute with S as defined in (5.7) is isometrically isomorphic to H^∞.

Theorem 5.2. Let $A = [a_{i\ell}]$ be an infinite matrix in $\mathcal{B}_S$. Then A is a Toeplitz matrix with $a_{i\ell} = a_{i-\ell,1}$ for $i \geq \ell$, and $a_{i\ell} = 0$ for $i < \ell$ as shown in (5.8). Furthermore, if $\phi(z)$ denotes the corresponding symbol of A as defined in (5.10), then $\phi(z) \in H^\infty$, and

$$(A\underline{\xi})_+(z) = \phi(z)\xi_+(z) \qquad \text{for all } \underline{\xi} \in l^2 .$$

Conversely, for any H^∞ function $\phi(z) := \sum_{i=1}^{\infty} b_{i1} z^{i-1}$, the matrix $A := [b_{i-\ell,1}]$, with $b_{m1} := 0$ for $m < 0$, is in $\mathcal{B}_S$. Moreover, $\|A\| = \|\phi(z)\|_{H^\infty}$.

To prove this theorem, it is sufficient to show that $\|A\| = \|\phi(z)\|_{H^\infty}$. In the following, the isometry of l^2 and H^2 is used repeatedly. We first have

$$\begin{aligned} \|A\underline{\xi}\|_{l^2} &= \|(A\underline{\xi})_+(z)\|_{H^2} \\ &= \|\phi(z)\xi_+(z)\|_{H^2} \\ &\leq \|\phi(z)\|_{H^\infty} \|\underline{\xi}\|_{l^2} , \end{aligned}$$

or

$$\|A\| = \sup_{\|\underline{\xi}\|_{l^2} \neq 0} \frac{\|A\underline{\xi}\|_{l^2}}{\|\underline{\xi}\|_{l^2}} \leq \|\phi(z)\|_{H^\infty} .$$

On the other hand, we also have

$$\begin{aligned}\|\phi(z)\xi_+(z)\|_{H^2} &= \|(A\underline{\xi})_+(z)\|_{H^2} \\ &= \|A\underline{\xi}\|_{l^2} \\ &\le \|A\|\,\|\underline{\xi}\|_{l^2} \\ &= \|A\|\,\|\xi_+(z)\|_{H^2},\end{aligned}$$

so that

$$\begin{aligned}\|\phi^2(z)\xi_+(z)\|_{H^2} &= \|\phi(z)(\phi(z)\xi_+(z))\|_{H^2} \\ &\le \|A\|\,\|\phi(z)\xi_+(z)\|_{H^2} \\ &\le \|A\|^2\|\xi_+(z)\|_{H^2},\end{aligned}$$

and in general,

$$\|\phi^k(z)\|_{H^2} = \sup_{\|\xi_+(z)\|_{H^2}\neq 0} \frac{\|\phi^k(z)\xi_+(z)\|_{H^2}}{\|\xi_+(z)\|_{H^2}} \le \|A\|^k .$$

Hence, we have

$$\left(\frac{1}{2\pi}\int_0^{2\pi} |\phi(e^{j\theta})|^{2k} d\theta\right)^{1/2k} \le \|A\|,$$

where $\phi(e^{j\theta})$ is the (almost everywhere) radial limit of $\phi(z)$. Since

$$\|\phi(z)\|_{H^\infty} = \lim_{p\to\infty} \|\phi(z)\|_{H^p} \tag{5.12}$$

(Problem 5.4), we obtain

$$\|\phi(z)\|_{H^\infty} \le \|A\| .$$

This completes the proof of the theorem.

5.1.3 Beurling's Theorem

In this section, we will introduce a very important theorem due to Beurling which essentially says that the closure of the linear span of all shifts of a given l^2 sequence is characterized by the inner factor of its corresponding H^2 function. We first need some preliminary lemmas.

Lemma 5.1. Let $\phi(z) = \Sigma_{i=1}^{\infty}\phi_i z^{i-1}$ be an inner function in H^∞ and S be the shift operator defined by (5.6). Then we have

(*a*) $\|\phi(S)\underline{\xi}\|_{l^2} = \|\underline{\xi}\|_{l^2}$ for all $\underline{\xi} \in l^2$;

(*b*) $\phi^*(S)\phi(S) = I$,

where $\phi^*(S)$ is the adjoint of the operator $\phi(S)$ and I is the identity operator; and

(*c*) $\phi(S)\phi^*(S)$ is an orthogonal projection onto the space $\phi(S)l^2$, namely, $\phi(S)\phi^*(S)\underline{\xi} = \underline{\xi}$ for all $\underline{\xi} \in \phi(S)l^2$.

The proof of this lemma is straightforward. First, we see that

$$\begin{aligned}\|\phi(S)\underline{\xi}\|_{l^2} &= \|(\phi(S)\underline{\xi})_+(z)\|_{H^2}\\ &= \|\phi(z)\xi_+(z)\|_{H^2}\\ &= \|\xi_+(z)\|_{H^2}\\ &= \|\underline{\xi}\|_{l^2}\,.\end{aligned}$$

Secondly, for all $\underline{\xi} \in l^2$, we have

$$\begin{aligned}&\|\phi^*(S)\phi(S)\underline{\xi} - \underline{\xi}\|_{l^2}^2\\ &= \sup_{\|\underline{\eta}\|_{l^2}\le 1} \langle \phi^*(S)\phi(S)\underline{\xi} - \underline{\xi}, \underline{\eta}\rangle_{l^2}\\ &= \sup_{\|\underline{\eta}\|_{l^2}\le 1} \{\langle \phi^*(S)\phi(S)\underline{\xi}, \underline{\eta}\rangle_{l^2} - \langle \underline{\xi}, \underline{\eta}\rangle_{l^2}\}\\ &= \sup_{\|\underline{\eta}\|_{l^2}\le 1} \{\langle \phi(S)\underline{\xi}, \phi(S)\underline{\eta}\rangle_{l^2} - \langle \underline{\xi}, \underline{\eta}\rangle_{l^2}\}\\ &= 0\,,\end{aligned}$$

where the last equality follows from the isometry as a result of part (a). Finally, for all $\underline{\xi} \in l^2$, it follows from part (b) that

$$\begin{aligned}&\|\phi(S)\phi^*(S)(\phi(S)\underline{\xi}) - \phi(S)\underline{\xi}\|_{l^2}\\ =&\|\phi(S)\underline{\xi} - \phi(S)\underline{\xi}\|_{l^2}\\ =&0\,,\end{aligned}$$

which implies that $\phi(S)\phi^*(S)$ is an orthogonal projection onto the space $\phi(S)l^2$.

The following lemma characterizes a Hankel matrix in terms of the shift operator S.

Lemma 5.2. Γ is a Hankel matrix if and only if $\Gamma S = S^*\Gamma$.

This can be proved by a direct calculation and hence is left to the reader as an Problem (Problem 5.5). As a consequence of this result, we have the following:

Lemma 5.3. Let Γ_1 and Γ_2 be two Hankel matrices such that $\Gamma_1\underline{\xi}^0 = \Gamma_2\underline{\xi}^0$ for some $\underline{\xi}^0 \neq 0$ in l^2. Then, $\Gamma_1\underline{\eta} = \Gamma_2\underline{\eta}$ for all $\underline{\eta}$ in the subspace $l^2_S(\underline{\xi}^0)$ of l^2 defined by

$$l^2_S(\underline{\xi}^0) = \mathrm{span}\{\ \underline{\xi}^0, S\underline{\xi}^0, S^2\underline{\xi}^0,\ \cdots\}\,.$$

The proof of this lemma is quite simple and is again left to the reader as an Problem (Problem 5.6).

We are now in a position to prove the following result:

Theorem 5.3. (Beurling)

Let $\underline{\xi} = [\ \xi_1\ \xi_2\ \xi_3\ \cdots\ ]^\top \in l^2$ and define $\xi_+(z) = \Sigma_{i=1}^{\infty} \xi_i z^{i-1}$. Decompose $\xi_+(z)$ as

$$\xi_+(z) = \xi_I(z)\xi_O(z)\,, \tag{5.13}$$

where $\xi_I(z)$ and $\xi_O(z)$ are the inner and outer factors of $\xi_+(z)$, respectively. Moreover, let

$$l_S^2(\underline{\xi}) = \mathrm{span}\{S^n\underline{\xi}: \quad n = 0, 1, \cdots\}\,, \tag{5.14}$$

and let $\overline{l_S^2(\underline{\xi})}$ denote its closure in l^2. Then,

$$\overline{l_S^2(\underline{\xi})} = \xi_I(S)l^2\,. \tag{5.15}$$

To prove the theorem, let T denote the isometric isomorphism from l^2 to H^2 defined by

$$T(\underline{\xi}) = \xi_+(z) = \xi_I(z)\xi_O(z)\,.$$

Then, for any $n = 0, 1, \cdots$,

$$T(S^n\underline{\xi}) = z^n\xi_+(z)\,,$$

(Problem 5.7). Consequently,

$$T(\overline{l_S^2(\underline{\xi})}) = \overline{\mathrm{span}\{z^n\xi_+(z):\ n = 0, 1, \cdots\}}\,,$$

which is the closure of the span$\{z^n\xi_+(z):\ n = 0, 1, \cdots\}$.

On the other hand, let $\mathcal{P}_n$ denote, as usual, the family of complex polynomials of z with degree no greater than n. Then, for any $f(z) \in H^2$, we have, by the Beurling approximation theorem (Theorem 3.1),

$$\begin{aligned} &\inf_{p\in\mathcal{P}_n} \|\xi_I(z)f(z) - p(z)\xi_+(z)\|_{H^2} \\ &= \inf_{p\in\mathcal{P}_n} \|\xi_I(z)(f(z) - p(z)\xi_O(z))\|_{H^2} \\ &= \inf_{p\in\mathcal{P}_n} \|f(z) - p(z)\xi_O(z)\|_{H^2} \to 0 \end{aligned}$$

as $n \to \infty$. This implies that

$$\xi_I(z)f(z) \in \overline{\mathrm{span}\{z^n\xi_+(z):\ n = 0, 1, \cdots\}}$$

so that

$$\xi_I(z)H^2 \subset \overline{\mathrm{span}\{z^n\xi_+(z):\ n = 0, 1, \cdots\}}\,.$$

Conversely, for any $f(z) \in \overline{\text{span}\{z^n \xi_+(z):\ n = 0, 1, \cdots\}}$, we have

$$\inf_{p \in \mathcal{P}_n} \|\xi_I(z) f(z) - p(z)\xi_+(z)\|_{H^2} \to 0$$

as $n \to \infty$, so that by Theorem 3.1, $f(z)$ is an inner function, or equivalently, $f(z) \in \xi_I(z) H^2$. Hence,

$$\overline{\text{span}\{z^n \xi_+(z):\ n = 0, 1, \cdots\}} \subset \xi_I(z) H^2 .$$

It then follows that

$$\xi_I(z) l^2 = \overline{l^2_S(\underline{\xi})},$$

which is the closure of the span$\{S^n \underline{\xi}:\ n = 0, 1, \cdots\}$, completing the proof of the theorem.

As an important consequence of Beurling's theorem, we will see that two Hankel operators are identical if they agree on some "outer sequence." The proof of this result is immediate (Problem 5.8):

Corollary 5.1. Let Γ_1 and Γ_2 be two Hankel matrices with $\Gamma_1 \underline{\xi} = \Gamma_2 \underline{\xi}$ for some $\underline{\xi} = [\ \xi_1\ \xi_2\ \xi_3\ \cdots\ ]^\top \in l^2$ such that $\xi_+(z) = \Sigma_{i=1}^{\infty} \xi_i z^{i-1}$ is an outer function. Then $\Gamma_1 \equiv \Gamma_2$.

5.1.4 Operator Norms of Hankel Matrices in Terms of Inner and Outer Factors

In this section, we will derive an explicit formula for the operator norm of a Hankel matrix Γ on l^2. This formula will be instrumental for determining a closed-form solution to the problem of best approximation of Γ from finite-rank Hankel operators.

We need some additional notations. For any function

$$\phi(z) = \sum_{i=1}^{\infty} \phi_i z^{i-1}, \tag{5.16}$$

in H^∞, we define

$$\bar{\phi}(z) = \sum_{i=1}^{\infty} \bar{\phi}_i z^{i-1} \tag{5.17}$$

and

$$\phi^*(S) = \sum_{i=1}^{\infty} \bar{\phi}_i (S^*)^{i-1}, \tag{5.18}$$

where S is the shift operator. Observe that

$$\bar{\phi}^*(S) = \sum_{i=1}^{\infty} \phi_i (S^*)^{i-1} . \tag{5.19}$$

We have the following results.

Lemma 5.4. Let Γ be a Hankel matrix and let $\phi(z) \in H^\infty$ be defined by (5.16). Then

$$\Gamma \phi(S) = \bar{\phi}^*(S) \Gamma . \tag{5.20}$$

We leave the proof to the reader as an exercise (Problem 5.9).

Lemma 5.5. Let Γ_1 and Γ_2 be two Hankel matrices satisfying

$$\Gamma_2 = \Gamma_1 \phi(S) ,$$

where $\phi(z) \in H^\infty$ is given by (5.16) such that $\|\phi\|_{H^\infty} \leq 1$. Then

$$\overline{\Gamma}_2 \Gamma_2 \leq \overline{\Gamma}_1 \Gamma_1 . \tag{5.21}$$

Here, we have used the standard notation that $A \leq B$ if $B - A$ is non-negative definite. Observe that

$$\|\bar{\phi}^*(S)\| = \|\bar{\phi}(S)\| = \|\bar{\phi}(z)\|_{H^\infty} \leq 1 ,$$

so that for any $\underline{\xi} \in l^2$ we have

$$\begin{aligned} &\langle \overline{\Gamma}_1 \Gamma_1 \underline{\xi}, \underline{\xi} \rangle_{l^2} - \langle \overline{\Gamma}_2 \Gamma_2 \underline{\xi}, \underline{\xi} \rangle_{l^2} \\ =&\|\Gamma_1 \underline{\xi}\|_{l^2}^2 - \|\Gamma_2 \underline{\xi}\|_{l^2}^2 \\ =&\|\Gamma_1 \underline{\xi}\|_{l^2}^2 - \|\Gamma_1 \phi(S) \underline{\xi}\|_{l^2}^2 \\ =&\|\Gamma_1 \underline{\xi}\|_{l^2}^2 - \|\bar{\phi}^*(S) \Gamma_1 \underline{\xi}\|_{l^2}^2 \geq 0 . \end{aligned}$$

Hence, (5.21) follows.

The following result characterizes the set of all vectors on which the norm of a Hankel matrix is attained. This result will provide an explicit formula for the operator norm and will be used frequently later.

Theorem 5.4. Let Γ be a Hankel matrix and $\underline{\xi} = [\, \xi_1 \ \xi_2 \ \xi_3 \ \cdots \,]^\top \in l^2$ such that

$$\|\Gamma \underline{\xi}\|_{l^2} = \|\Gamma\| \ \|\underline{\xi}\|_{l^2} .$$

Denote by $\xi_I(z)$ the inner factor of the function $\xi_+(z) = \sum_{i=1}^{\infty} \xi_i z^{i-1}$ and by $\eta_I(z)$ the inner factor of the function $\eta_+(z) = \overline{(\Gamma \underline{\xi})_+(\bar{z})}$. Moreover, let $\gamma_I(z)$ be any inner function that satisfies $(\xi_I(z)\eta_I(z)/\gamma_I(z)) \in H^\infty$. Then

$$\|\Gamma\| = \frac{\|\Gamma\gamma_I(S)\xi_I^*(S)\underline{\xi}\|_{l^2}}{\|\gamma_I(S)\xi_I^*(S)\underline{\xi}\|_{l^2}}, \tag{5.22}$$

where $\xi_I^*(S)$ is the adjoint of the operator $\xi_I(S)$. In particular,

$$\|\Gamma\| = \frac{\|\Gamma\underline{\xi}^0\|_{l^2}}{\|\underline{\xi}^0\|_{l^2}}, \tag{5.23}$$

where $\underline{\xi}^0 = [\ \xi_1^0\ \xi_2^0\ \xi_3^0\ \cdots\ ]^\top \in l^2$ with $\xi_+^0(z) = \sum_{i=1}^\infty \xi_i^0 z^{i-1}$ being the outer factor of $\xi_+(z)$.

To prove this theorem, we first observe that $\eta_I(z)$ is the inner factor of $\overline{(\Gamma\underline{\xi})_+(\bar{z})}$, and hence $\bar{\eta}_I(z)$ [see (5.17) for its definition] is the inner factor of $(\Gamma\underline{\xi})_+(z)$ (Problem 5.10). It follows from Theorem 5.3 (with $\underline{\xi}$ replaced by $\Gamma\underline{\xi}$) that $\Gamma\underline{\xi} \in \bar{\eta}_I(S)l^2$. Hence, from Lemma 5.1(b), we have $\bar{\eta}_I^*(S)\Gamma\underline{\xi} \in l^2$, so that it follows from Lemma 5.1(a) and (c) that

$$\|\bar{\eta}_I^*(S)\Gamma\underline{\xi}\|_{l^2} = \|\bar{\eta}_I(S)\bar{\eta}_I^*(S)\Gamma\underline{\xi}\|_{l^2} = \|\Gamma\underline{\xi}\|_{l^2}.$$

Consequently, from Lemmas 5.2 and 5.1 and under the assumption in the statement of the theorem, we have

$$\begin{aligned}\|\Gamma\eta_I(S)\underline{\xi}\|_{l^2} &= \|\bar{\eta}_I^*(S)\Gamma\underline{\xi}\|_{l^2}\\ &= \|\Gamma\underline{\xi}\|_{l^2}\\ &= \|\Gamma\|\ \|\underline{\xi}\|_{l^2}\\ &= \|\Gamma\|\ \|\eta_I(S)\underline{\xi}\|_{l^2}\end{aligned}$$

which implies that the operator norm $\|\Gamma\|$ of Γ is attained on the subspace $\eta_I(S)l^2$.

Moreover, for any inner function $\gamma_I(z)$ with

$$\beta(z) := \frac{\xi_I(z)\eta_I(z)}{\gamma_I(z)} \in H^\infty,$$

$\beta(z)$ is an inner function and satisfies

$$\xi_I(S)\eta_I(S) = \gamma_I(S)\beta(S).$$

It also follows from Lemma 5.1 (b) that

$$\gamma_I^*(S)\xi_I(S)\eta_I(S) = \beta(S).$$

Note that $\Gamma\beta(S)$ is a Hankel matrix (Problem 5.9a). Using Lemmas 5.5 and 5.1, as well as Problems 5.11 and 5.12, we obtain, for $\underline{\xi} \in \xi_I(S)l^2$,

$$\begin{aligned}\|\Gamma\gamma_I(S)\underline{\xi}_I^*(S)\underline{\xi}\|_{l^2} &\geq \|\Gamma\beta(S)\gamma_I(S)\xi_I^*(S)\underline{\xi}\|_{l^2}\\ &= \|\Gamma\gamma_I^*(S)\xi_I(S)\eta_I(S)\gamma_I(S)\xi_I^*(S)\underline{\xi}\|_{l^2}\\ &= \|\Gamma\gamma_I^*(S)\gamma_I(S)\eta_I(S)\xi_I(S)\xi_I^*(S)\underline{\xi}\|_{l^2}\\ &= \|\Gamma\eta_I(S)\underline{\xi}\|_{l^2}\\ &= \|\Gamma\|\ \|\eta_I(S)\underline{\xi}\|_{l^2}\\ &= \|\Gamma\|\ \|\gamma_I(S)\beta(S)\xi_I^*(S)\underline{\xi}\|_{l^2}\\ &= \|\Gamma\|\ \|\xi_I^*(S)\underline{\xi}\|_{l^2}\\ &= \|\Gamma\|\ \|\gamma_I(S)\xi_I^*(S)\underline{\xi}\|_{l^2}\\ &\geq \|\Gamma\gamma_I(S)\xi_I^*(S)\underline{\xi}\|_{l^2}\,.\end{aligned}$$

Hence, (5.22) follows.

In particular, for $\underline{\xi} \in \xi_I(S)l^2$, it follows from Lemma 5.1(a) that

$$\|\gamma_I(S)\xi_I^*(S)\underline{\xi}\|_{l^2} = \|\xi_I^*(S)\underline{\xi}\|_{l^2}\,,$$

so that by setting $\underline{\xi}^0 = \xi_I^*(S)\underline{\xi}$ with $\underline{\xi} \in \xi_I(S)l^2$ we have

$$\begin{aligned}\|\Gamma\underline{\xi}^0\|_{l^2} &= \|\Gamma\xi_I^*(S)\underline{\xi}\|_{l^2}\\ &= \|\Gamma\gamma_I(S)\xi_I^*(S)\underline{\xi}\|_{l^2}\\ &= \|\Gamma\|\ \|\gamma_I(S)\xi_I^*(S)\underline{\xi}\|_{l^2}\\ &= \|\Gamma\|\ \|\xi_I^*(S)\underline{\xi}\|_{l^2}\\ &= \|\Gamma\|\ \|\underline{\xi}^0\|_{l^2}\,.\end{aligned}$$

Note that $\xi_I(S)\underline{\xi}^0 = \xi_I(S)\xi_I^*(S)\underline{\xi} = \underline{\xi}$ for all $\underline{\xi} \in \xi_I(S)l^2$, so that $\xi_+(z) = \xi_I(z)\xi_+^0(z)$, or $\xi_+^0(z) = \xi_O(z)$ is an outer function. This completes the proof of the theorem.

5.1.5 Properties of the Norm of Hankel Matrices

In the previous section, we expressed the operator norm of a Hankel matrix in terms of the outer and inner factors of a nontrivial l^2 sequence. In this section, we will further characterize the set on which the operator norm of the Hankel matrix is attained, and obtain some properties of the norm under consideration, which is important to the proof of the general AAK theorem later in this chapter.

Let Γ be a bounded Hankel matrix defined on l^2, and set

$$L_\Gamma = \{\underline{\xi} \in l^2:\ \ \|\Gamma\underline{\xi}\|_{l^2} = \|\Gamma\|\ \|\underline{\xi}\|_{l^2}\}\,. \tag{5.24}$$

It can be easily verified that $\underline{\xi} \in L_\Gamma$ if and only if $\overline{\Gamma}\Gamma\underline{\xi} = \|\Gamma\|^2\underline{\xi}$, so that (Problem 5.13)

$$L_\Gamma = \{\underline{\xi} \in l^2:\ \ \overline{\Gamma}\Gamma\underline{\xi} = \|\Gamma\|^2\underline{\xi}\}\,. \tag{5.25}$$

It is now clear that L_Γ is a subspace of l^2, see, again, Problem 5.13. Denote, as usual, the family of polynomials of z with degree no greater than n by $\mathcal{P}_n$. Then, we have the following result which characterizes the linear space L_Γ when it is finite-dimensional. Another characterization theorem for an infinite-dimensional linear space L_Γ will be given later in Theorem 5.7.

Theorem 5.5. Suppose that the linear space L_Γ defined by (5.25) has finite dimension n. Then there exists a vector $\underline{\xi} \in L_\Gamma$ such that

$$L_\Gamma = \{p(S)(S^*)^{n-1}\underline{\xi}:\ p(z) \in \mathcal{P}_{n-1}\}, \tag{5.26}$$

where S^* is the adjoint of the shift operator S.

To determine such a vector $\underline{\xi}$ in L_Γ, we first observe that since $\dim(L_\Gamma) = n < \infty$, there exists a vector $\underline{\xi}$ in L_Γ such that

$$\underline{\xi} = [\,0\ \cdots\ 0\ \xi_k\ \xi_{k+1}\ \cdots\,]^\top,$$

where $k \geq n$ and $\xi_k \neq 0$ (Problem 5.14). Set

$$\xi_+(z) = \sum_{i=k}^{\infty} \xi_i z^{i-1} = z^{k-1} \sum_{i=1}^{\infty} \xi_{k+i-1} z^{i-1} := z^{k-1}\xi_I(z)\xi_o(z). \tag{5.27}$$

Then, $\underline{\xi} = S^{k-1}\xi_I(S)\underline{\xi}_o$ for some vector $\underline{\xi}_o \in l^2$ corresponding to the outer factor $\xi_o(z)$ of $\xi_+(z)$. We will show that $k = n$. Indeed, it can be verified by using Theorem 5.4 that the norm of the Hankel matrix Γ is attained on the set (Problem 5.15)

$$V = \{S^i\xi_I(S)\xi_I^*(S)(S^*)^{k-1}\underline{\xi}:\quad 0 \leq i \leq k-1\}. \tag{5.28}$$

Moreover, since

$$(S^*)^{k-1}\underline{\xi} = (S^*)^{k-1}(S^{k-1})\xi_I(S)\underline{\xi}_o = \xi_I(S)\underline{\xi}_o \in \xi_I(S)l^2,$$

it follows from Lemma 5.1 that

$$V = \{S^i(S^*)^{k-1}\underline{\xi}:\quad 0 \leq i \leq k-1\}. \tag{5.29}$$

However, for such a vector $\underline{\xi} \in l^2$, the vectors in V are linearly independent (Problem 5.16), and hence $k \leq n$, so that $k = n$ as claimed.

Now, we have n linearly independent vectors $\{S^i(S^*)^{n-1}\underline{\xi}\}_{i=0}^{n-1}$ on which the operator norm of Γ is attained, and we already know that L_Γ is an n-dimensional linear vector space. This yields

$$\begin{aligned} L_\Gamma &= \operatorname{span}\{S^i(S^*)^{n-1}\underline{\xi}:\quad 0 \leq i \leq n-1\} \\ &= \{p(S)(S^*)^{n-1}\underline{\xi}:\quad p(z) \in \mathcal{P}_{n-1}\} \end{aligned}$$

and completes the proof of the theorem.

To derive the next result, we need the following lemma.

Lemma 5.6. Let $f(z) \in H^\infty$ be such that $1/f(z) \in H^\infty$, and let S be the shift operator. Then $f(S)$ and its adjoint operator $f^*(S)$ are both invertible on l^2, namely, $f^{-1}(S)$ and $[f^*(S)]^{-1}$ are both bounded linear operators on l^2.

In view of the open mapping theorem, it is sufficient to show that $f(S)$ is a one-to-one mapping from l^2 onto l^2. We will then prove the boundedness of $[f^*(S)]^{-1}$ from the existence of $f^{-1}(S)$.

For any $\underline{\xi} \in l^2$, since $1/f(z) \in H^\infty$ we have $\eta_+(z) := \xi_+(z)/f(z) \in H^2$. Hence, if we denote by $\underline{\eta}$ the vector in l^2 corresponding to the function $\eta_+(z) \in H^2$, we have

$$(f(S)\underline{\eta})_+(z) = f(z)\eta_+(z) = \xi_+(z),$$

so that $f(S)\underline{\eta} = \underline{\xi}$. This implies that $f(S)$ maps l^2 onto itself.

Now, suppose that $f(S)\underline{\xi}^1 = f(S)\underline{\xi}^2$. Then we have $f(z)\xi_+^1(z) = f(z)\xi_+^2(z)$, so that $\xi_+^1(z) = \xi_+^2(z)$, or equivalently $\underline{\xi}^1 = \underline{\xi}^2$. Hence, $f(S)$ is also a one-to-one mapping. It follows that $f^{-1}(S)$ exists and is bounded from l^2 to l^2.

Let $g(S) = f^{-1}(S)$. Then we have

$$\langle g^*(S)f^*(S)\underline{\xi}, \underline{\eta}\rangle_{l^2} = \langle \underline{\xi}, f(S)g(S)\underline{\eta}\rangle_{l^2} = \langle \underline{\xi}, \underline{\eta}\rangle_{l^2}.$$

This implies that $f^*(S)$ is invertible and

$$[f^*(S)]^{-1} = g^*(S) = [f^{-1}(S)]^*,$$

completing the proof of the lemma.

Recall from Sect. 3.1.2 that an inner function $\phi(z)$ can be factorized as

$$\phi(z) = \prod \frac{z - z_i}{1 - \bar{z}_i z}\psi(z),$$

where $\psi(z)$ is a zero-free singular inner function. Here, if $\psi(z)$ is a nontrivial constant and the Blaschke product is a finite product, then the number of zeros of $\phi(z)$, counting multiplicities, is called the *degree* of $\phi(z)$. Otherwise, we say that the degree of $\phi(z)$ is infinite.

In the following, we also recall a notation from functional analysis: The *kernel* (or *null space*) of an operator A on l^2 is defined by

$$\operatorname{Ker}(A) = \{\underline{\xi} \in l^2 : \quad A\underline{\xi} = 0\}. \tag{5.30}$$

We can now state and prove the following useful result:

Theorem 5.6. Let $\phi(z)$ be an inner function of the form

$$\phi(z) = \prod_{i=1}^{n} \frac{z - z_i}{1 - \bar{z}_i z}.$$

Then

$$\mathrm{Ker}(\phi^*(S)) = (\phi(S)l^2)^{\perp} \tag{5.31}$$

and

$$\dim(\mathrm{Ker}(\phi^*(S)) = n. \tag{5.32}$$

We first note that (5.31) follows immediately from the identity

$$\langle \phi^*(S)\underline{\xi}, \underline{\eta} \rangle_{l^2} = \langle \underline{\xi}, \phi(S)\underline{\eta} \rangle_{l^2}.$$

To prove (5.32), let $\phi(z) = \Pi_{i=1}^{n} \phi_i(z)$ with

$$\phi_i(z) = \frac{z - z_i}{1 - \bar{z}_i z}, \qquad i = 1, 2, \ldots, n.$$

We first consider $\phi_1(z) = d_1(z)(z - z_1)$ with $d_1(z) = 1/(1 - \bar{z}_1 z)$ by noting that

$$\phi_1^*(S) = d_1^*(S)(S^* - \bar{z}_1 I),$$

where I is the identity operator on l^2. By Lemma 5.6, we see that $d_1^*(S)$ is invertible, so that for any $\underline{\xi} \in l^2$, we have $\phi_1^*(S)\underline{\xi} = 0$ if and only if $(S^* - \bar{z}_1 I)\underline{\xi} = 0$.

Let $(S^* - \bar{z}_1 I)\underline{\xi} = 0$ where $\underline{\xi} \in l^2$ and $\underline{\xi} \neq 0$. Then we have $\underline{\xi} = c[\ 1\ \bar{z}_1\ \bar{z}_1^2\ \cdots\ ]^{\top}$ for some nonzero constant c (Problem 5.17). Set $\underline{\eta}_1 = [\ 1\ \bar{z}_1\ \bar{z}_1^2\ \cdots\ ]^{\top}$. Then it is clear that

$$\mathrm{Ker}(\phi_1^*(S)) = \mathrm{span}\{\underline{\eta}_1\}.$$

Since we have $\|\phi_1(S)\underline{\xi}\|_{l^2} = \|\underline{\xi}\|_{l^2}$ for all $\underline{\xi} \in l^2$, it follows that $\phi_1(S)l^2$ is a closed subspace of l^2. Hence, in view of (5.31), we have $\underline{\eta}_1 \in (\phi_1(S)l^2)^{\perp}$ so that

$$\begin{aligned} l^2 &= (\phi_1(S)l^2)^{\perp} \oplus (\phi_1(S)l^2) \\ &= \mathrm{span}\{\underline{\eta}_1\} \oplus (\phi_1(S)l^2). \end{aligned}$$

We then consider $\phi_2(z)$ and set $\phi_2(z) = d_2(z)(z - z_2)$ with $d_2(z) = 1/(1 - \bar{z}_2 z)$, and similarly obtain

$$l^2 = \mathrm{span}\{\underline{\eta}_2\} \oplus (\phi_2(S)l^2).$$

Moreover, observe that since $\underline{\eta}_1 \in (\phi_1(S)l^2)^{\perp}$ we have

$$\langle \phi_2(S)\underline{\eta}_1, \phi_2(S)\phi_1(S)\underline{\xi}\rangle_{l^2} = \langle \underline{\eta}_1, \phi_1(S)\underline{\xi}\rangle_{l^2} = 0$$

for all $\underline{\xi} \in l^2$, so that

$$l^2 = \text{span}\{\underline{\eta}_2\} \oplus \text{span}\{\phi_2(S)\underline{\eta}_1\} \oplus (\phi_2(S)\phi_1(S)l^2).$$

Hence, in general we have

$$\begin{aligned} l^2 =& \text{span}\{\underline{\eta}_n\} \oplus \text{span}\{\phi_n(S)\underline{\eta}_{n-1}\} \oplus \\ & \cdots \oplus \text{span}\{\phi_n(S)\cdots\phi_2(S)\underline{\eta}_1\} \oplus (\phi_n(S)\cdots\phi_1(S)l^2), \end{aligned}$$

where

$$\underline{\eta}_i = [\, 1 \ \bar{z}_i \ \bar{z}_i^2 \ \cdots \,]^\top, \qquad i = 1, 2, \ldots, n.$$

This implies that

$$\begin{aligned} & \text{Ker}(\phi^*(S)) \\ &= (\phi(S)l^2)^\perp = (\phi_n(S)\cdots\phi_1(S)l^2)^\perp \\ &= \text{span}\{\underline{\eta}_n\} \oplus \text{span}\{\phi_n(S)\underline{\eta}_{n-1}\} \oplus \cdots \oplus \text{span}\{\phi_n(S)\cdots\phi_2(S)\underline{\eta}_1\}, \end{aligned}$$

and so (5.32) follows. The proof of the theorem is then completed.

Now, in addition to Theorem 5.5, we will give a characterization theorem for the linear space L_Γ defined by either (5.24) or (5.25) when it is not necessarily finite-dimensional.

Theorem 5.7. Let L_Γ be the linear space defined by (5.24) and let $\underline{\xi} = [\, \xi_1 \ \xi_2 \ \cdots \,]^\top$ be a vector in L_Γ. Moreover, let $\xi_I(z)$ denote the inner factor of the function $\xi_+(z) = \sum_{i=1}^\infty \xi_i z^{i-1}$, and suppose that the dimension of the space L_Γ is greater than the (finite) degree of $\xi_I(z)$. Then, the norm of the Hankel matrix Γ which defines L_Γ is attained on the subspace $\xi_I(S)l^2$ of l^2.

Let $\underline{\xi}$ be a vector in L_Γ such that its first m components are equal to zero but its $(m+1)$st component is nonzero. Decompose the corresponding $\xi_+(z)$ as

$$\xi_+(z) = z^m \xi_I(z)\xi_o(z),$$

where $\xi_I(z)$ is an inner function that does not vanish at 0 and $\xi_o(z)$ is the outer factor of $\xi_+(z)$. Write $\xi_o(z) = \sum_{i=1}^\infty \xi_{oi} z^{-i}$ and set $\underline{\xi}_o = [\, \xi_{o1} \ \xi_{o2} \ \cdots \,]^\top$. Then, it follows from Theorem 5.4 that the norm of the Hankel matrix Γ is attained at

$$S^i(S^*)^m \tilde{\xi}_I^*(S)\underline{\xi} = S^i \underline{\xi}_o, \qquad 0 \le i \le m,$$

which implies that the norm of Γ is attained at all vectors in the linear span of

$$\{p(S)\underline{\xi}_o : \quad p(z) \in \mathcal{P}_m\}.$$

Write

$$\tilde{\xi}_I(z) = \prod_{i=1}^{n} \frac{z - z_i}{1 - \bar{z}_i z} = \tilde{\psi}(z) \prod_{i=1}^{n} (z - z_i).$$

Then, by Lemma 5.6 we have $\prod_{i=1}^{n}(z - z_i) = \tilde{\psi}^{-1}(z)\tilde{\xi}_I(z)$, so that the norm of Γ is attained on the subspace $\xi_I(z)l^2$ of l^2. This completes the proof of the theorem.

The following result can be easily verified (Problem 5.18). Recall from (5.17) the notation of $\bar{\xi}_I(z)$ which is obtained from $\xi_I(z)$.

Lemma 5.7. Let Γ be a Hankel matrix, L_Γ be defined by (5.24) and $\underline{\xi}$ be any vector in L_Γ with inner and outer factorization $\xi_+(z) = \xi_I(z)\xi_o(z)$. Moreover, let $\tilde{\Gamma}$ be another Hankel matrix such that

$$\Gamma\underline{\eta} = \tilde{\Gamma}\underline{\eta} \qquad \text{for all } \underline{\eta} \in \xi_I(S)l^2 .$$

Then, the range of $\Gamma - \tilde{\Gamma}$ is orthogonal to the subspace $\bar{\xi}_I(S)l^2$ in the sense that

$$\langle(\Gamma - \tilde{\Gamma})\underline{\xi}, \bar{\xi}_I(S)\underline{\eta}\rangle = 0$$

for all $\underline{\xi} \in \xi_I(S)l^2$ and all $\underline{\eta} \in \bar{\xi}_I(S)l^2$.

We are now in a position to prove the following minimality result on the norm of a Hankel operator.

Theorem 5.8. Suppose that the norm of a Hankel matrix Γ is attained on the subspace $\xi_I(S)l^2$ for some vector $\underline{\xi} \in L_\Gamma$ where L_Γ is defined by (5.24), and suppose that Λ is any bounded Hankel matrix such that $\Lambda\underline{\xi} = 0$ for all $\underline{\xi} \in \xi_I(S)l^2$. Then, we have

$$\|\Gamma + \Lambda\| \geq \|\Gamma\|, \tag{5.33}$$

where the equality holds if and only if $\Lambda \equiv 0$.

To prove the theorem, let $\underline{\xi} \in L_\Gamma$ be so chosen that $\underline{\tilde{\xi}} := \xi_I(S)\underline{\xi}$ satisfies $\|\Gamma\underline{\tilde{\xi}}\|_{l^2} = \|\Gamma\| \, \|\underline{\tilde{\xi}}\|_{l^2}$. Then, it follows from Lemmas 5.4 and 5.1 that

$$\|\Gamma\xi_I(S)\underline{\xi}\|_{l^2} = \|\bar{\xi}_I^*(S)\Gamma\underline{\xi}\|_{l^2} = \|\Gamma\underline{\xi}\|_{l^2} = \|\Gamma\| \, \|\underline{\xi}\|_{l^2},$$

and hence $\|\Gamma\| \, \|\underline{\tilde{\xi}}\|_{l^2} = \|\Gamma\| \, \|\underline{\xi}\|_{l^2}$. Now, set

$$\tilde{\xi}_+(z) = \xi_I(z)\xi_I(z)\xi_o(z).$$

Then, it follows from Theorem 5.4 that the norm of Γ is attained both at the vector

$$\xi_I^*(S)\xi_I^*(S)\tilde{\underline{\xi}} = \underline{\xi}_O$$

and the vector

$$\xi_I(S)(\xi_I^*(S)\xi_I^*(S)\tilde{\underline{\xi}}) = \xi_I(S)\xi_O\,.$$

Hence, we have

$$\|\bar{\xi}_I^*(S)\Gamma\underline{\xi}_O\|_{l^2} = \|\Gamma\xi_I(S)\underline{\xi}_O\|_{l^2} = \|\Gamma\|\;\|\xi_I(S)\underline{\xi}_O\|_{l^2} = \|\Gamma\|\;\|\underline{\xi}_O\|_{l^2}\,,$$

which yields $\Gamma\underline{\xi}_O \in \bar{\xi}_I(S)l^2$. Moreover, if we define $\tilde{\Gamma} = \Gamma - \Lambda$, then we have $\|\tilde{\Gamma}\underline{\xi}\|_{l^2} = \|\Gamma\underline{\xi}\|_{l^2}$ for all $\xi_I(S)l^2$, so that by Lemma 5.7, we have

$$\langle(\Gamma - \tilde{\Gamma})\underline{\xi}, \Gamma\underline{\xi}_O\rangle = 0\,.$$

Consequently,

$$\|(\Gamma + \Lambda)\underline{\xi}_O\|_{l^2}^2 = \|\Gamma\underline{\xi}_O\|_{l^2}^2 + \|\Lambda\underline{\xi}_O\|_{l^2}^2 \geq \|\Gamma\|^2\|\underline{\xi}_O\|_{l^2}^2\,.$$

This implies that $\|\Gamma+\Lambda\| \geq \|\Gamma\|$. Since $\xi_O(z)$ is an outer function, it follows from Corollary 5.1 that $\Lambda\xi_O = 0$ implies $\Lambda \equiv 0$. This completes the proof of the theorem.

5.2 Uniqueness of Schmidt Pairs

In this section, we study the uniqueness property of Schmidt pairs corresponding to any s-number of a general bounded Hankel matrix.

5.2.1 Uniqueness of Ratios of Schmidt Pairs

Let Γ be a bounded Hankel matrix and s any nonzero s-number of Γ with a corresponding Schmidt pair $(\underline{\xi}, \underline{\eta})$ in l^2. Recall from (4.25) that

$$\begin{cases} \Gamma\underline{\xi} = s\underline{\eta}\,, \\ \overline{\Gamma}\underline{\eta} = s\underline{\xi}\,. \end{cases} \tag{5.34}$$

Of course, to a given s-number there corresponds more than one Schmidt pairs. However, we have the following uniqueness result:

Theorem 5.9. Let $(\underline{\xi}^1, \underline{\eta}^1)$ and $(\underline{\xi}^2, \underline{\eta}^2)$ be two Schmidt pairs corresponding to the same s-number s of a bounded Hankel matrix Γ. Then, we have

$$\xi_+^1(z)\overline{\xi_+^2(z)} = \eta_-^1(z)\overline{\eta_-^2(z)}$$

on the unit circle $|z| = 1$.

It follows from direct calculations that, for $|z| = 1$, we have

$$\begin{aligned}\xi_+^1(z)\overline{\xi_+^2(z)} &= \left(\sum_{i=1}^{\infty} \xi_i^1 z^{i-1}\right)\overline{\left(\sum_{i=1}^{\infty} \xi_i^2 z^{i-1}\right)} \\ &= \sum_{k=0}^{\infty}\left(\sum_{i=1}^{\infty} \xi_{i+k}^1 \overline{\xi_i^2}\right) z^k + \sum_{k=1}^{\infty}\left(\sum_{i=1}^{\infty} \xi_i^1 \overline{\xi_{i+k}^2}\right) z^{-k}\end{aligned}$$

and that

$$\begin{aligned}\eta_-^1(z)\overline{\eta_-^2(z)} &= \left(\sum_{i=1}^{\infty} \eta_i^1 z^{-i}\right)\overline{\left(\sum_{i=1}^{\infty} \eta_i^2 z^{-i}\right)} \\ &= \sum_{k=0}^{\infty}\left(\sum_{i=1}^{\infty} \eta_i^1 \overline{\eta_{i+k}^2}\right) z^k + \sum_{k=1}^{\infty}\left(\sum_{i=1}^{\infty} \eta_{i+k}^1 \overline{\eta_i^2}\right) z^{-k},\end{aligned}$$

(Problem 5.19). Hence, it is sufficient to verify

$$\sum_{i=1}^{\infty} \xi_{i+k}^1 \overline{\xi_i^2} = \sum_{i=1}^{\infty} \eta_i^1 \overline{\eta_{i+k}^2} \tag{5.35}$$

and

$$\sum_{i=1}^{\infty} \xi_i^1 \overline{\xi_{i+k}^2} = \sum_{i=1}^{\infty} \eta_{i+k}^1 \overline{\eta_i^2}\,. \tag{5.36}$$

This can be accomplished as follows: First, observe that

$$\begin{aligned}\langle (S^*)^k \underline{\xi}^1, \underline{\xi}^2\rangle &= \langle (S^*)^k (s^{-1}\overline{\Gamma}\underline{\eta}^1), \underline{\xi}^2\rangle \\ &= s^{-1}\langle \overline{\Gamma} S^k \underline{\eta}^1, \underline{\xi}^2\rangle \\ &= \langle S^k \underline{\eta}^1, s^{-1}\Gamma \underline{\xi}^2\rangle \\ &= \langle S^k \underline{\eta}^1, \underline{\eta}^2\rangle,\end{aligned}$$

where

$$(S^*)^k \underline{\xi}^1 = [\ \xi_{k+1}^1\ \xi_{k+2}^1\ \cdots\ ]^\top$$

and

$$S^k \underline{\eta}^1 = [\ \underbrace{0\ \cdots\ 0}_{k}\ \eta_1^1\ \eta_2^1\ \cdots\ ]^\top,$$

so that

$$\langle (S^*)^k \underline{\xi}^1, \underline{\xi}^2\rangle = \sum_{i=1}^{\infty} \xi_{i+k}^1 \overline{\xi_i^2}$$

and

$$\langle S^k \underline{\eta}^1, \underline{\eta}^2 \rangle = \sum_{i=1}^{\infty} \eta_i^1 \overline{\eta_{i+k}^2}\,.$$

Hence, (5.35) is verified. The second identity (5.36) can be verified in a similar manner (Problem 5.20). This completes the proof of the theorem.

As a consequence of this theorem, we have the following uniqueness result:

Corollary 5.2. Let $\mathcal{S}$ be the collection of all Schmidt pairs corresponding to the same positive s-number of a bounded Hankel matrix Γ. Then, the function $\phi_s(z) := \eta_-(z)/\xi_+(z)$ does not depend on the choice of the Schmidt pair $(\underline{\xi}, \underline{\eta})$ over the set $\mathcal{S}$. Moreover, $|\phi_s(z)| \equiv 1$ on the unit circle $|z| = 1$.

Let $(\underline{\xi}^1, \underline{\eta}^1)$ and $(\underline{\xi}^2, \underline{\eta}^2)$ be any two Schmidt pairs in $\mathcal{S}$. Then, by Theorem 5.9, we have $\xi_+^1(z)\overline{\xi_+^2(z)} = \eta_-^1(z)\overline{\eta_-^2(z)}$, or

$$\frac{\eta_-^1(z)}{\xi_+^1(z)} = \overline{\left(\frac{\xi_+^2(z)}{\eta_-^2(z)}\right)}$$

for $|z| = 1$. In particular, setting $(\underline{\xi}^2, \underline{\eta}^2) = (\underline{\xi}^1, \underline{\eta}^1)$ yields $|\xi_+^1(z)|^2 = |\eta_-^1(z)|^2$, or $|\eta_-^1(z)/\xi_+^1(z)| = 1$ so that $\overline{(\xi_+^1(z)/\eta_-^1(z))} = \eta_-^1(z)/\xi_+^1(z)$, for $|z| = 1$. This yields

$$\frac{\eta_-^1(z)}{\xi_+^1(z)} = \frac{\eta_-^2(z)}{\xi_+^2(z)}$$

on $|z| = 1$ and hence in $|z| < 1$ by analyticity. Moreover, $|\phi_s(z)| = |\eta_-^2(z)/\xi_+^2(z)| = 1$ on $|z| = 1$, completing the proof of the corollary.

5.2.2 Hankel Operators Generated by Schmidt Pairs

Let s be a positive s-number of a bounded Hankel matrix Γ, and $\mathcal{S}$ the collection of all Schmidt pairs corresponding to s. Moreover, let $\phi_s(z)$ be the function defined as in Corollary 5.2, namely, $\phi_s(z) = \eta_-(z)/\xi_+(z)$ with $(\underline{\xi}, \underline{\eta}) \in \mathcal{S}$. Then, $\phi_s(z)$ defines another bounded Hankel matrix

$$\Gamma_s := \Gamma(s\phi_s(z)) = \Gamma(s\eta_-(z)/\xi_+(z))\,, \tag{5.37}$$

which is uniquely determined by the s-number s. The following interesting result is very useful.

Lemma 5.8. Let $(\underline{\xi}, \underline{\eta})$ be a Schmidt pair of Γ corresponding to an s-number s of Γ and Γ_s be defined as in (5.37). Then, s is also an s-number of Γ_s with Schmidt pair $(\underline{\xi}, \underline{\eta})$ of Γ_s, corresponding to this s-number s.

Since the statement of this theorem is trivial for $s = 0$, we only consider $s > 0$. As usual, let $\eta_-(z) = \sum_{i=1}^{\infty} \eta_i z^{-i}$. Then, for each $i = 1, 2, \ldots$, we have, on the unit circle $z = e^{j\theta}$,

$$\begin{aligned} s\eta_i &= s\frac{1}{2\pi}\int_0^{2\pi} z^i \eta_-(z) d\theta \\ &= s\frac{1}{2\pi}\int_0^{2\pi} z^i \phi_s(z)\xi_+(z) d\theta \\ &= s\frac{1}{2\pi}\int_0^{2\pi} z^i \phi_s(z)\sum_{k=1}^{\infty} \xi_k z^{k-1} d\theta \\ &= \sum_{k=1}^{\infty} \xi_k \left(s\frac{1}{2\pi}\int_0^{2\pi} \phi_s(z) z^{k+i-1} d\theta \right) \\ &= [\Gamma(s\phi_s(z))\underline{\xi}]_i \,, \end{aligned}$$

where $[\Gamma(s\phi_s(z))\underline{\xi}]_i$ denotes the ith component of the vector $\Gamma(s\phi_s(z))\underline{\xi}$, and $\Gamma(s\phi_s(z))$ denotes the Hankel matrix corresponding to the function $s\phi_s(z)$. Hence, we have

$$s\underline{\eta} = \Gamma_s \underline{\xi} \,.$$

In the above derivation, the interchange of summation and integration is allowed based on the fact that $\sum_{i=1}^{\infty} \xi_i z^{i-1}$ converges uniformly in $H^2(|z| = 1)$. Similarly, one can show that (Problem 5.21)

$$s\underline{\xi} = \overline{\Gamma}_s \underline{\eta} \,.$$

This implies that s is an s-number of Γ_s and that (ξ, η) is also a Schmidt pair of Γ_s corresponding to s, completing the proof of the lemma.

As to the matrix Γ_s defined above, the s-number s is its largest s-number. This result is a simple consequence of Nehari's theorem. More precisely, we have the following:

Lemma 5.9. Let $(\underline{\xi}, \underline{\eta})$ be a Schmidt pair of a bounded Hankel matrix Γ corresponding to an s-number $s > 0$, and let $\Gamma_s = \Gamma(s\eta_-(z)/\xi_+(z))$. Then,

$$\|\Gamma_s\| = s \,.$$

On one hand, we have $\|\Gamma_s\underline{\xi}\|_{l^2}^2 = \langle \overline{\Gamma}_s\Gamma_s\underline{\xi}, \underline{\xi}\rangle = s^2\|\underline{\xi}\|_{l^2}^2$, so that $\|\Gamma_s\| \geq s$. On the other hand, from the Nehari theorem (Theorem 4.1), we have

$$\begin{aligned} \|\Gamma_s\| &= s\|\Gamma(\eta_-(z)/\xi_+(z))\| \\ &= s \inf_{g\in H^\infty} \|\eta_-(z)/\xi_+(z) - g(z)\|_{L^\infty} \\ &\leq s\|\eta_-(z)/\xi_+(z)\|_{L^\infty} = s \,. \end{aligned}$$

Hence, we have $\|\Gamma_s\| = s$.

5.3 The Greatest Common Divisor: The Inner Function $\xi_I^0(z)$

In this section, we investigate the basic properties of an inner function $\xi_I^0(z)$ which is the greatest common divisor of all inner factors corresponding to null sequences of $\Gamma - \Gamma_s$. These important properties are of essential importance in completing the proof of AAK's general theorem in the next section.

5.3.1 Basic Properties of the Inner Function $\xi_I^0(z)$

Let Γ be a bounded Hankel matrix defined on l^2 and $(\underline{\xi}, \underline{\eta})$ a Schmidt pair of Γ corresponding to an s-number $s > 0$. For any $\underline{\xi} \in l^2$, we may factorize $\xi_+(z) = \sum_{i=1}^{\infty} \xi_i z^{i-1}$, where $\underline{\xi} = [\ \xi_1\ \xi_2\ \cdots\ ]^\top \in l^2$, into the canonical form $\xi_+(z) = \xi_I(z)\xi_O(z)$, where $\xi_I(z)$ and $\xi_O(z)$ are inner and outer functions, respectively. We will see that $\xi_I(z)$ indeed plays an important role in the proof of the AAK theorem.

Let Γ_s be the Hankel matrix corresponding to the function $s\eta_-(z)/\xi_+(z)$, and consider the null space of $\Gamma - \Gamma_s$ defined by

$$\mathcal{S}_0 = \{\underline{\xi} \in l^2 \colon \ \Gamma\underline{\xi} = \Gamma_s\underline{\xi}\}\,. \tag{5.38}$$

Then, it is easy to see that $\mathcal{S}_0$ is an S-invariant subspace of l^2 in the sense that $\underline{\xi} \in \mathcal{S}_0$ implies that $S\underline{\xi} \in \mathcal{S}_0$, where S is the shift operator defined by (5.6) (Problem 5.22). In addition, in view of Lemma 5.3, it can be shown that there exists some $\underline{\xi}^0 \in \mathcal{S}_0$ such that $\mathcal{S}_0 = \overline{l_S^2(\underline{\xi}^0)}$, the closure of the space $l_S^2(\underline{\xi}^0)$ defined in Lemma 5.3 (Problem 5.22). Note that the inner factor $\xi_I^0(z)$ of $\xi_+^0(z)$ is the greatest common divisor of all $\xi_I(z)$ where $\underline{\xi} \in \mathcal{S}_0$, as we will see in the following theorem. Hence, by Beurling's theorem (Theorem 5.3), we have

$$\mathcal{S}_0 = \xi_I^0(S)l^2\,. \tag{5.39}$$

Moreover, $\xi_I^0(z)$ has the following important property:

Theorem 5.10. Let Γ be a bounded Hankel matrix and s a positive s-number of Γ. Denote by $\mathcal{S}$ the collection of all Schmidt pairs of Γ corresponding to this s-number s. Then there exists a pair $(\underline{\xi}^0, \underline{\eta}^0) \in \mathcal{S}$ such that $\underline{\xi}^0$ defines the inner function $\xi_I^0(z)$ in (5.39). Furthermore, $\xi_I^0(z)$ is the greatest common divisor of all $\xi_I(z)$ with $\underline{\xi} \in \mathcal{S}_0$.

To prove the theorem, we first recall from Lemma 5.8 that for an arbitrary pair $(\underline{\xi}, \underline{\eta})$ in $\mathcal{S}$, we have $\Gamma\underline{\xi} = s\underline{\eta} = \Gamma_s\underline{\xi}$, where Γ_s is defined as in (5.37), so that $\underline{\xi} \in \mathcal{S}_0$. Conversely, for any $\underline{\xi} \in \mathcal{S}_0$, by defining $\underline{\eta} = s^{-1}\Gamma\underline{\xi}$, it is clear that $(\xi, \eta) \in \mathcal{S}$. That is, there is a one-to-one correspondence between $\mathcal{S}$ and $\mathcal{S}_0$. By (5.39), it is clear that for any $\underline{\xi} \in \mathcal{S}_0$, we have

$$\xi_+(z) = \xi_I^0(z)\tilde{\xi}_I(z)\xi_o(z), \tag{5.40}$$

where $\tilde{\xi}_I(z)$ is an inner function. This implies that $\xi_I^0(z)$ is a common divisor of all $\xi_I(z)$, where $\underline{\xi} \in \mathcal{S}_0$. To see that $\xi_I^0(z)$ is the greatest common divisor, we will show that $\xi_I^0(z)$ is the inner factor of some $\xi_+^0(z)$ where $\underline{\xi}^0 \in \mathcal{S}_0$.

Let $(\underline{\xi}, \underline{\eta})$ be an arbitrary Schmidt pair of Γ corresponding to Γ, that is, $(\underline{\xi}, \underline{\eta}) \in \mathcal{S}$. We will show that $\underline{\xi}^0 := \xi_I^0(S)\underline{\xi}_o \in \mathcal{S}_0$. Since it is clear that $\xi_+^0(z) = \xi_I^0(z)(\xi_o)_+(z)$, where $(\xi_o)_+(z)$ is the outer factor of $\xi_+^0(z)$, the proof of the theorem is then completed. Now, since the norm of Γ_s is attained at $\underline{\xi}$, by Theorem 5.4 it follows that this norm is also attained at the vector $\underline{\xi}_I^0(S)\underline{\xi}$. On the other hand, it can be easily verified that $\|\overline{\Gamma}_s\| \, \|\underline{\eta}\|_{l^2} = s\|\underline{\eta}\|_{l^2}$ (Problem 5.23), so that from $\|\overline{\Gamma}_s\| = \|\Gamma_s\| = s$ we have $\|\underline{\xi}\|_{l^2} = \|\underline{\eta}\|_{l^2}$.

Since $\overline{\Gamma}_s\underline{\eta} = s\underline{\xi} = s\xi_I^0(S)\tilde{\xi}_I(S)\underline{\xi}_o$ where $\xi_I^0(z)\tilde{\xi}_I(z) = \xi_I(z)$ is the inner factor of $\xi_+(z)$, see (5.40), we have

$$\tilde{\xi}_I^*(S)\overline{\Gamma}_s\underline{\eta} = s\xi_I^0(S)\underline{\xi}_o$$

or, by Lemma 5.4,

$$\overline{\Gamma}_s(\overline{\tilde{\xi}}_I(S)\underline{\eta}) = s(\xi_I^0(S)\underline{\xi}_o), \tag{5.41}$$

where $\overline{\tilde{\xi}}_I(z)$ is defined from $\tilde{\xi}$ by (5.17). Moreover, we have

$$\|\overline{\tilde{\xi}}_I(S)\underline{\eta}\|_{l^2} = \|\underline{\eta}\|_{l^2} = \|\underline{\xi}\|_{l^2} = \|\underline{\xi}_o\|_{l^2} = \|\xi_I^0(S)\underline{\xi}_o\|_{l^2}.$$

This, together with (5.41), implies that the norm of $\overline{\Gamma}_s$ is attained at $\overline{\tilde{\xi}}_I(S)\underline{\eta}$. Hence, $\Gamma_s\overline{\Gamma}_s(\overline{\tilde{\xi}}_I(s)\underline{\eta}) = s^2\overline{\tilde{\xi}}_I(S)\underline{\eta}$, or

$$\Gamma_s(\xi_I^0(S)\underline{\xi}_o) = s(\overline{\tilde{\xi}}(S)\underline{\eta}),$$

which, together with (5.41), implies that $(\underline{\xi}^0, \underline{\eta}^0) := (\xi_I^0(S)\underline{\xi}_o, \overline{\tilde{\xi}}_I(S)\underline{\eta})$ is a Schmidt pair of the Hankel matrix Γ_s corresponding to the same s-number s.

Now, since $\underline{\xi}_o \in l^2$ we have, by (5.39),

$$\underline{\xi}^0 = \xi_I^0(S)\underline{\xi}_o \in \xi_I^0(S)l^2 = \mathcal{S}_0. \tag{5.42}$$

This completes the proof of the theorem.

The following result follows from the above proof.

Corollary 5.3. Let Γ be a bounded Hankel matrix and $\underline{\xi} \in l^2$ such that $\|\Gamma\underline{\xi}\|_{l^2} = \|\Gamma\| \, \|\underline{\xi}\|_{l^2}$. Then, there exists an $\underline{\eta} \in l^2$ with $\|\underline{\eta}\|_{l^2} = \|\underline{\xi}\|_{l^2}$ such that $(\underline{\xi}, \underline{\eta})$ is a Schmidt pair of Γ corresponding to the s-number $s_1 := \|\Gamma\|$.

Obviously, if $(\underline{\xi}, \underline{\eta})$ is a Schmidt pair of Γ corresponding to s_1, then $(\underline{\bar{\xi}}, \underline{\bar{\eta}})$ is also a Schmidt pair of Γ corresponding to s_1 (Problem 5.24a). Consequently, we have the following result where, as usual, $\bar{\eta}_+(z) := \sum_{i=1}^{\infty} \bar{\eta}_i z^{i-1}$.

Corollary 5.4. Let Γ be a bounded Hankel matrix and s a positive s-number of Γ. Also, let $\mathcal{S}$ be the set of Schmidt pairs corresponding to s and $\mathcal{M}$ the set of all functions $\bar{\eta}_+(z)$ with $\underline{\eta} \in \mathcal{S}$. Then, there exists a $(\underline{\xi}^0, \underline{\eta}^0) \in \mathcal{S}$ such that the inner factor $\xi_I^0(z)$ of $\xi_+^0(z)$ is the greatest common divisor of all $\bar{\eta}_+(z)$ in $\mathcal{M}$.

Of course, it is clear that $\bar{\eta}_+(z) = \overline{z\eta_-(z)}$ on the unit circle $|z| = 1$.

5.3.2 Relations Between Dimensions and Degrees

Analogous to the subspace L_Γ of l^2 defined by (5.24), we consider the subspace

$$\begin{aligned} L_{\Gamma_s} &= \{\underline{\xi} \in l^2 \colon \|\Gamma_s\underline{\xi}\|_{l^2} = \|\Gamma_s\| \, \|\underline{\xi}\|_{l^2}\} \\ &= \{\underline{\xi} \in l^2 \colon \|\Gamma_s\underline{\xi}\|_{l^2} = s\|\underline{\xi}\|_{l^2}\}, \end{aligned} \tag{5.43}$$

where $\Gamma_s := \Gamma(s\eta_-(z)/\xi_+(z))$. Here, Lemma 5.8 has been used. Recall that $\mathcal{S}$ is the set of Schmidt pairs of Γ corresponding to s and that the inner factor of $\xi_+^0(z)$ is denoted by $\xi_I^0(z)$. Then we have the following result:

Theorem 5.11. Suppose that $\dim(L_{\Gamma_s}) = k, \dim(\mathcal{S}) = \ell$, and degree $(\xi_I^0(z)) = n$. Then, we have $k \geq 2n + \ell$.

If $\ell = \infty$, then since any Schmidt pair $(\underline{\xi}, \underline{\eta})$ in $\mathcal{S}$ is also a Schmidt pair of Γ_s corresponding to the s-number $\|\Gamma_s\|$, so that $\underline{\xi} \in L_{\Gamma_s}$, we have $k = \infty$. On the other hand, suppose that $n = \infty$. Let us write

$$\xi_I^0(z) = c \prod_{i=1}^{k} \frac{z - z_i}{1 - \bar{z}_i z} \psi(z),$$

where k may be either finite or infinite, $|c| = 1$, and

$$\psi(z) = \exp\left(-\frac{1}{2\pi} \int_0^{2\pi} \frac{e^{jt} + z}{e^{jt} - z} d\mu(z) \right)$$

for a finite positive Borel measure μ (which is singular with respect to the Lebesgue measure) on the unit circle (Sect. 3.1.2). For any integer $m > 1$, write $\psi(z) = [\psi_m(z)]^m$, where

$$\psi_m(z) := \exp\left(-\frac{1}{2\pi m}\int_0^{2\pi}\frac{e^{jt}+z}{e^{jt}-z}d\mu(z)\right),$$

which is also an inner function. It then follows from Theorem 5.4 that there is a vector $\tilde{\xi} \in l^2$ with $\tilde{\xi}_+(z)$ being an outer function such that the norm of Γ_s is attained at $\tilde{\xi}$. Consequently, it follows from Theorem 5.7 that the norm of Γ_s is also attained at $\xi_+^0(S)\tilde{\xi}$. Moreover, by Theorem 5.4, we see that the norm of Γ_s is attained at the vectors $\{\beta_m^i\}$ satisfying

$$(\beta_m^i)_+(z) = \exp\left(\frac{i}{2\pi m}\int_0^{2\pi}\frac{e^{jt}+z}{e^{jt}-z}d\mu(z)\right)\xi_+^0(z)\tilde{\xi}_+(z)$$

for $0 \leq i \leq m$ where $m > 1$. Since the vectors $\{\beta_m^i\}$ are linearly independent and m is arbitrary, we have $k = \infty$. If the inner factor of $\xi_+^0(z)$ is an infinite Blaschke product, it can be shown that $k = \infty$, by using similar arguments.

Now, suppose that both ℓ and n are finite. Consider the expression $\xi_+^0(z) = z^m\phi(z)$ where $\phi(z)$ is a finite Blaschke product with $\phi(0) \neq 0$. Let $(\underline{\xi}^1, \underline{\eta}^1), \ldots, (\underline{\xi}^\ell, \underline{\eta}^\ell)$ be ℓ linearly independent pairs in $\mathcal{S}$. Then, since $\xi_+^0(z)$ is a common divisor of all $\xi_+^i(z), i = 1, \ldots, \ell$ (Theoerm 5.10), each $\xi_+^i(z)$ has the factor $z^m, i = 1, \ldots, \ell$, so that

$$\underline{\xi}^i = [\,0 \;\cdots\; 0 \;\xi_{m+1}^i \;\xi_{m+2}^i \;\cdots\,]^\top, \qquad i = 1, 2, \ldots, \ell.$$

It follows from the linear independence of $\{(\underline{\xi}^i, \underline{\eta}^i)\}_{i=1}^\ell$ that there exists a vector pair $(\underline{\xi}^*, \underline{\eta}^*)$ in $\mathcal{S}$ such that

$$\underline{\xi}^* = [\,0 \;\cdots\; 0 \;\xi_{m+\ell}^* \;\xi_{m+\ell+1}^* \;\cdots\,]^\top,$$

which gives

$$\xi_+^*(z) = z^{m+\ell+1}\phi(z)\xi_I^*(z)\xi_O^*(z) = z^{\ell-1}\xi_I^0(z)\xi_I^*(z)\xi_O^*(z)$$

for some inner and outer functions $\xi_I^*(z)$ and $\xi_O^*(z)$. Moreover, we have

$$(\overline{\Gamma_s\underline{\xi}^*})_+(z) = s\bar{\eta}_+^*(z),$$

and by Corollary 5.4, $\xi_I^0(z)$ is also a divisor of $\bar{\eta}_+^*(z)$. Hence, the conditions in Theorem 5.4 are satisfied, so that with $\gamma_I(z) = \xi_I^0(z)/\xi_I^*(z)$ in the theorem we may conclude that the norm of Γ_s is also attained at the vector $S^{\ell-1}[\xi_I^0(S)]^2\xi_O^*$.

On the other hand, since $\dim(L_{\Gamma_s}) = k < \infty$, it follows from Theorem 5.5 that there exists a vector $\underline{\xi}^\# \in L_{\Gamma_s}$ such that

$$L_{\Gamma_s} = \{p(S)(S^*)^{k-1}\underline{\xi}^\#:\;\; p(z) \in \mathcal{P}_{k-1}\},$$

where $((S^*)^{k-1}\underline{\xi}^\#)_+(z) = z^{-k+1}\xi_+^\#(z)$ is an outer function.

Hence, there is a polynomial $p_0(z) \in \mathcal{P}_{k-1}$ such that

$$p_0(S)(S^*)^{k-1}\underline{\xi}^{\#} = S^{\ell-1}[\xi_I^0(S)]^2\underline{\xi}_o^*,$$

or

$$p_0(z)((S^*)^{k-1}\underline{\xi}^{\#})_+(z) = S^{\ell-1}[\xi_I^0(z)]^2\xi_o^*(z).$$

This implies that $k \geq 2n+\ell$, completing the proof of the theorem.

The following consequence can be verified by noting that $\Gamma\xi_I^0(S) = \Gamma_s\xi_I^0(S)$ [Problem 5.25(*a*)].

Corollary 5.5. $\|\Gamma\xi_I^0(S)\| = s$ and $\dim(L_{\Gamma\xi_I^0(S)}) \geq n+\ell$.

We leave the proof to the reader as an exercise (Problem 5.24b).

5.3.3 Relations Between $\xi_I^0(z)$ and s-Numbers

In this section, we derive some very important results which describe the relationship between s-numbers and the inner function $\xi_I^0(z)$ studied previously. These results are also needed in the proof of AAK's main result in the next section. We first have the following:

Theorem 5.12. Let Γ be a bounded Hankel matrix with s-numbers $s_1 \geq s_2 \geq \ldots$. Suppose that $s_m > s_{m+1} = \cdots = s_{m+r} > s_{m+r+1}$ for some m and r with $m \geq 1$ and $r \geq 1$. Moreover, let $(\underline{\xi}^0, \underline{\eta}^0)$ be a Schmidt pair of Γ corresponding to the s-number $s := s_{m+1}$, with $\xi_+^0(z) = \xi_I^0(z)\xi_o^0(z)$. Then, the degree of $\xi_I^0(z)$ is equal to m.

To prove the theorem, let the degree of $\xi_I^0(z)$ be equal to n. It follows from Corollary 5.5 (see also Problem 5.25) that $\Gamma\xi_I^0(S) = \Gamma_{s_{m+1}}\xi_I^0(S)$, where $\Gamma_{s_{m+1}} := \Gamma(s_{m+1}\eta_-(z)/\xi_+(z))$ in which $(\underline{\xi}, \underline{\eta})$ is a Schmidt pair of Γ corresponding to s_{m+1}. Moreover, we have $\|\Gamma\xi_I^0(S)\| = \|\Gamma_{s_{m+1}}\xi_I^0(S)\| = s_{m+1}$, and $\dim(L_{\Gamma\xi_I^0(S)}) \geq n+r$, where $L_{\Gamma\xi_I^0(S)}$ is defined by (5.24). Consequently, we have

$$\tilde{s}_1 = \tilde{s}_2 = \cdots = \tilde{s}_{n+r} = s_{m+1},$$

where $\{\tilde{s}_i\}_{i=1}^\infty$ are the s-numbers of the bounded Hankel matrix $\Gamma\xi_I^0(S)$. It follows from Lemma 5.5 that $(\xi_I^0(S))^*\overline{\Gamma}\Gamma\xi_I^0(S) \leq \overline{\Gamma}\Gamma$, so that

$$\tilde{s}_k \leq s_k, \qquad k = 1, 2, \cdots.$$

Hence, we have

$$s_{m+r+1} < s_{m+r} = s_{m+1} = \tilde{s}_{n+r} \leq s_{n+r},$$

which yields $m \geq n$.

On the other hand, it follows from Theorem 5.6 that

$$\dim(\xi_I^0(S)l^2)^\perp = \dim(\mathrm{Ker}(\xi_I^0(S))^*) = n\,.$$

Since $(\Gamma - \Gamma_{s_{m+1}})\xi_I^0(S)l^2 = 0$ by Problem 5.25a, we have rank $(\Gamma - \Gamma_{s_{m+1}}) \leq n$, so that (Problem 5.26)

$$s_{m+1} = \|\Gamma_{s_{m+1}}\| = \|\Gamma - (\Gamma - \Gamma_{s_{m+1}})\| \geq s_{n+1}$$

This implies that $k \leq n$, completing the proof of the theorem.

The following result is immediate.

Corollary 5.6. Let Γ be a bounded Hankel matrix with s-numbers $s_1 \geq s_2 \geq \cdots$. Suppose that $s_m > s_{m+1} = \cdots = s_\infty$. Moreover, let $(\underline{\xi}^0, \underline{\eta}^0)$ be a Schmidt pair of Γ corresponding to the s-number $s := s_{m+1}$, with $\xi_+^0(z) = \xi_I^0(z)\xi_O^0(z)$. Then, the degree of $\xi_I^0(z)$ is not greater than k. Furthermore, if $s_{m+1} < s_\infty$, then $\xi_I^0(z)$ has infinite degree.

The first assertion is clear from the second part of the proof of the above theorem. If the degree of $\xi_I^0(z)$ is equal to $n < \infty$, then for any m such that $s_m > s_{m+1}$, we must have

$$s_{m+1} = \|\Gamma_{s_{m+1}}\| = \|\Gamma - (\Gamma - \Gamma_{s_{m+1}})\| \geq s_{n+1} \geq s_\infty$$

as has been shown in the second part of the above proof, so that $s_{m+1} < s_\infty$ which is impossible.

The following result on a finite-rank Hankel matrix will be useful later.

Theorem 5.13. Let Γ be a Hankel matrix of rank m. Then, there exists an inner function $\phi(z)$ of degree m such that $\Gamma\phi(S) = 0$.

Let

$$\Gamma = \begin{bmatrix} \gamma_1 & \gamma_2 & \cdots \\ \gamma_2 & \cdots & \cdots \\ \cdots & \cdots & \cdots \end{bmatrix} \quad \text{and} \quad \underline{\gamma}_i = \begin{bmatrix} \gamma_i \\ \gamma_{i+1} \\ \vdots \end{bmatrix}, \qquad i = 1, 2, \cdots.$$

Since rank $(\Gamma) = m$, it follows from Kronecker's theorem (see also Theorem 2.9) that

$$\sum_{i=1}^{\infty} \gamma_i z^{-i} = \frac{p_1 z^{m-1} + \cdots + p_m}{z^m - c_m z^{m-1} - \cdots - c_1} = \frac{p(z)}{\prod_{i=1}^m (z - z_i)},$$

where $|z_i| < 1, i = 1, \ldots, m$, and $\{c_i\}_{i=1}^m$ satisfies

$$\underline{\gamma}_{m+1} = \sum_{i=1}^{m} c_{m-i+1}\underline{\gamma}_{m-i+1}\,.$$

Define

$$\phi(z) = \prod_{i=1}^{m} \frac{z - z_i}{1 - \bar{z}_i z} := \phi_1(z)\phi_2(z)$$

with $\phi_1(z) = \prod_{i=1}^{m} 1/(1 - \bar{z}_i z)$ and $\phi_2(z) = \Pi_{i=1}^{m}(z - z_i)$. Then, it follows from Lemma 5.4 that

$$\Gamma\phi(S) = \Gamma\phi_1(S)\phi_2(S) = \hat{\phi}_1^*(S)\Gamma\phi_2(S),$$

where $\hat{\phi}_1^*(S)$ is invertible by Lemma 5.5. Hence, $\Gamma\phi(S) = 0$ if and only if $\Gamma\phi_2(S) = 0$. Furthermore, since $\Gamma\phi_2(S)$ is a Hankel matrix, we have $\Gamma\phi_2(S) = 0$ if and only if $\Gamma\phi_2(S)\mathbf{e}_1 = 0$ where $\mathbf{e}_1 = [\,1\ 0\ 0\ \cdots\,]^\top$. Finally, since

$$\begin{aligned}
\Gamma\phi_2(S)\mathbf{e}_1 &= \Gamma\Big(\prod_{i=1}^{m}(S - z_i I)\Big)\mathbf{e}_1 \\
&= \Gamma(S^m - c_m S^{m-1} - \cdots - c_1 I)\mathbf{e}_1 \\
&= \Gamma[-c_1\ \cdots\ -c_m\ 1\ 0\ 0\ \cdots]^\top \\
&= -\sum_{i=1}^{m} c_{m-i+1}\gamma_{m-i+1} + \gamma_{m+1} \\
&= 0,
\end{aligned}$$

it follows that $\Gamma\phi_2(S) = 0$. This completes the proof of the theorem.

5.4 AAK's Main Theorem on Best Hankel-Norm Approximation

We are now in a position to present a proof of AAK's main result on optimal Hankel-norm approximation. When the given Hankel matrix Γ is a finite-rank operator on l^2, this result was stated in Theorem 4.6 and has been proved in Sect. 4.3.2. We will now extend Theorem 4.6 to any general bounded Hankel matrix Γ. To state the theorem, let us recall the notation

$$\mathcal{R}_{\tilde{m}}^{s} = \Big\{ r_n(z):\ \ r_n(z) = \frac{p_1 z^{n-1} + \cdots + p_n}{z^n + q_1 z^{n-1} + \cdots + q_n} \qquad \text{all poles of } r_n(z) \text{ lie in } |z| < 1, n \le \tilde{m} \Big\}, \tag{5.44}$$

see (4.6). Let $f(z)$ be a given function on $L^\infty = L^\infty(|z| = 1)$ and Γ_f be its corresponding bounded operator on l^2. Consider the extremal problem:

$$\|f - \hat{r}_{\tilde{m}}\|_\Gamma = \inf_{r \in \mathcal{R}_{\tilde{m}}^{s}} \|f - r\|_\Gamma\,. \tag{5.45}$$

Then, we can state AAK's general result as follows:

Theorem 5.14. (Adamjan, Arov, and Krein)

Let $f(z)$ be a given function in L^∞ and Γ_f denote the corresponding bounded operator with s-numbers

$$s_1 \geq \cdots \geq s_m > s_{m+1} = \cdots = s_{m+k} > s_{m+k+1} \geq \cdots .$$

Also let $(\underline{\xi}^{m+1}, \underline{\eta}^{m+1})$ with

$$\underline{\xi}^{m+1} = [\ \xi_1^{m+1}\ \xi_2^{m+1}\ \cdots\ ]^\top \qquad \text{and} \qquad \underline{\eta}^{m+1} = [\ \eta_1^{m+1}\ \eta_2^{m+1}\ \cdots\ ]^\top$$

be a Schmidt pair of Γ_f corresponding to s_{m+1}. Then, the extremal problem stated in (5.45) with $\tilde{m} = m + k - 1$ has a unique solution which is given by the singular part $\hat{r}_m(z)$ of the function

$$h(z) = f(z) - s_{m+1} \frac{\eta_-^{m+1}(z)}{\xi_+^{m+1}(z)} \tag{5.46}$$

where

$$\xi_+^{m+1}(z) = \sum_{i=1}^{\infty} \xi_i^{m+1} z^{i-1}$$

and

$$\eta_-^{m+1}(z) = \sum_{i=1}^{\infty} \eta_i^{m+1} z^{-i} .$$

Moreover,

$$\|f - \hat{r}_m\|_\Gamma = \inf_{r \in \mathcal{R}^s_{m+k-1}} \|f - r\|_\Gamma = s_{m+1} . \tag{5.47}$$

Note that the result is independent of the multiplicity k of the s-number s_{m+1}. Also, recall that the extremal problem (5.45) (with $\tilde{m} = m + k - 1$) is equivalent to the problem of determining a $\Gamma_{\hat{r}_m}$ in $G^{[m+k-1]}$ such that

$$\|\Gamma_f - \Gamma_{\hat{r}_m}\| = \inf_{\Gamma_r \in G^{[m+k-1]}} \|\Gamma - \Gamma_r\| , \tag{5.48}$$

where $G^{[\cdot]}$ was already defined in (5.1). It has been shown in Theorem 5.1 that this problem is solvable, namely, the Hankel matrix $\Gamma_{\hat{r}_m}$ exists. Hence, what is left to show is that the Hankel matrix $\Gamma_{\hat{r}_m}$ is explicitly given by the singular part of the function $h(z)$ defined in (5.46), and hence $\|\Gamma_f - \Gamma_{\hat{r}_m}\| = s_{m+1}$.

Although several preliminary results have already been obtained in the previous sections, the proof of Theorem 5.14 is still quite long. Hence, it is better to give an outline of the proof before going into any details. We will prove the theorem by considering three cases:

Case 1: $m = 0$. In this case, we will define $s_0 = \infty$ and show that $\|\Gamma_f - \Gamma_r\| = s_1$ for some $\Gamma_r \in G^{[k-1]}$ if and only if $\Gamma_r = 0$. Then, since $\Gamma_{\hat{r}_0} := \Gamma_f - \Gamma_{s_1}$ satisfies $\|\Gamma_f - \Gamma_{\hat{r}_0}\| = s_1$, we will show that $\Gamma_{\hat{r}_0} = 0$, so that $\Gamma_{\hat{r}_0} \in G^{[k-1]}$.

Case 2: $m > 0$ and $s_m > s_{m+1} = \cdots = s_{m+k} > s_\infty \geq 0$. In this case, since $\Gamma_{\hat{r}_m} := \Gamma - \Gamma_{s_{m+1}}$ satisfies $\|\Gamma - \hat{\Gamma}_{\hat{r}_m}\| = s_{m+1}$, we will show that $\Gamma_{\hat{r}_m}$ has rank $m+1$, and then show that $\Gamma_{\hat{r}_m}$ is the unique element in $G^{[m+k-1]}$ which has this property.

Case 3: $m > 0$ and $s_m > s_{m+1} = \cdots = s_\infty \geq 0$. In this case, we will also show that $\Gamma_{\hat{r}_m} := \Gamma - \Gamma_{s_{m+1}}$ satisfies $\|\Gamma - \hat{\Gamma}_{\hat{r}_m}\| = s_{m+1}$.

In the meantime, we must keep in mind that the Hankel matrix $\Gamma_{\hat{r}_m}$ is explicitly given by the singular part of the function $h(z)$ defined by (5.46).

For convenience, we simplify the notations by setting $\Gamma = \Gamma_f, \Gamma' = \Gamma_{\hat{r}_m}$, and $\Lambda = \Gamma_r$ in the following presentation.

5.4.1 Proof of AAK's Main Theorem: Case 1

Let $m = 0$. In this case, we define $s_0 = \infty$.

We first show that $\|\Gamma - \Lambda\| = s_1$ for some $\Lambda \in G^{[k-1]}$ if and only if $\Lambda = 0$. One direction is trivial. Suppose that $\|\Gamma - \Lambda\| = s_1$ for some $\Lambda \in G^{[k-1]}$. Then, since $m = 0$, we have $\|\Gamma\| = s_1 = s_2 = \cdots = s_k$. Let

$$L_\Gamma = \{\underline{\xi} \in l^2 \colon \|\Gamma\underline{\xi}\| = \|\Gamma\|\underline{\xi}\}\,.$$

Then, $\dim(L_\Gamma) \geq k$. On the other hand, since $\Lambda \in G^{[k-1]}$, there is an inner function $\phi(z)$ of degree less than or equal to $k-1$ such that by Theorem 5.6 $\Lambda\phi(S) = 0$, that is, Λ vanishes on $\phi(S)l^2$. It then follows from Theorem 5.7 that the norm of Γ is attained on the subspace $\phi(S)l^2$, so that Theorem 5.8 implies that $\|\Gamma + \Lambda\| \geq \|\Gamma\|$, where the equality holds if and only if $\Lambda \equiv 0$. Hence, $\|\Gamma - \Lambda\| = s_1 = \|\Gamma\|$, yielding $\Lambda = 0$.

We next show that the matrix $\tilde{\Gamma} := \Gamma - \Gamma_{s_1}$ is in fact the zero matrix, where Γ_{s_1} is the Hankel matrix of the function $s_1\eta_-^1(z)/\xi_+^1(z)$. Note that $\tilde{\Gamma}$ is in $G^{[k-1]}$ and satisfies $\|\Gamma - \tilde{\Gamma}\| = \|\Gamma_{s_1}\| = s_1$. From the definition of $\tilde{\Gamma}$, we see that as a solution to the optimal Hankel-norm approximation problem (5.46) in the case $m = 0$, the matrix $\tilde{\Gamma} = \Gamma_{\hat{r}_0}$ is given by the singular part of the function $h(z)$ defined by (5.46), namely,

$$h(z) = f(z) - s_1\frac{\eta_-^1(z)}{\xi_+^1(z)}\,.$$

To show that $\tilde{\Gamma} = 0$, we need the following lemma:

Lemma 5.10. There exists a Schmidt pair $(\underline{\xi}, \underline{\eta})$ of Γ corresponding to the s-number $s_1 = \|\Gamma\|$ such that $\xi_+(z)$ is an outer function.

To prove this lemma, let $(\underline{\xi}^0, \underline{\eta}^0)$ be an arbitrary Schmidt pair of Γ corresponding to $s_1 = \|\Gamma\|$. Then, we have

$$\Gamma\underline{\xi}^0 = \|\Gamma\|\underline{\eta}^0, \quad \overline{\Gamma}\underline{\eta}^0 = \|\Gamma\|\underline{\xi}^0, \quad \text{and} \quad \|\underline{\xi}^0\|_{l^2} = \|\underline{\eta}^0\|_{l^2}.$$

Factorize $\xi^0_+(z)$ into $\xi^0_+(z) = \xi^0_I(z)\xi^0_O(z)$, where $\xi^0_I(z) := \sum_{i=0}^{\infty} \alpha_i z^i$ and $\xi^0_O(z) := \sum_{i=0}^{\infty} \beta_i z^i$ are inner and outer functions, respectively, and then define

$$\underline{\xi} = [\ \beta_1\ \beta_2\ \cdots\ ]^\top \quad \text{and} \quad \underline{\eta} = \hat{\xi}^0_I(S)\underline{\eta}^0,$$

where $\hat{\xi}^0_I(z) := \sum_{i=0}^{\infty} \bar{\alpha}_i z^i$ as in (5.17). Then, we may conclude that $(\underline{\xi}, \underline{\eta})$ is also a Schmidt pair of Γ corresponding to $s_1 = \|\Gamma\|$. Indeed, since $\overline{\Gamma}\underline{\eta}^0 = \|\Gamma\|\underline{\xi}^0 = \|\Gamma\|\xi^0_I(S)\underline{\xi}$, we have $(\xi^0_I(S))^*\overline{\Gamma}\underline{\eta}^0 = \|\Gamma\|\underline{\xi}$, or $\overline{\Gamma}\hat{\xi}^0_I(S)\underline{\eta}^0 = \|\Gamma\|\underline{\xi}$, by Lemma 5.4, so that

$$\overline{\Gamma}\underline{\eta} = \|\Gamma\|\underline{\xi}. \tag{5.49}$$

On the other hand, since

$$\|\overline{\Gamma}\underline{\eta}\|_{l^2} = \|\Gamma\|\ \|\underline{\xi}\|_{l^2} = \|\Gamma\|\ \|\underline{\xi}^0\|_{l^2} = \|\Gamma\|\ \|\underline{\eta}^0\|_{l^2} = \|\overline{\Gamma}\|\ \|\underline{\eta}\|_{l^2},$$

which implies that the norm of $\overline{\Gamma}$ is attained at $\underline{\eta}$, and consequently, we have $\Gamma\overline{\Gamma}\underline{\eta} = \|\overline{\Gamma}\|^2\underline{\eta}$, so that by (5.49),

$$\Gamma\underline{\xi} = \|\Gamma\|\underline{\eta}.$$

This, together with (5.49), implies that $(\underline{\xi}, \underline{\eta})$ is also a Schmidt pair of Γ corresponding to $s_1 = \|\Gamma\|$, completing the proof of the lemma.

It then follows immediately that

$$\Gamma\underline{\xi} = \|\Gamma\|\underline{\eta} = s_1\underline{\eta} = \Gamma_{s_1}\underline{\xi},$$

where $\underline{\xi}$ is the l^2 sequence that yields the outer function $\xi_+(z)$, so that by Corollary 5.1 $\Gamma \equiv \Gamma_{s_1}$. This implies that $\Gamma' = \Gamma - \Gamma_{s_1} = 0$, completing the proof for this special case.

5.4.2 Proof of AAK's Main Theorem: Case 2

Let $m > 0$ and $s_m > s_{m+1} = \cdots = s_{m+k} > s_\infty \geq 0$.

We first observe that the Hankel matrix $\Gamma - \Gamma_{s_{m+1}}$ satisfies

$$\|\Gamma - (\Gamma - \Gamma_{s_{m+1}})\| = \|\Gamma_{s_{m+1}}\| = s_{m+1}.$$

We may conclude that rank $(\Gamma - \Gamma_{s_{m+1}}) = m + 1$. Indeed, since by (5.38) and (5.39) we have

$$S_0 = \{\underline{\xi} \in l^2 \colon \ \Gamma\underline{\xi} = \Gamma_{s_{m+1}}\underline{\xi}\} = \xi^0_I(S)l^2,$$

where $\xi_I^0(z)$ is the inner factor of $\xi_+^0(z)$ corresponding to a vector $\underline{\xi}^0$ in $\mathcal{S}_0$ such that the degree of $\xi_I^0(z)$ is equal to $m+1$ (Theorem 5.12), $l^2 = (\xi_I^0(S)l^2)^\perp \oplus (\xi_I^0(S)l^2)$, and $\dim((\xi_I^0(S)l^2)^\perp) = m+1$ (Theorem 5.6). Hence, we have

$$\operatorname{rank}(\Gamma - \Gamma_{s_{m+1}}) \le m+1\,,$$

(Problem 5.26). On the other hand, from $s_{m+1} = \|\Gamma - (\Gamma - \Gamma_{s_{m+1}})\| < s_m$, we have

$$\operatorname{rank}(\Gamma - \Gamma_{s_{m+1}}) \ge m+1\,,$$

and consequently, rank $(\Gamma - \Gamma_{s_{m+1}}) = m+1$.

What is left to show is the uniqueness of the Hankel matrix $\Gamma' = \Gamma - \Gamma_{s_{m+1}}$. Suppose that there is another Hankel matrix $\Gamma^0 \in G^{[m+k-1]}$ that satisfies $\|\Gamma - \Gamma^0\| = s_{m+1}$. Then, since the degree of $\xi_I^0(z)$ is $m+1$ and there exist k linearly independent Schmidt pairs of Γ corresponding to s_{m+1}, we have

$$\dim(L_{\Gamma_{s_{m+1}}}) \ge 2m+k+1$$

by Theorem 5.11, where

$$L_{\Gamma_{s_{m+1}}} = \{\underline{\xi} \in l^2\colon \quad \|\Gamma_{s_{m+1}}\underline{\xi}\|_{l^2} = s_{m+1}\|\underline{\xi}\|_{l^2}\}\,.$$

On the other hand, we have

$$\operatorname{rank}((\Gamma - \Gamma_{s_{m+1}}) - \Gamma^0) \le (m+1) + (m+k-1) = 2m+k\,,$$

so that there is an inner function $\phi(z)$ of degree less than or equal to $2m+k$ such that

$$((\Gamma - \Gamma_{s_{m+1}}) - \Gamma^0)\phi(S) = 0\,,$$

(Problem 5.27); that is, the matrix $(\Gamma - \Gamma_{s_{m+1}}) - \Gamma^0$ vanishes on $\phi(S)l^2$. Moreover, it follows from Theorem 5.7 that the norm of $\Gamma_{s_{m+1}}$ is attained on $\phi(S)l^2$, so that by Theorem 5.8 we have

$$s_{m+1} = \|\Gamma - \Gamma^0\| = \|\Gamma_{s_{m+1}} + (\Gamma - \Gamma_{s_{m+1}}) - \Gamma^0\| \ge \|\Gamma_{s_{m+1}}\| = s_{m+1}\,.$$

This implies that $(\Gamma - \Gamma_{s_{m+1}}) - \Gamma^0 = 0$, or $\Gamma^0 = \Gamma - \Gamma_{s_{m+1}}$, completing the proof for this case.

5.4.3 Proof of AAK's Main Theorem: Case 3

Let $m > 0$ and $s_m > s_{m+1} = \cdots = s_\infty \ge 0$.

We first assume that Γ is a compact operator, so that $s_{m+1} = \cdots = s_\infty = 0$. In this case, we have

$$\Gamma = \sum_{i=1}^{m} s_i \underline{\eta}_i \underline{\xi}_i^* ,$$

where $(\underline{\xi}_i, \underline{\eta}_i)$ is a Schmidt pair of Γ corresponding to the s-number s_i, and $\underline{\xi}_i^* = \bar{\underline{\xi}}_i^\top$. Hence, rank $(\Gamma) \le m$. Since $\Gamma_{s_{m+1}} = \Gamma(s_{m+1}\eta_-^{m+1}(z)/\xi_+^{m+1}(z)) = 0$, by defining $\Lambda = \Gamma - \Gamma_{s_{m+1}} (= \Gamma)$, we have $\|\Gamma - \Lambda\| = 0 = s_{m+1}$, and $\Lambda = \Gamma - \Gamma_{s_{m+1}}$ is unique.

We then consider a bounded Hankel operator Γ which may not be compact. From Theorem 5.1, we first observe that there exists a $\Lambda \in G^{[n]}$ for some $n > m$ such that $\|\Gamma - \Lambda\|$ attains its infimum on the set of all Hankel matrices whose ranks are larger than m. Following the same basic idea, it can be verified that this Hankel matrix Λ satisfies $\|\Gamma - \Lambda\| = s_{m+1}$ (Problem 5.29). On the other hand, by (5.38) and (5.39), we have

$$\mathcal{S}_0 = \{\underline{\xi} \in l^2\colon \quad \Gamma\underline{\xi} = \Gamma_{s_{m+1}}\underline{\xi}\} = \xi_I^0(S)l^2 ,$$

where $\xi_I^0(z)$ is the inner factor of the function $\xi^0(z)$ corresponding to a vector $\underline{\xi}^0$ in $\mathcal{S}_0$, and the degree of $\xi_I^0(z)$ is equal to k for some $k \ge m+1$ (Problem 5.30). Hence, the rank of $(\Gamma - \Gamma_{s_{m+1}})$ is less than or equal to k, and consequently,

$$\text{rank}((\Gamma - \Gamma_{s_{m+1}}) - \Lambda) \le k + n .$$

It follows that there exists an inner function $\phi(z)$ of degree less than or equal to $k + n$ such that (Problem 5.31a)

$$((\Gamma - \Gamma_{s_{m+1}}) - \Lambda)\phi(S) = 0 .$$

that is, the matrix $(\Gamma - \Gamma_{s_{m+1}}) - \Lambda$ vanishes on $\phi(S)l^2$. Observe that since $s_m > s_{m+1} = \cdots = s_\infty \ge 0$ we have $\dim(L_{\Gamma_{s_{m+1}}}) = \infty$, so that the norm of $\Gamma_{s_{m+1}}$ is attained on $\phi(S)l^2$ (Problem 5.31b). Hence, it follows from Theorem 5.8 that

$$s_{m+1} = \|\Gamma - \Lambda\| = \|\Gamma_{s_{m+1}} + (\Gamma - \Gamma_{s_{m+1}}) - \Lambda\| \ge \|\Gamma_{s_{m+1}}\| = s_{m+1} .$$

This yields $\Lambda = \Gamma - \Gamma_{s_{m+1}}$, completing the proof of the AAK Theorem.

Problems

Problem 5.1. Let $\{A_n\}$ be a sequence of (finite or infinite) matrices whose ranks are at most equal to $k, 0 \le k < \infty$, and suppose that

$$A = \lim_{n\to\infty} A_n ,$$

where convergence in the operator norm is considered. Prove that A has rank at most k.

Problem 5.2. Let $\{A_n\}$ be a sequence of bounded infinite matrices on l^2 which converges weakly to some bounded matrix A in the sense that for each $\ell = 1, 2, \ldots$, the ℓth columns of A_n converge to the ℓth column of A weakly. Show that for all $\underline{\xi}, \underline{\eta} \in l^2$,

$$\langle A_n \underline{\xi}, \underline{\eta} \rangle_{l^2} \to \langle A \underline{\xi}, \underline{\eta} \rangle_{l^2}$$

as $n \to \infty$. Also, show that for any bounded matrix B

$$\|B - A\| \leq \overline{\lim_{n \to \infty}} \|B - A_n\|.$$

Problem 5.3. Let S be the shift operator defined by $S(x_1, x_2, \cdots) = (0, x_1, x_2, \cdots)$. Show that

$$S = \begin{bmatrix} 0 & 0 & 0 & \cdots \\ 1 & 0 & 0 & \cdots \\ 0 & 1 & 0 & \cdots \\ 0 & 0 & 1 & \cdots \\ \cdots & \cdots & \cdots & \cdots \end{bmatrix}$$

and that for any sequence $\{a_i\}$ in l^2, $A_n := \sum_{i=1}^{n} a_i S^{i-1}$ converges in the operator norm to some bounded infinite matrix A on l^2. Also, verify that

$$A = \begin{bmatrix} a_1 & 0 & 0 & \cdots \\ a_2 & a_1 & 0 & \cdots \\ a_3 & a_2 & a_1 & \cdots \\ \cdots & \cdots & \cdots & \cdots \end{bmatrix}.$$

Problem 5.4. Verify the identity (5.12).

Problem 5.5. Show that Γ is an infinite Hankel matrix if and only if $\Gamma S = S^* \Gamma$ where S is the shift operator and S^* is the adjoint of S. Note that (see Example 4.6)

$$S^* = \begin{bmatrix} 0 & 1 & 0 & 0 & \cdots \\ 0 & 0 & 1 & 0 & \cdots \\ 0 & 0 & 0 & 1 & \cdots \\ 0 & 0 & 0 & 0 & \cdots \\ \cdots & \cdots & \cdots & \cdots & \cdots \end{bmatrix}.$$

Problem 5.6. Let Γ_1 and Γ_2 be two Hankel matrices such that $\Gamma_1 \underline{\xi}^0 = \Gamma_2 \underline{\xi}^0$ for some $\underline{\xi}^0 \neq 0$ in l^2. Show that $\Gamma_1 \underline{\eta} = \Gamma_2 \underline{\eta}$ for all $\underline{\eta}$ in the subspace $l^2_S(\underline{\xi}^0)$ of l^2 defined by

$$l_S^2(\underline{\xi}^0) = \text{span}\{\underline{\xi}^0, S\underline{\xi}^0, S^2\underline{\xi}^0, \ldots\}.$$

[Hint: Apply Problem 5.5.]

Problem 5.7. Let T be an isometric isomorphism from l^2 onto H^2 defined by

$$T(\underline{\xi}) = \xi_+(z),$$

where $\underline{\xi} = [\ \xi_1\ \xi_2\ \cdots\ ]^\top \in l^2$ and $\xi_+(z) = \Sigma_{i=1}^{\infty}\xi_i z^{i-1}$. Show that for any $n \geq 0$,

$$T(S^n\underline{\xi}) = z^n\xi_+(z),$$

where S is the shift operator.

Problem 5.8. Use Beurling's theorem to show that if Γ_1 and Γ_2 are two Hankel matrices with $\Gamma_1\underline{\xi} = \Gamma_2\underline{\xi}$ for a $\underline{\xi} = [\ \xi_1\ \xi_2\ \xi_3\ \cdots\ ]^\top \in l^2$ such that $\xi_+(z) = \Sigma_{i=1}^{\infty}\xi_i z^{i-1}$ is an outer function, then $\Gamma_1 \equiv \Gamma_2$.

Problem 5.9. Let Γ be a Hankel matrix, $\phi(z) \in H^\infty$ be given by (5.16), and $\bar{\phi}(z)$ be defined by (5.17). Moreover, let S be the shift operator. Show that

(a) $S^*(\Gamma\phi(S)) = (\Gamma\phi(S))S$, so that it follows from Lemma 5.2 that $\Gamma\phi(S)$ is a Hankel matrix;
(b) $\overline{\Gamma}\bar{\phi}(S)$ is a Hankel matrix, so that $\overline{(\overline{\Gamma}\bar{\phi}(S))} = (\Gamma\bar{\phi}(S))^*$; and
(c) $\Gamma\phi(S) = \bar{\phi}^*(S)\Gamma$.

Problem 5.10. Let $\phi_I(z)$ be the inner factor of a function $\phi_+(z)$ and write

$$\phi_I(z) = \sum_{i=1}^{\infty}\phi_i z^{i-1}.$$

Show that $\bar{\phi}_I(z) := \sum_{i=1}^{\infty}\bar{\phi}_i z^{i-1}$ is the inner factor of $\overline{\phi_+(\bar{z})}$.

Problem 5.11. Let $\alpha(z)$ and $\beta(z)$ be two polynomials defined by $\alpha(z) = \Sigma_{i=1}^{n}\alpha_i z^{i-1}$ and $\beta(z) = \Sigma_{i=1}^{n}\beta_i z^{i-1}$, respectively, and let A be an arbitrary constant square matrix. Verify that $\alpha(A)\beta(A) = \beta(A)\alpha(A)$.

Problem 5.12. Let $\xi_I(z)$ and $\gamma_I(z)$ be two inner functions. Show that for all sequences $\mathbf{x} \in \xi_I(S)l^2$ where S is the shift operator, we have

$$\xi_I^*(S)\gamma_I(S)\mathbf{x} = \gamma_I(S)\xi_I^*(S)\mathbf{x}.$$

[Hint: Lemma 5.3 and Problem 5.11 may be useful.]

Problem 5.13. Let Γ be a bounded Hankel matrix on l^2 and define

$$L_\Gamma = \{\underline{\xi} \in l^2: \quad \|\Gamma\underline{\xi}\|_{l^2} = \|\Gamma\| \, \|\underline{\xi}\|_{l^2}\}.$$

Verify that

$$L_\Gamma = \{\underline{\xi} \in l^2: \quad \overline{\Gamma}\Gamma\underline{\xi} = \|\Gamma\|^2\underline{\xi}\}$$

and that L_Γ is a subspace of l^2.

Problem 5.14. Let L be a linear vector space consisting of infinite-dimensional vectors and suppose that L has finite dimension n. Show that there exists a vector $\underline{\xi}$ in L such that

$$\underline{\xi} = [\,0 \ \cdots \ 0 \ \xi_k \ \xi_{k+1} \ \cdots\,]^\top$$

with $k \geq n$ and $\xi_k \neq 0$.

Problem 5.15. Apply Theorem 5.4 to show that the norm of a bounded Hankel matrix Γ on l^2 is attained on the set

$$V = \{S^i \xi_I(S)\xi_I^*(S)(S^*)^{k-1}\underline{\xi}: \quad 0 \leq i \leq k-1\}$$

where the linear vector space L_Γ defined by (5.24) has finite dimension n, and where $\xi_I(z)$ is defined in (5.27) by a vector $\underline{\xi} = [\,0 \ \cdots \ 0 \ \xi_k \ \xi_{k+1} \ \cdots\,]^\top$ in L_Γ with $k \geq n$ and $\xi_k \neq 0$.

Problem 5.16. Let k be a fixed positive integer. Suppose that

$$\underline{\xi} = [\,0 \ \cdots \ 0 \ \xi_k \ \xi_{k+1} \ \ldots\,]^\top \in l^2$$

with $\xi_k \neq 0$. Show that $\{S^i(S^*)^{k-1}\underline{\xi}\}_{i=0}^{k-1}$ is a linearly independent set.

Problem 5.17. Let S^* be the adjoint of the shift operator S, I the identity operator on l^2, and α an arbitrary constant. Show that $(S^* - \alpha I)\underline{\xi} = 0$ with $\underline{\xi} \neq 0$ implies that

$$\underline{\xi} = c[\,1 \ \alpha \ \alpha^2 \ \cdots\,]^\top$$

for some nonzero constant c.

Problem 5.18. Prove Lemma 5.7. [Hint: Use Lemma 5.5.]

Problem 5.19. Verify that for $|z| = 1$

$$(a) \quad \left(\sum_{i=1}^{\infty} \xi_i^1 z^{i-1}\right)\overline{\left(\sum_{i=1}^{\infty} \xi_i^2 z^{i-1}\right)} = \sum_{k=0}^{\infty}\left(\sum_{i=1}^{\infty} \xi_{i+k}^1 \overline{\xi_i^2}\right) z^k + \sum_{k=1}^{\infty}\left(\sum_{i=1}^{\infty} \xi_i^1 \overline{\xi_{i+k}^2}\right) z^{-k},$$

and

$$(b)\quad \left(\sum_{i=1}^{\infty}\eta_i^1 z^{-i}\right)\overline{\left(\sum_{i=1}^{\infty}\eta_i^2 z^{-i}\right)} = \sum_{k=0}^{\infty}\left(\sum_{i=1}^{\infty}\eta_i^1\overline{\eta_{i+k}^2}\right)z^k$$

$$+\sum_{k=1}^{\infty}\left(\sum_{i=1}^{\infty}\eta_{i+k}^1\overline{\eta_i^2}\right)z^{-k}.$$

Problem 5.20. Establish the identity in (5.36) by imitating the proof of (5.35).

Problem 5.21. Let $(\underline{\xi},\underline{\eta})$ be a Schmidt pair of a bounded infinite Hankel matrix Γ on l^2 corresponding to a positive s-number s. Show that

$$s\underline{\xi} = \overline{\Gamma}_s\underline{\eta},$$

where $\Gamma_s := \Gamma(s\eta_-(z)/\xi_+(z))$ is the Hankel matrix corresponding to the function $s\eta_-(z)/\xi_+(z)$.

Problem 5.22. Verify that the space $\mathcal{S}_0$ defined in (5.38) is an S-invariant subspace of l^2 in the sense that $\underline{\xi} \in \mathcal{S}_0$ implies that $S\underline{\xi} \in \mathcal{S}_0$, where S is the shift operator defined in (5.6). Moreover, show that there exists a $\underline{\xi}^0 \in \mathcal{S}_0$ such that $\mathcal{S}_0 = \overline{l^2_S(\underline{\xi}^0)}$, the closure of the space $l^2_S(\underline{\xi}^0)$ defined in Lemma 5.3.

Problem 5.23. Let $(\underline{\xi},\underline{\eta})$ be a Schmidt pair of a bounded Hankel matrix Γ corresponding to a positive s-number s. Show that $\|\Gamma\|\;\|\underline{\xi}\|_{l^2} = s\|\underline{\xi}\|_{l^2}$ and $\|\overline{\Gamma}\|\;\|\underline{\eta}\|_{l^2} = s\|\underline{\eta}\|_{l^2}$.

Problem 5.24. (a) Let $(\underline{\xi},\underline{\eta})$ be a Schmidt pair of a bounded Hankel matrix Γ corresponding to a positive s-number s. Show that $(\underline{\bar{\xi}},\underline{\bar{\eta}})$ is also a Schmidt pair of $\overline{\Gamma}$ corresponding to the same s-number s. (b) Let $\bar{\eta}_+(z) = \Sigma_{i=1}^{\infty}\bar{\eta}_i z^{i-1}$. Show that $\bar{\eta}_+(z) = z\eta_-(z)$ for all z on the unit circle $|z| = 1$.

Problem 5.25. Let Γ be a bounded Hankel matrix and $(\underline{\xi},\underline{\eta})$ a Schmidt pair of Γ corresponding to a positive s-number s. Let $\Gamma_s = \overline{\Gamma}(s\eta_-(z)/\xi_+(z))$ and let $(\underline{\xi}^0,\underline{\eta}^0)$ be any Schmidt pair of Γ corresponding to the same s, with $\xi_+^0(z) = \xi_I^0(z)\xi_O^0(z)$. Moreover, as usual let S be the shift operator.

(a) Show that $\Gamma\xi_I^0(S) = \Gamma_s\xi_I^0(S)$.

(b) Show that $\|\Gamma\xi_I^0(S)\| = s$ and $\dim(L_{\Gamma\xi_I^0(S)}) \geq n + \ell$, where $L_{\Gamma\xi_I^0(S)}$ is defined by (5.24), $\ell = \dim(\mathcal{S})$, with $\mathcal{S}$ denoting the set of all Schmidt pairs of Γ corresponding to s, and n the degree of the inner function $\xi_I^0(z)$. This completes the proof of Corollary 5.5.

Problem 5.26. Let Γ be a bounded Hankel matrix with s-numbers $s_1 \geq s_2 \geq \cdots$, and let Γ_m be another Hankel matrix of rank no greater than $m \geq 0$. Show that

$$\|\Gamma - \Gamma_m\| \geq s_{m+1} .$$

Problem 5.27. Let S be the shift operator defined in (5.6) and $\mathcal{N} = \phi(S)l^2$ be the null space of the Hankel matrix Γ, where $\phi(z)$ is an inner function of degree k. Write $l^2 = (\phi(S)l^2)^{\perp} \oplus (\phi(S)l^2)$, and suppose that $\dim((\phi(S)l^2)^{\perp}) = k$. Show that rank $(\Gamma) \leq k$.

Problem 5.28. Let Γ be a Hankel matrix of rank k. Show that there exists an inner function $\phi(z)$ of degree less than or equal to k such that

$$\Gamma\phi(S) \equiv 0 ,$$

where S is the shift operator.

Problem 5.29. Let Γ be a bounded Hankel matrix with s-numbers $s_1 \geq \cdots \geq s_m > s_{m+1} = \cdots = s_\infty \geq 0$, and let $G^{[n]}$ denote the set of Hankel matrices whose ranks are less than or equal to n. Show that there exists a $\Lambda \in G^{[n]}$ for some $n > m$ such that $\|\Gamma - \Lambda\| = s_{m+1}$.

Problem 5.30. Let Γ be a bounded Hankel matrix with s-numbers $s_1 \geq \cdots \geq s_m > s_{m+1} = \cdots = s_\infty \geq 0$, and let

$$\mathcal{S}_0 = \{\underline{\xi} \in l^2 \colon \quad \Gamma\underline{\xi} = \Gamma_{s_{m+1}}\underline{\xi}\} .$$

Moreover, let $\underline{\xi}^0 = [\ \xi_1^0\ \xi_2^0\ \cdots\ ]^\top$ be an arbitrary but fixed vector in $\mathcal{S}_0$ and define $\xi_+^0(z) = \sum_{i=1}^{\infty} \xi_i^0 z^{i-1}$ with the canonical inner-outer factorization $\xi_+^0(z) = \xi_I^0(z)\xi_O^0(z)$. Show that degree $(\xi_I^0(z)) \geq m+1$.

Problem 5.31. Let Γ be a bounded Hankel matrix with s-numbers $s_1 \geq \cdots \geq s_m > s_{m+1} = \cdots = s_\infty > 0$ and let $\underline{\xi}^0 \in \mathcal{S}_0 = \{\underline{\xi} \in l^2 \colon \ \Gamma\underline{\xi} = \Gamma_{s_{m+1}}\underline{\xi}\}$ with the corresponding inner factor $\xi_I^0(z)$ as defined in Problem 5.30. Moreover, let $\Lambda \in G^{[n]}$ for some $n > m$ such that $\|\Gamma - \Lambda\| = s_{m+1}$. Suppose that the degree of $\xi_I^0(z)$ is equal to k.

(*a*) Show that there exists an inner function $\phi(z)$ of degree less than or equal to $k+n$ such that

$$((\Gamma - \Gamma_{s_{m+1}}) - \Lambda)\phi(S) = 0\,.$$

(*b*) Show that the norm of the Hankel matrix $\Gamma_{s_{m+1}}$ is attained on the subspace $\phi(S)l^2$ of l^2.

6. H^∞-Optimization and System Reduction for MIMO Systems

In this chapter, we study the general theory of H^∞-optimization and problems on systems reduction for discrete-time MIMO linear systems. In Chap. 4, we have discussed the so-called AAK approach for SISO systems. The matrix-valued analog of AAK's theorem in the study of MIMO systems differs from the SISO setting in that the Hankel matrix Γ_H corresponding to the matrix-valued proper rational transfer function

$$H(z) = H_0 + H_1 z^{-1} + H_2 z^{-2} + \cdots$$

is a (finite-rank) infinite block-Hankel matrix of the form

$$\Gamma_H = \begin{bmatrix} H_1 & H_2 & H_3 & \cdots \\ H_2 & H_3 & \cdots & \cdots \\ H_3 & \cdots & \cdots & \cdots \\ \cdots & \cdots & \cdots & \cdots \end{bmatrix},$$

see (2.15) and (2.23). This representation will obviously cause new difficulties, especially in the computational aspect. Theoretically, the matrix-valued analog of the AAK theorem has also been established, at least in principle, by the same authors AAK [1978]. In addition, an earlier paper applying this theory to systems reduction by Kung and Lin [1981] is worth mentioning. For continuous-time linear systems, there are quite a few research papers related to the matrix-valued setting of the AAK theorem including, for example, the elegant balanced realization approach of Glover [1984].

6.1 Balanced Realization of MIMO Linear Systems

The derivations throughout this chapter will depend on the technique of balanced realization which is an elegant and interesting technique in its own right. This technique can be traced back to Laub [1980], Moore [1981], and Pernebo and Silverman [1979], etc. The idea of balanced realization is quite simple but turns out to be useful in the study of the matrix-valued analog of the AAK theory for both continuous-time and discrete-time MIMO systems. In the spirit of this text, we will only discuss discrete-time systems.

Analogous formulations for continuous-time systems can be found in Glover [1984].

6.1.1 Lyapunov's Equations

Consider the discrete-time MIMO linear time-invariant system

$$\begin{cases} \mathbf{x}_{n+1} = A\mathbf{x}_n + B\mathbf{u}_n \,, \\ \mathbf{v}_n = C\mathbf{x}_n + D\mathbf{u}_n \,, \end{cases} \tag{6.1}$$

which has been reviewed in some detail in Chap. 2. Here, A, B, C, and D are respectively $n \times n, n \times p, q \times n$, and $q \times p$ constant matrices, and in general $1 \le p, q \le n$. We also need the concept of controllability and observability Gramians for this linear system from linear systems theory as follows: For any linear system with a state-space description given by (6.1), the controllability Gramian P and observability Gramian Q of the system are defined respectively by

$$P = \sum_{k=0}^{\infty} A^k BB^\top (A^\top)^k \tag{6.2}$$

and

$$Q = \sum_{k=0}^{\infty} (A^\top)^k C^\top C A^k \,. \tag{6.3}$$

It is easily seen that if these two infinite sums converge (to constant matrices), then P and Q satisfy, respectively, the following so-called *matrix Lyapunov equations*:

$$P - APA^\top = BB^\top \,, \tag{6.4}$$

$$Q - A^\top Q A = C^\top C \,. \tag{6.5}$$

However, the infinite sums (6.2) and (6.3) may not converge in general, as can be seen in the following simple example:

Example 6.1.. Consider a linear system described by

$$A = \begin{bmatrix} 0 & 1 \\ 1 & 0 \end{bmatrix}, \quad B = \begin{bmatrix} 1 \\ 0 \end{bmatrix}, \quad C = [1 \;\; 0], \quad \text{and} \quad D = 0 \,.$$

It can be easily verified that

$$\sum_{k=0}^{2N-1} A^k BB^\top (A^\top)^k = \sum_{k=0}^{2N-1} (A^\top)^k C^\top C A^k = \begin{bmatrix} N & 0 \\ 0 & N \end{bmatrix},$$

which diverges as $N \to \infty$, so that (6.2) and (6.3) are not well-defined and hence both (6.4) and (6.5) do not hold. In fact, a direct calculation also shows that in this example the matrix Lyapunov equations (6.4) and (6.5) have no solution.

Note that this simple linear system is a very nice one in the sense that it is both completely controllable and observable. Indeed, it is clear that

$$M_{AB} = [B \ \ AB] = \begin{bmatrix} 1 & 0 \\ 0 & 1 \end{bmatrix} \quad \text{and} \quad N_{CA} = \begin{bmatrix} C \\ CA \end{bmatrix} = \begin{bmatrix} 1 & 0 \\ 0 & 1 \end{bmatrix}.$$

It is also clear that a sufficient condition for the infinite sums (6.2) and (6.3) to converge is that the operator norm $\|A\| < 1$, where $\|A\| = \max_{|\mathbf{x}|_2=1} |A\mathbf{x}|_2$ with $|\mathbf{x}|_2$ denoting the length of the vector $\mathbf{x}$. This means that the linear system is stable [see the remark following Theorem 2.3 in Chap. 2]. This condition is obviously not necessary in general, of course, as can be easily seen from the special situation where either $B = 0$ or $C = 0$.

It is also clear that if the infinite sums (6.2) and (6.3) converge, then both P and Q are symmetric and non-negative definite (and we will use the standard notation $P \geq 0$ and $Q \geq 0$). The importance of the controllability Gramian P is that if P is positive definite (and we will use the notation $P > 0$), then the linear system is completely controllable, which in turn means that the controllability matrix

$$M_{AB} = [\, B \ \ AB \ \ \cdots \ \ A^{n-1}B \,]$$

has full rank, namely, $\text{rank} M_{AB} = n$, where A is an $n \times n$ matrix, see (2.8). Conversely, if the linear system is stable and completely controllable, then the Lyapunov equation (6.4) has a positive definite solution P given by (6.2). Observe, however, that controllability alone does not imply the positive definiteness of P, as has been seen from Example 6.1 above.

Similarly, the positive definiteness of the observability Gramian Q characterizes the observability of the stable linear system, which in turn is equivalent to the observability matrix

$$N_{CA} = \begin{bmatrix} C \\ CA \\ \vdots \\ CA^{n-1} \end{bmatrix}$$

having full rank, see (2.9). These two facts can be simply verified by using the relationships

$$P = [\, B \ \ AB \ \ \cdots \ \ A^{n-1}B \ \ \cdots \,] \begin{bmatrix} B^\top \\ B^\top A^\top \\ \vdots \\ B^\top (A^{n-1})^\top \\ \vdots \end{bmatrix}$$

and

$$Q = [\; C^\top A^\top C^\top \;\cdots\; (A^{n-1})^\top C^\top \;\cdots\;] \begin{bmatrix} C \\ CA \\ \vdots \\ CA^{n-1} \\ \vdots \end{bmatrix} .$$

We leave the details as an exercise (Problem 6.3).

The above discussion may be summarized as follows:

Theorem 6.1. Suppose that the linear system described by (6.1) is stable. Then, the matrix Lyapunov equations (6.4) and (6.5) have solutions $P > 0$ and $Q > 0$ given, respectively, by (6.2) and (6.3) if and only if the system is both completely controllable and observable.

We remark that, as has been seen from Chaps. 4 and 5, in the study of optimal Hankel-norm approximation problems we only consider stable linear systems. Hence, the above theorem provides a useful result for further investigation of this subject.

6.1.2 Balanced Realizations

To introduce the notion of balanced realizations, let us assume that the linear system (6.1) is stable and is both completely controllable and observable. Hence, from the above discussion, both the controllability Gramian P and observability Gramian Q are symmetric and positive definite. In addition, recall from Theorem 2.6 that the state-space description (6.1) is a minimal realization of the system.

As usual, let $Q^{1/2}$ denote the positive square root of Q, that is, $Q^{1/2} > 0$ and $Q = (Q^{1/2})^\top Q^{1/2}$. Then, $Q^{1/2}P(Q^{1/2})^\top$ is also symmetric and positive definite, and hence can be diagonalized by using a unitary matrix U, namely,

$$Q^{1/2}P(Q^{1/2})^\top = U\Sigma^2 U^\top ,$$

where $\Sigma = \mathrm{diag}[\sigma_1, \cdots, \sigma_n]$ with $\sigma_1 \geq \cdots \geq \sigma_n > 0$ and n is the dimension of the matrix P, which is the same as that of the system matrix A. Let T be a nonsingular transformation defined by

$$T = \Sigma^{-1/2} U^\top Q^{1/2} , \tag{6.6}$$

where $\Sigma^{-1/2} := (\Sigma^{-1})^{1/2} = (\Sigma^{1/2})^{-1}$. Then, we have

$$TPT^\top = \Sigma^{-1/2} U^\top Q^{1/2} P (Q^{1/2})^\top U \Sigma^{-1/2} = \Sigma$$

and

$$T^{-\top} Q T^{-1} = \Sigma^{1/2} U^\top (Q^{-1/2})^\top (Q^{1/2})^\top Q^{1/2} Q^{-1/2} U \Sigma^{1/2} = \Sigma .$$

Here and throughout, we will use the notation

$$T^{-\top} := (T^{-1})^\top = (T^\top)^{-1}.$$

As we will see below, this implies that under the nonsingular transform T, the controllability and observability Gramians are equal and both diagonal and, in fact, are identical. To see this, first note that under the nonsingular transform $T : \mathbf{x}_n \to \mathbf{y}_n$ the state-space description (6.1) of the linear system becomes $(\tilde{A}, \tilde{B}, \tilde{C}, \tilde{D})$ with

$$\tilde{A} = TAT^{-1}, \quad \tilde{B} = TB, \quad \tilde{C} = CT^{-1}, \quad \text{and} \quad \tilde{D} = D. \tag{6.7}$$

Hence, from (6.2) and (6.3), the controllability and observability Gramians of $(\tilde{A}, \tilde{B}, \tilde{C}, \tilde{D})$ are given by $\tilde{P} = TPT^\top$ and $\tilde{Q} = T^{-\top}QT^{-1}$ with $\tilde{P} = \tilde{Q} = \Sigma$ (Problem 6.4). The state-space description $(\tilde{A}, \tilde{B}, \tilde{C}, \tilde{D})$ is called a *balanced realization* of the linear system described by (A, B, C, D).

The following result is very useful:

Theorem 6.2. Let (A, B, C, D) be a minimal realization of a stable linear system described by (6.1) and let $H(z)$ be its corresponding transfer matrix. If the system has a balanced realization with the controllability and observability Gramians $P = Q = \Sigma$, where Σ is a diagonal matrix, then there exist two unitary matrices U and V such that the Hankel matrix Γ_H associated with $H(z)$ has the diagonalization

$$\Gamma_H = U \Sigma V^\top.$$

Note that this factorization, called the *singular value decomposition* of Γ_H, implies that the matrix Σ consists of all of the s-numbers of the Hankel matrix Γ_H.

We leave the proof of this useful result to the reader as an exercise (Problem 6.5).

Example 6.2. Consider a linear system described by

$$A = \begin{bmatrix} 0 & \frac{1}{2} \\ a & 0 \end{bmatrix} \ (|a| < 2), \quad B = \begin{bmatrix} 1 \\ 0 \end{bmatrix}, \quad C = [1 \ \ 0], \quad \text{and} \quad D = 0.$$

Since

$$P = Q = \begin{bmatrix} \frac{4}{4-a^2} & 0 \\ 0 & \frac{4a^2}{4-a^2} \end{bmatrix},$$

we have

$$Q^{1/2} P (Q^{1/2})^\top = \begin{bmatrix} \left(\frac{4}{4-a^2}\right)^2 & 0 \\ 0 & \left(\frac{4a^2}{4-a^2}\right)^2 \end{bmatrix} := \Sigma^2,$$

so that $U = I$ and

$$T = \Sigma^{-1/2} U^\top Q^{1/2} = I .$$

Consequently, $\tilde{A} = A, \tilde{B} = B, \tilde{C} = C, \tilde{D} = D$, and

$$\tilde{P} = \tilde{Q} = \Sigma = \begin{bmatrix} \frac{4}{4-a^2} & 0 \\ 0 & \frac{4a^2}{4-a^2} \end{bmatrix} = P = Q .$$

This implies that the original system is already a balanced realization for any value of a with $|a| < 2$.

It will be seen in the following that the balanced realization technique plays an important role in the study of optimal Hankel-norm approximation for MIMO linear systems.

6.2 Matrix-Valued All-Pass Transfer Functions

We first point out that the discussion given below in this section is independent of the balanced realization that we studied previously. Hence, the linear system under consideration here need not be stable. We only require that the system be a minimal realization, which in turn means that the linear system is both completely controllable and observable (Theorem 2.6). The balanced realization technique will be needed later in this chapter.

We now introduce the concept of all-pass transfer matrices. A transfer matrix $H(z) = D + C(zI - A)^{-1}B$ is said to be *all-pass* if $H(z)$ is a square matrix satisfying

$$H(z)H^*(\bar{z}^{-1}) \equiv \sigma^2 I \tag{6.8}$$

for some positive constant σ, where the identity holds for all z in the extended complex z-plane $\hat{C}$ and, as usual, $H^* = \bar{H}^\top$ denotes the adjoint of H. By setting $\tilde{D} = D/\sigma$, $\tilde{A} = A$, $\tilde{B} = B/\sqrt{\sigma}$, and $\tilde{C} = C/\sqrt{\sigma}$, we may assume, without loss of generality, that $\sigma = 1$.

Clearly, the transfer function $H(z) = 1/z$ is both stable and all-pass. However, as remarked above, a linear system with an all-pass transfer matrix is not necessarily stable. For example, the transfer function $H(z) = (z + \mathrm{j})/(1 - \mathrm{j}z)$ is all-pass but has a pole at $z = -\mathrm{j}$.

It will be seen that there is a very tight relationship between an all-pass transfer matrix and some of the Lyapunov equations. We first establish the following result:

Lemma 6.1. Let (A, B, C, D) be a minimal realization of a discrete-time linear system and let

$$H(z) = D + C(zI - A)^{-1}B \tag{6.9}$$

be its corresponding transfer matrix. Then, $H(z)$ is all-pass in the sense that

$$H(z)H^*(\bar{z}^{-1}) = I \qquad \text{for all } z \in \hat{\mathcal{C}}, \tag{6.10}$$

where $\hat{\mathcal{C}}$ is the extended complex plane, if and only if the matrix D is square and nonsingular and there exists a nonsingular symmetric matrix T such that

$$\begin{cases} A^\top TB = C^\top D, \\ CT^{-1}A^\top = DB^\top, \\ A^\top TA - A^\top TBD^{-1}C = T, \\ D^\top D - B^\top TB = I. \end{cases} \tag{6.11}$$

We remark that the linear system studied in Example 6.1 is both completely controllable and observable, and so is a minimal realization. However, its transfer function $H(z) = z/(z^2 - 1)$ is not all-pass. This is simply due to the fact that the matrix $D = 0$ is singular.

Before proving the lemma, we remark that if $\{\lambda_i\}_{i=1}^n$ is the set of eigenvalues of the matrix A then $\{s + \lambda_i\}_{i=1}^n$ is the set of eigenvalues of the matrix

$$A_s := sI + A. \tag{6.12}$$

Hence, if A is singular, there always exist $s \in \hat{\mathcal{C}}$, with $s \to 0$, such that the matrices A_s are nonsingular. Moreover, if we let

$$H_s(z) = D + C(zI - A_s)^{-1}B, \tag{6.13}$$

then, since $zI - A_s = (z - s)I - A$, the transfer matrix $H(z)$ is all-pass if and only if $H_s(z)$ is all-pass. Observe also that there is no inversion for the matrix A in the statements of the lemma. This implies that we may verify the lemma for the matrix $H_s(z)$ instead of the matrix $H(z)$, and then let $s \to 0$ to complete the proof. In other words, we may assume, without loss of generality, that A is nonsingular in the following proof.

Suppose that $H(z)$ is all-pass. Then, it follows that $H(z)$ is square and

$$H^*(\bar{z}^{-1}) = H^{-1}(z) = [D + C(zI - A)^{-1}B]^{-1}.$$

Since $H(z)H^*(\bar{z}^{-1}) = I$ in the extended complex plane $\hat{\mathcal{C}}$, we see that D is square and D^{-1} exists, so that by the matrix inversion lemma, see, for example, Chui and Chen [1987], we have

$$H^*(\bar{z}^{-1}) = D^{-1} - D^{-1}C[zI - (A - BD^{-1}C)]^{-1}BD^{-1}. \tag{6.14}$$

On the other hand, we also have

$$\begin{aligned}
H^*(\bar{z}^{-1}) &= \bar{H}^\top(\bar{z}^{-1}) \\
&= D^\top + B^\top(z^{-1}I - A^\top)^{-1}C^\top \\
&= D^\top + B^\top z(I - zA^\top)^{-1}C^\top \\
&= D^\top + B^\top[-A^{-\top}(I - zA^\top) + A^{-\top}](I - zA^\top)^{-1}C^\top \\
&= (D^\top - B^\top A^{-\top}C^\top) - B^\top A^{-\top}(zI - A^{-\top})^{-1}A^{-\top}C^\top, \quad (6.15)
\end{aligned}$$

where $A^{-\top} := (A^{-1})^\top = (A^\top)^{-1}$. Hence, by combining (6.15) with (6.14), it follows from the linear systems theory, which can be found from Chui and Chen [1989], that there exists a nonsingular transformation T that relates the above two identities in the sense that

$$\begin{cases}
BD^{-1} = T^{-1}A^{-\top}C^\top, \\
D^{-1}C = B^\top A^{-\top}T, \\
A - BD^{-1}C = T^{-1}A^{-\top}T, \\
D^{-1} = D^\top - B^\top A^{-\top}C^\top,
\end{cases} \tag{6.16}$$

see also Problem 6.7, or equivalently,

$$\begin{cases}
A^\top TB = C^\top D, \\
CT^{-1}A^\top = DB^\top, \\
A^\top TA - A^\top TBD^{-1}C = T, \\
D^\top D - B^\top TB = I,
\end{cases} \tag{6.17}$$

where the last identity follows from the first and the last equations in (6.16). That is, we have verified (6.11).

What is left to show in this direction is that T is symmetric. In view of (6.16), it is sufficient to show that

$$\begin{cases}
BD^{-1} = T^{-\top}A^{-\top}C^\top, \\
D^{-1}C = B^\top A^{-\top}T^\top, \\
A - BD^{-1}C = T^{-\top}A^{-\top}T^\top.
\end{cases} \tag{6.18}$$

The reason is that (A, B, C, D) is a minimal realization so that T is unique (Problem 6.7). To verify the first equation in (6.18), we apply the second equation in (6.16) to conclude that

$$B^\top = D^{-1}CT^{-1}A^\top,$$

so that by incorporating with the last equation in (6.16), we have

$$BD^{-1} = AT^{-\top}C^\top D^{-\top}D^{-1} = AT^{-\top}C^\top(I - D^{-\top}B^\top A^{-\top}C^\top).$$

It is now clear that the first equation in (6.18) follows if we can show that

$$AT^{-\top}C^\top(I - D^{-\top}B^\top A^{-\top}C^\top) = T^{-\top}A^{-\top}C^\top,$$

or

$$AT^{-\top}(I - C^\top D^{-\top} B^\top A^{-\top})C^\top = T^{-\top} A^{-\top} C^\top .$$

In order to do so, it is sufficient to verify that

$$AT^{-\top}(I - C^\top D^{-\top} B^\top A^{-\top}) = T^{-\top} A^{-\top} . \tag{6.19}$$

Writing

$$\begin{aligned} T^{-\top}A^{-\top} &= (T^{-\top}A^{-\top} - AT^{-\top}) + AT^{-\top} \\ &= AT^{-\top}(T^\top A^{-1} T^{-\top} - A^\top)A^{-\top} + AT^{-\top} \end{aligned}$$

and then applying the third equation in (6.16), we have

$$\begin{aligned} T^{-\top}A^{-\top} &= AT^{-\top}(-C^{-\top} D^{-\top} B^\top)A^{-\top} + AT^{-\top} \\ &= AT^{-\top}(I - C^\top D^{-\top} B^\top A^{-\top}), \end{aligned}$$

which is (6.19). Hence, the first equation in (6.18) is verified.

The second equation in (6.17) can be derived in a similar manner (Problem 6.8). Finally, the last equation in (6.18) may be verified as follows: Starting from the third equation in (6.16), we have

$$\begin{aligned} A^\top &= T(A - BD^{-1}C)^{-1}T^{-1} \\ &= T(A^{-1} + A^{-1}B(D - CA^{-1}B)^{-1}CA^{-1})T^{-1} \\ &= T(A^{-1} + A^{-1}BD^\top CA^{-1})T^{-1} \\ &= TA^{-1}T^{-1} + (TA^{-1}B)D^\top(CA^{-1}T^{-1}), \end{aligned}$$

where the matrix inversion lemma and the last equation in (6.16) have been applied. Substituting the first two equations in (6.18) that have just been verified into this identity, we obtain

$$A^\top = TA^{-1}T^{-1} + (C^\top D^{-\top})D^\top(D^{-\top}B^\top),$$

which is the last equation in (6.18). This completes the proof of the claim that T is symmetric, and hence one direction of the lemma is verified.

The converse of the lemma follows immediately from direct calculation. Indeed, by (6.15), (6.11), and the matrix inversion lemma, we have

$$
\begin{aligned}
&H(z)H^*(\bar{z}^{-1}) \\
=&[D + C(zI - A)^{-1}B][(D^\top - B^\top A^{-\top} C^\top) \\
&\quad - B^\top A^{-\top}(zI - A^{-\top})^{-1} A^{-\top} C^\top] \\
=&[D + C(zI - A)^{-1}B][D^\top D - B^\top A^{-\top} C^\top D \\
&\quad - B^\top A^{-\top}(zI - A^{-\top})^{-1} A^{-\top} C^\top D] D^{-1} \\
=&[D + C(zI - A)^{-1}B][(I + B^\top T B) - B^\top T B \\
&\quad - D^{-1} C T^{-1}(zI - A^{-\top})^{-1} T B] D^{-1} \\
=&[D + C(zI - A)^{-1}B][D^{-1} - D^{-1} C T^{-1}(zI - A^{-\top})^{-1} T B D^{-1}] \\
=&[D + C(zI - A)^{-1}B][D^{-1} - D^{-1} C(zI - T^{-1} A^{-\top} T)^{-1} B D^{-1}] \\
=&[D + C(zI - A)^{-1}B][D^{-1} - D^{-1} C(zI - A + BD^{-1}C)^{-1} B D^{-1}] \\
=&[D + C(zI - A)^{-1}B][D + C(zI - A)^{-1}B]^{-1} \\
=&I\,.
\end{aligned}
$$

This completes the proof of the lemma.

The following theorem indicates a tight relationship of an all-pass transfer matrix with two Lyapunov equations. Note that, as pointed out at the beginning of this section, since no balanced realization is needed in this section, the linear system under consideration does not have to be stable.

Theorem 6.3. Let (A, B, C, D) be a minimal realization of a discrete-time linear system. Then, the corresponding transfer matrix

$$H(z) = D + C(zI - A)^{-1}B$$

is all-pass if and only if the matrix D is square and nonsingular and there exist two symmetric nonsingular matrices P and Q that satisfy

(a) $P - APA^\top = BB^\top$,
(b) $Q - A^\top QA = C^\top C$,
(c) $PQ = \sigma^2 I$ for some constant $\sigma > 0$, and
(d) $D^\top D + B^\top QB = \sigma^2 I$.

We first remark that the two matrices P and Q described in this theorem may not be the same as the controllability and observability Gramians defined respectively in the previous section, namely, they may not satisfy (6.2) and (6.3). In fact, since the system is not necessarily stable, the infinite sums (6.2) and (6.3) may not even converge.

Again, in the following proof we assume, without loss of generality, that $\sigma = 1$.

First, suppose that $H(z)$ is all-pass. Then, it follows from Lemma 6.1 that there exists a nonsingular symmetric matrix T that satisfies (6.11). Let

$$P = -T^{-1} \quad \text{and} \quad Q = -T\,.$$

Then, both P and Q are symmetric, nonsingular, and satisfy (c) (with $\sigma = 1$). Moreover, from the last equation in (6.11) we have (d) (with $\sigma = 1$). By substituting the first equation in (6.11) into the third equation there, we have

$$A^\top TA - C^\top C = T\,,$$

which is (b). To verify (a), the perturbation technique applied in the proof of Lemma 6.1 is again useful. In other words, there always exist values of $s \to 0$ such that the matrices $A_s = sI + A$ are all nonsingular. So, we may again assume, without loss of generality, that the matrix A is nonsingular. Now, it follows from the third equation in (6.11) that

$$A^\top TAT^{-1}A^\top - A^\top TBD^{-1}CT^{-1}A^\top = A^\top\,.$$

Hence, substituting the second equation in (6.11) into this equation yields

$$A^\top TAT^{-1}A^\top - A^\top TBB^\top = A^\top\,,$$

or

$$A^\top T(AT^{-1}A^\top - BB^\top - T^{-1}) = 0\,.$$

Since $A^\top T$ is nonsingular, we have

$$AT^{-1}A^\top - BB^\top - T^{-1} = 0\,,$$

which is (a) with $P = -T^{-1}$. This completes the proof of one direction of the theorem.

To prove the converse, in view of Lemma 6.1, it is sufficient to verify that the conditions (a)-(d) together imply (6.11) for some symmetric nonsingular matrix T. To do so, we may actually select $T := -Q$. Then, T is symmetric and nonsingular. Moreover, with $\sigma = 1$ in (c), we have $P = Q^{-1} = -T^{-1}$. The last equation in (6.11) follows immediately from (d). Note that (a) and (b) become, respectively,

$$-T^{-1} + AT^{-1}A^\top = BB^\top \tag{6.20}$$

and

$$-T + A^\top TA = C^\top C\,. \tag{6.21}$$

We proceed next to verify the second equality in (6.11). First observe that since D is a nonsingular matrix that satisfies (d), the matrix $\tilde{D} := UD$ also satisfies (d), where U is an arbitrary unitary, that is,

$$\tilde{D}^\top \tilde{D} + B^\top QB = I\,. \tag{6.22}$$

Substituting this into (6.20), we have

$$B(\tilde{D}^\top \tilde{D} - B^\top TB)B^\top = -T^{-1} + AT^{-1}A^\top ,$$

or

$$(\tilde{D}B^\top)^\top(\tilde{D}B^\top) = (BB^\top)T(BB^\top) - T^{-1} + AT^{-1}A^\top .$$

By applying (6.20) again, we also obtain

$$\begin{aligned}
&(\tilde{D}B^\top)^\top(\tilde{D}B^\top)\\
=&(-T^{-1} + AT^{-1}A^\top)T(-T^{-1} + AT^{-1}A^\top) - T^{-1} + AT^{-1}A^\top\\
=&- AT^{-1}A^\top + AT^{-1}(A^\top TA)T^{-1}A^\top .
\end{aligned}$$

Furthermore, substituting (6.21) into this equation yields

$$\begin{aligned}
&(\tilde{D}B^\top)^\top(\tilde{D}B^\top)\\
=&- AT^{-1}A^\top + AT^{-1}(T + C^\top C)T^{-1}A^\top\\
=&(CT^{-1}A^\top)^\top(CT^{-1}A^\top) .
\end{aligned}$$

Hence, it follows that

$$\tilde{U}\tilde{D}B^\top = CT^{-1}A^\top ,$$

for some unitary matrix $\tilde{U}$. Now, set $U = \tilde{U}^\top$. Then, we obtain $DB^\top = CT^{-1}A^\top$, which is the second equation in (6.11).

To derive the first equation in (6.11), we start with the second equation that has just been verified and obtain

$$CT^{-1}A^\top T^\top A = DB^\top T^\top A .$$

Taking the transposes of both sides yields

$$A^\top TAT^{-1}C^\top = A^\top TBD^\top$$

or

$$A^\top TAT^{-1}C^\top D = A^\top TB(D^\top D) .$$

In addition, substituting the last equation in (6.11) into the above equation and then applying (6.20), we also have

$$\begin{aligned}
&A^\top TAT^{-1}C^\top D\\
&= A^\top TB(I + B^\top TB)\\
&= A^\top TB + A^\top T(BB^\top)TB\\
&= A^\top TB + A^\top T(-T^{-1} + AT^{-1}A^\top)TB\\
&= A^\top TAT^{-1}A^\top TB .
\end{aligned}$$

Now, since T is nonsingular it follows from (6.21) that $A^\top TA = T + C^\top C$ must also be nonsingular, so that

$$C^\top D = A^\top TB \,. \tag{6.23}$$

This is the first equation in (6.11).

Finally, to verify the third equation of (6.11), we note from (6.23) that $C^\top = A^\top TBD^{-1}$, so that (6.21) gives

$$T = A^\top TA - C^\top C = A^\top TA - A^\top TBD^{-1}C \,,$$

as expected. This completes the proof of the theorem.

6.3 Optimal Hankel-Norm Approximation for MIMO Systems

Let $H(z)$ be the transfer matrix of a discrete-time MIMO system and let Γ_H denote its corresponding infinite block-Hankel matrix. This section is devoted to establishing the analogous results of Nehari and AAK under the assumption that Γ_H has finite rank.

6.3.1 Preliminary Results

Recall from Sect. 4.2.2 that the s-numbers of a Hankel matrix Γ are defined to be the corresponding eigenvalues of the positive square root of the symmetric non-negative definite matrix $\bar{\Gamma}\Gamma$. To indicate the correspondence of this block-Hankel matrix Γ to the transfer matrix $H(z) = D + C(zI - A)^{-1}B$, we will, as usual, use the more precise notation Γ_H for Γ. For a given block-Hankel matrix Γ_H, let $\{s_i^2\}$ be the set of eigenvalues of the matrix $\bar{\Gamma}_H\Gamma_H$, and let $\{\underline{\xi}_i\}$ be the corresponding set of eigenvectors. That is,

$$\bar{\Gamma}_H\Gamma_H\underline{\xi}_i = s_i^2\underline{\xi}_i \,, \qquad i = 1, 2, \ldots \,.$$

First, we need some new notations. Let l^2 denote the space of square-summable vector-valued sequences, namely,

$$l^2 = \left\{ \{\mathbf{x}_k\}_{k=-\infty}^{\infty} : \quad \sum_{k=-\infty}^{\infty} \mathbf{x}_k^\top \bar{\mathbf{x}}_k < \infty \right\}. \tag{6.24}$$

It is clear that l^2 is a Hilbert space with inner product defined by

$$\langle\{\mathbf{x}_k\}, \{\mathbf{y}_k\}\rangle_{l^2} = \sum_{k=-\infty}^{\infty} \mathbf{x}_k^\top \bar{\mathbf{y}}_k \,. \tag{6.25}$$

In addition, the $\mathcal{L}^2$ function space isometric to l^2 is the space of all vector-valued z-transforms

$$X(z) := \sum_{k=-\infty}^{\infty} \mathbf{x}_k z^{-k},$$

where $\{\mathbf{x}_k\} \in l^2$, with inner product defined by

$$\langle X, Y \rangle_{\mathcal{L}^2} = \frac{1}{2\pi} \int_{|z|=1} X^\top(z) Y(z) \frac{dz}{\mathrm{j}z}.$$

The isometry between l^2 and $\mathcal{L}^2$ is given by the identity

$$\langle \{\mathbf{x}_k\}, \{\mathbf{y}_k\} \rangle_{l^2} = \langle X, Y \rangle_{\mathcal{L}^2}. \tag{6.26}$$

Moreover, for any sequence $\{\mathbf{u}_k\}_{k=-\infty}^{\infty}$ in l^2, we define two corresponding truncated sequences $\{\mathbf{u}_k^-\}$ and $\{\mathbf{u}_k^+\}$ of $\{\mathbf{u}_k\}$ by

$$\mathbf{u}_k^- = \begin{cases} \mathbf{u}_k, & k \le 0, \\ 0, & k > 0, \end{cases} \tag{6.27}$$

and

$$\mathbf{u}_k^+ = \begin{cases} 0, & k \le 0, \\ \mathbf{u}_k, & k > 0, \end{cases} \tag{6.28}$$

and their corresponding z-transforms $U_a(z)$ and $U_s(z)$ by

$$U_a(z) := \sum_{k=-\infty}^{\infty} \mathbf{u}_k^- z^{-k} = \sum_{k=0}^{\infty} \mathbf{u}_{-k} z^{k}, \tag{6.29}$$

and

$$U_s(z) := \sum_{k=-\infty}^{\infty} \mathbf{u}_k^+ z^{-k} = \sum_{k=1}^{\infty} \mathbf{u}_k z^{-k}. \tag{6.30}$$

Then, we have the following result which holds for not necessarily stable linear systems.

Lemma 6.2. Let $H(z)$ be the transfer matrix associated with a minimal realization (A, B, C, D) of a linear system, and $\{\mathbf{v}_k\}$ the system output corresponding to the system input $\{\mathbf{u}_k^-\} = \{\mathbf{u}_k\}$, with $\mathbf{u}_k = 0$ for $k > 0$. Also, let $\{\mathbf{v}_k^+\}$ and $V_s(z)$ denote the truncation and its corresponding one-sided z-transform related to $\{\mathbf{v}_k\}$ as defined in (6.28) and (6.30), respectively. Then,

$$\|H\|_\Gamma = \sup_{\{\mathbf{u}_k^-\}} \frac{\|\{\mathbf{v}_k^+\}\|_{l^2}}{\|\{\mathbf{u}_k^-\}\|_{l^2}} = \sup_{U_a} \frac{\|V_s\|_{\mathcal{L}^2}}{\|U_a\|_{\mathcal{L}^2}}. \tag{6.31}$$

This result says that the Hankel norm of a transfer matrix of a linear system gives the l^2-gain from the past inputs to the future outputs.

To prove the lemma, consider the discrete-time linear system

$$\begin{cases} \mathbf{x}_{k+1} = A\mathbf{x}_k + B\mathbf{u}_k \,, \\ \mathbf{v}_k = C\mathbf{x}_k + D\mathbf{u}_k \,, \end{cases}$$

and set

$$\widetilde{M}_{AB} = [\, B \;\; AB \;\; \cdots \;\; A^n B \; \cdots \,]$$

and

$$\widetilde{N}_{CA} = \begin{bmatrix} C \\ CA \\ \vdots \\ CA^n \\ \vdots \end{bmatrix} .$$

Then, for an input $\{\mathbf{u}_k\}_{k=-\infty}^{\infty}$ with $\mathbf{u}_k = 0$ for $k > 0$, we have

$$\mathbf{x}_1 = A\mathbf{x}_0 + B\mathbf{u}_0 = A(A\mathbf{x}_{-1} + B\mathbf{u}_{-1}) + B\mathbf{u}_0 = \cdots ,$$

so that

$$\mathbf{x}_1 = \widetilde{M}_{AB} \begin{bmatrix} \mathbf{u}_0 \\ \mathbf{u}_{-1} \\ \vdots \end{bmatrix} .$$

Similarly, for the truncated output $\{\mathbf{v}_k^+\}$, we have

$$\begin{bmatrix} \mathbf{v}_1^+ \\ \mathbf{v}_2^+ \\ \vdots \end{bmatrix} = \widetilde{N}_{CA}\mathbf{x}_1 .$$

Hence,

$$\begin{bmatrix} \mathbf{v}_1^+ \\ \mathbf{v}_2^+ \\ \vdots \end{bmatrix} = \widetilde{N}_{CA}\widetilde{M}_{AB} \begin{bmatrix} \mathbf{u}_0 \\ \mathbf{u}_{-1} \\ \vdots \end{bmatrix} ,$$

and it follows that the operator norm of $\widetilde{N}_{CA}\widetilde{M}_{AB}$ is given by

$$\|\widetilde{N}_{CA}\widetilde{M}_{AB}\| = \sup_{\{\mathbf{u}_k^-\}} \frac{\|\{\mathbf{v}_k^+\}\|_{l^2}}{\|\{\mathbf{u}_k^-\}\|_{l^2}} .$$

Next, recall from (2.20) that the Hankel matrix Γ_H associated with the transfer matrix $H(z)$ is given by

$$\Gamma_H = \tilde{N}_{CA}\widetilde{M}_{AB} \,,$$

so that

$$\bar{\Gamma}_H \Gamma_H = \bar{\Gamma}_H^\top \Gamma_H = \widetilde{M}_{AB}^\top \tilde{N}_{CA}^\top \tilde{N}_{CA}\widetilde{M}_{AB} \,.$$

Consequently, if we again use the notation $\lambda_{max}(M)$ to denote the maximum eigenvalue of a non-negative definite matrix M, then it follows from (6.26) that

$$\begin{aligned} \|H\|_\Gamma &= \lambda_{max}^{1/2}(\bar{\Gamma}_H \Gamma_H) \\ &= \lambda_{max}^{1/2}(\widetilde{M}_{AB}^\top \tilde{N}_{CA}^\top \tilde{N}_{CA}\widetilde{M}_{AB}) \\ &= \|\tilde{N}_{CA}\widetilde{M}_{AB}\| \\ &= \sup_{\{\mathbf{u}_k^-\}} \frac{\|\{\mathbf{v}_k^+\}\|_{l^2}}{\|\{\mathbf{u}_k^-\}\|_{l^2}} \\ &= \sup_{U_a} \frac{\|V_s\|_{\mathcal{L}^2}}{\|U_a\|_{\mathcal{L}^2}} \,. \end{aligned}$$

This completes the proof of the lemma.

Similar to the $\mathcal{L}^2$ space defined above, we now introduce the $\mathcal{L}^\infty_{q\times p}$ space of $q \times p$ matrix-valued bounded measurable functions on the unit circle $|z| = 1$ by

$$\mathcal{L}^\infty_{q\times p} = \{A_{q\times p}(z): \quad s_1(A_{q\times p}(z)) < \infty \quad \text{for all } z \text{ with } |z| = 1\} \tag{6.32}$$

with the supremum norm

$$\|A\|_{\mathcal{L}^\infty_{q\times p}} = ess \sup_{|z|=1} s_1(A(z)) \,, \tag{6.33}$$

where $s_1(A(z))$ denotes, as usual, the largest s-number of the matrix $A(z)$ for each fixed z. As in the scalar-valued setting in Chap. 1, we decompose each function $H(z)$ in $\mathcal{L}^\infty_{q\times p}$ as the sum of its singular and analytic parts, namely,

$$H(z) = H_s(z) + H_a(z) \,, \tag{6.34}$$

with $H_s(z) \in H^\infty_{q\times p,s}$ and $H_a(z) \in H^\infty_{q\times p,a}$, where

$$H^\infty_{q\times p,s} = \{A_{q\times p}(z) \in \mathcal{L}^\infty_{q\times p}: \quad A_{q\times p}(z) \text{ is analytic on } |z| \geq 1\}$$

and

$$H^\infty_{q\times p,a} = \{A_{q\times p}(z) \in \mathcal{L}^\infty_{q\times p}: \quad A_{q\times p}(z) \text{ is analytic on } |z| \leq 1\} \,.$$

Note that $H^\infty_{q\times p,s}$ contains all $q \times p$ stable rational transfer matrices.

To establish the matrix-valued extension of Nehari's theorem, we first need the following result:

Lemma 6.3. Let $H(z) \in H^\infty_{q\times p,s}$. Then, we have

$$\|H\|_\Gamma \le \|H - F\|_{\mathcal{L}^\infty_{q\times p}} \tag{6.35}$$

for all $F(z) \in H^\infty_{q\times p,a}$.

To prove the lemma, let $\{\mathbf{u}_k\}_{k=-\infty}^0 \in l^2$ be an input sequence of the system. Define, as before,

$$\mathbf{u}_k^- = \begin{cases} \mathbf{u}_k, & k \le 0, \\ 0, & k > 0, \end{cases}$$

$$U_a(z) = \sum_{k=-\infty}^{0} \mathbf{u}_k^- z^{-k} = \sum_{k=0}^{\infty} \mathbf{u}_{-k} z^k,$$

$$V(z) := H(z)U_a(z) := \sum_{k=-\infty}^{\infty} \mathbf{v}_k z^{-k}, \tag{6.36}$$

$$\mathbf{v}_k^+ = \begin{cases} 0, & k \le 0, \\ \mathbf{v}_k, & k > 0, \end{cases}$$

and

$$V_s(z) = \sum_{k=1}^{\infty} \mathbf{v}_k z^{-k}.$$

Then, since

$$\langle \{\mathbf{v}_k^+\}, \{\mathbf{v}_k^+\} \rangle_{l^2} = \langle \{\mathbf{v}_k^+\}, \{\mathbf{v}_k\} \rangle_{l^2},$$

it follows from the isometry (6.26) that

$$\|V_s\|^2_{\mathcal{L}^2} = \langle V_s(z), V(z) \rangle_{\mathcal{L}^2} = \langle V_s(z), H(z)U_a(z) \rangle_{\mathcal{L}^2}.$$

Now, let $F(z)$ be arbitrarily chosen from $H^\infty_{q\times p,a}$. Then, since $F(z)U_a(z)$ is in $\mathcal{L}^2$ and is analytic in $|z| \le 1$, we have

$$\langle V_s(z), F(z)U_a(z) \rangle_{\mathcal{L}^2} = 0.$$

Consequently,

$$\begin{aligned} \|V_s\|^2_{\mathcal{L}^2} &= \langle V_s(z), (H(z) - F(z))U_a(z) \rangle_{\mathcal{L}^2} \\ &\le \|V_s\|_{\mathcal{L}^2} \|(H - F)U_a\|_{\mathcal{L}^2} \\ &\le \|V_s\|_{\mathcal{L}^2} \|H - F\|_{\mathcal{L}^\infty_{q\times p}} \|U_a\|_{\mathcal{L}^2}, \end{aligned}$$

so that

$$\|V_s\|_{\mathcal{L}^2} \le \|H - F\|_{\mathcal{L}^\infty_{q\times p}} \|U_a\|_{\mathcal{L}^2} . \tag{6.37}$$

This holds for any input-output relationship $V(z) = H(z)U_a(z)$ defined in (6.36). Hence, (6.35) follows from Lemma 6.2 and (6.37).

6.3.2 Matrix-Valued Extensions of the Nehari and AAK Theorems

The fundamental result of Nehari (Theorem 4.1) was the starting point of the beautiful theory of optimal Hankel-norm approximation. As we have already seen in Chaps. 4 and 5, the AAK theory can be considered as a generalization of the Nehari theorem. In this section, we will state the matrix-valued extensions of the Nehari and AAK theorems when the transfer function to be approximated is derived from a finite-dimensional MIMO system. In other words, we do not consider the full extension of scalar-valued Hankel-norm approximation to the matrix-valued setting, but restrict our attention to approximation of matrix-valued stable proper rational functions. Since the matrix-valued extension of the AAK theorem includes the corresponding matrix-valued version of the Nehari theorem as a special case, it is sufficient to establish the matrix-valued analog of the AAK theorem. Details of this proof will be presented in Sect. 6.3.3.

Throughout this section, A, B, C, D will denote $n \times n, n \times p, q \times n, q \times q$ constant matrices, respectively, with $1 \le q \le p \le n$. To simplify our presentation, we will always assume, in the remainder of this chapter, that (A, B, C, D) represents a stable minimal realization of a discrete-time MIMO linear system. The transfer matrix of this system is given by

$$H(z) = D + C(zI - A)^{-1}B . \tag{6.38}$$

Since the matrix-valued transfer function $H(z)$ defined in (6.38) is in $H^\infty_{q\times p,s}$, an application of Lemma 6.3 immediately yields

$$\|H\|_\Gamma \le \inf_{F\in H^\infty_{q\times p,a}} \|H - F\|_{\mathcal{L}^\infty_{q\times p}} ,$$

where the $\mathcal{L}^\infty_{q\times p}$ norm was defined in (6.33). The matrix-valued extension of Nehari's theorem for the transfer matrix $H(z)$ in (6.38) says that the Hankel norm of $H(z)$ is indeed the magnitude of the error of approximation in $\mathcal{L}^\infty_{q\times p}$ of $H(z)$ by bounded analytic matrix-valued functions.

In this situation, we can even determine a bounded analytic matrix-valued function $\hat{F}(z)$ that best approximates $H(z)$ in $\mathcal{L}^\infty_{q\times p}$. To state this result and its AAK extension for MIMO systems, we need the notion of McMillan degree of a matrix-valued rational function.

Let $M(z)$ be an $m \times n$ matrix-valued strictly proper rational (transfer) function, where we assume, without loss of generality, that all entries of $M(z)$ are in coprime form. Then, $M(z)$ can be rewritten as

$$M(z) = \frac{1}{d(z)} N(z),$$

where $d(z)$ is the least common multiple of the denominators of the entries of $M(z)$ with leading coefficient 1, and $N(z) = d(z)M(z)$ is a polynomial matrix. Consider the polynomial matrix $N(z)$. It can be proved that there exist two nonsingular polynomial matrices $U_1(z)$ and $U_2(z)$ such that

$$d(z)M(z) = N(z) = U_1(z) \begin{bmatrix} n_1(z) & \cdots & 0 & 0 \\ \vdots & \ddots & \vdots & \vdots \\ 0 & \cdots & n_{p'}(z) & 0 \\ 0 & \cdots & 0 & 0 \end{bmatrix} U_2(z),$$

where $p' \le \min(m, n)$ and $n_i(z)$ divides $n_j(z)$ for each $i \le j$ (Problem 6.9). Note that although the two polynomial matrices $U_1(z)$ and $U_2(z)$ are in general not unique, the diagonal matrix is uniquely determined by $N(z)$. $U_1(z)$ and $U_2(z)$ are usually called *unimodular* matrices and the diagonal matrix is called the *Smith form* of $M(z)$. Next, write

$$M(z) = U_1(z) \begin{bmatrix} \frac{n_1(z)}{d(z)} & \cdots & 0 & 0 \\ \vdots & \ddots & \vdots & \vdots \\ 0 & \cdots & \frac{n_{p'}(z)}{d(z)} & 0 \\ 0 & \cdots & 0 & 0 \end{bmatrix} U_2(z), \tag{6.39}$$

and reduce all entries of the diagonal matrix to coprime forms. Then, the diagonal matrix is called the *Smith-McMillan form* of $M(z)$. The so-called *McMillan degree* of $M(z)$ is then defined to be the sum of the degrees of the denominators in the Smith-McMillan form of the matrix.

For a general rational (transfer) matrix $M(z)$, we decompose it as

$$M(z) = M_s(z) + M_a(z),$$

where $M_s(z)$ is strictly proper and $M_a(z)$ is a polynomial matrix. Then, the McMillan degree of $M(z)$ is defined to be the sum of the McMillan degrees of $M_s(z)$ and $M_a(z^{-1})$.

As a historical remark, we would like to mention that the McMillan degree is the minimum number of inductors and capacitors required to realize a so-called passive impedance matrix $M(z)$, first discovered by McMillan [1952]. Later, Kalman [1965,1966] discovered that this degree is also the minimum number of integrators required in constructing a controllable and observable realization of a proper rational transfer matrix $M(z)$.

Example 6.3. Consider the strictly proper rational matrix $M(z) = N(z)/d(z)$ with

$$N(z) = \begin{bmatrix} z & z(z+1)^2 \\ -z(z+1)^2 & -z(z+1)^2 \end{bmatrix} \quad \text{and} \quad d(z) = (z+1)^2(z+2)^2.$$

By following the above procedure, it can be verified that the Smith-McMillan form of the matrix is

$$M(z) = \begin{bmatrix} 1 & 0 \\ -(z+1)^2 & 1 \end{bmatrix} \begin{bmatrix} \frac{z}{(z+1)^2(z+2)^2} & 0 \\ 0 & \frac{z^2}{z+2} \end{bmatrix} \begin{bmatrix} 1 & (z+1)^2 \\ 0 & 1 \end{bmatrix} .$$

In this example, the McMillan degree of the matrix $M(z)$ is 5.

It should be pointed out that even if $M(z)$ is strictly proper, its Smith-McMillan form may not be so, as has been seen from this example. This discrepancy is due to the unimodular matrices which may introduce additional poles and zeros.

More importantly, it should be noted that for MIMO linear systems, there is no analog of the Kronecker theorem, so that finding the rank of the block-Hankel matrix associated with its transfer matrix is not simple in general. However, it turns out that calculating the McMillan degree of the transfer matrix is not difficult and this degree is equal to the rank (or, the number of the nonzero s-numbers) of its corresponding block-Hankel matrix (Problem 6.10), which is, in turn, the same as the dimension of its minimal realization. These facts are consistent with the analogous results for SISO systems. Hence, for convenience the matrix-valued version of the AAK Theorem will be formulated in terms of the McMillan degree of the given rational transfer matrix.

Now we are in a position to state the following two theorems. To simplify the notation in the derivation, we will assume, by replacing $H(z)$ with $\left[\begin{smallmatrix} H(z) \\ 0 \end{smallmatrix}\right]$ if necessary, that $H(z)$ is a $p \times p$ (square) matrix of McMillan degree n although a slightly more general result may be obtained.

Theorem 6.4. Let $H(z) \in H^\infty_{p\times p,s}$ be the transfer matrix of a stable minimal realization of a discrete-time MIMO linear system defined in (6.38). Then

$$\|H\|_\Gamma = \inf_{F\in H^\infty_{p\times p,a}} \|H - F\|_{\mathcal{L}^\infty_{p\times p}} . \tag{6.40}$$

Furthermore, there exists an $\hat{F}(z)$ in $H^\infty_{p\times p,a}$ such that

$$\|H\|_\Gamma = \|H - \hat{F}\|_{\mathcal{L}^\infty_{q\times p}} . \tag{6.41}$$

To describe $\hat{F}(z)$, assume that the s-numbers of $H(z)$ are

$$s_1 = \cdots = s_r > s_{r+1} \geq \cdots \geq s_n > 0$$

and that $H(z)$ has a balanced realization $(\tilde{A}, \tilde{B}, \tilde{C}, \tilde{D})$ of dimensions $n \times n, n \times p, p \times n$, and $p \times p$, respectively, with controllability and observability Gramians $P = Q = \Sigma$. Set

$$\Sigma = \begin{bmatrix} s_1 I & 0 \\ 0 & \Sigma_0 \end{bmatrix} \tag{6.42}$$

with $\Sigma_0 = \operatorname{diag}\{s_{r+1}, \cdots, s_n\}$. Then $\hat{F}(z) \in H^\infty_{p\times p,a}$ can be formulated as the analytic part $\hat{F}(z) = F_a(z)$ of the following matrix-valued rational function:

$$F(z) = F_a(z) + F_s(z) = \hat{D} + \hat{C}(zI - \hat{A})^{-1}\hat{B}\,, \tag{6.43}$$

where

$$\begin{cases} \hat{A} = [M - s_1^2 I]^{-1}[\Sigma_0^2 - M]\,, \\ \hat{B} = [M - s_1^2 I]^{-1}[\Sigma_0[0\ I](I+\tilde{A})^{-1}\tilde{B} \\ \qquad\qquad + [0\ s_1 I](I+\tilde{A})^{-\top}\tilde{C}^\top U]\,, \\ \hat{C} = \left[\tilde{C}(I+\tilde{A})^{-1}\begin{bmatrix}0\\I\end{bmatrix}\Sigma_0 + s_1 U\tilde{B}^\top(I+\tilde{A})^{-\top}\begin{bmatrix}0\\I\end{bmatrix}\right](I+\hat{A}), \\ \hat{D} = \tilde{D} - \tilde{C}(I+\tilde{A})^{-1}\tilde{B} - s_1 U + \hat{C}(I+\hat{A})^{-1}\hat{B}, \end{cases} \tag{6.44}$$

in which

$$\begin{aligned} M =& s_1^2[0\ I](I+\tilde{A})^{-\top}\begin{bmatrix}0\\I\end{bmatrix} + \Sigma_0[0\ I](I+\tilde{A})^{-1}\begin{bmatrix}0\\I\end{bmatrix}\Sigma_0 \\ &+ s_1[0\ I](I+\tilde{A})^{-\top}\tilde{C}^\top U\tilde{B}^\top(I+\tilde{A})^{-\top}\begin{bmatrix}0\\I\end{bmatrix}, \end{aligned} \tag{6.45}$$

and U is a unitary matrix satisfying

$$\tilde{C}(I+\tilde{A})^{-1}\begin{bmatrix}I\\0\end{bmatrix} = -U\tilde{B}^\top(I+\tilde{A})^{-\top}\begin{bmatrix}I\\0\end{bmatrix}. \tag{6.46}$$

Note that since $\hat{C}(zI-\hat{A})^{-1}\hat{B}$ in (6.43) may not be strictly proper in general, the analytic part $F_a(z)$ in (6.43) is usually not given by the single constant matrix $\hat{D}$.

The above result can be considered as the matrix-valued version of the Nehari theorem.

Next, to state the matrix-valued version of the AAK theorem, we set, for $k = 0, 1, \cdots$,

$$\mathcal{G}_k = \{\ G(z) \in H^\infty_{p\times p,s}:\quad G(z) \text{ has McMillan degree } k\ \}. \tag{6.47}$$

Theorem 6.5. Let (A, B, C, D) be a stable minimal realization of a discrete-time MIMO linear system and $H(z)$ the corresponding rational transfer matrix of McMillan degree n. Also, let

$$s_1 \geq \cdots \geq s_k > s_{k+1} = \cdots = s_{k+r} > s_{k+r+1} \geq \cdots \geq s_n > 0 \tag{6.48}$$

be the s-numbers of $H(z)$. Then there exists a $\hat{G}(z) \in \mathcal{G}_k$ such that

$$\|H - \hat{G}\|_\Gamma = \inf_{G\in\mathcal{G}_k} \|H - G\|_\Gamma = s_{k+1}\,. \tag{6.49}$$

To describe $\hat{G}(z)$, assume that $H(z)$ has a balanced realization $(\tilde{A}, \tilde{B}, \tilde{C}, \tilde{D})$. Then the matrix-valued rational function $\hat{H}(z)$ to be defined in (6.51) satisfies

$$[H(z) - \hat{H}(z)][H(\bar{z}^{-1}) - \hat{H}(\bar{z}^{-1})]^* = s_{k+1}^2 I\,, \tag{6.50}$$

and $\hat{G}(z)$ can be formulated as the singular part of $\hat{H}(z)$, where

$$\hat{H}(z) = \hat{D} + \hat{C}(zI - \hat{A})^{-1}\hat{B} \tag{6.51}$$

with

$$\begin{cases} \hat{A} = [M - s_{k+1}^2 I]^{-1}[\Sigma_0^2 - M]\,, \\ \hat{B} = [M - s_{k+1}^2 I]^{-1}[\Sigma_0[0\ I](I + \tilde{A})^{-1}\tilde{B} \\ \qquad + [0\ s_{k+1}I](I + \tilde{A})^{-\top}\tilde{C}^\top U]\,, \\ \hat{C} = \left[\tilde{C}(I + \tilde{A})^{-1}\begin{bmatrix}0\\I\end{bmatrix}\Sigma_0 + s_{k+1}U\tilde{B}^\top(I + \tilde{A})^{-\top}\begin{bmatrix}0\\I\end{bmatrix}\right](I + \hat{A})\,, \\ \hat{D} = \tilde{D} - \tilde{C}(I + \tilde{A})^{-1}\tilde{B} - s_{k+1}U + \hat{C}(I + \hat{A})^{-1}\hat{B}\,, \end{cases} \tag{6.52}$$

in which Σ_0 is defined in (6.42),

$$\begin{aligned} M =& s_{k+1}^2[0\ I](I + \tilde{A})^{-\top}\begin{bmatrix}0\\I\end{bmatrix} + \Sigma_0[0\ I](I + \tilde{A})^{-1}\begin{bmatrix}0\\I\end{bmatrix}\Sigma_0 \\ &+ s_{k+1}[0\ I](I + \tilde{A})^{-\top}\tilde{C}^\top U\tilde{B}^\top(I + \tilde{A})^{-\top}\begin{bmatrix}0\\I\end{bmatrix}, \end{aligned} \tag{6.53}$$

and U is a unitary matrix satisfying

$$\tilde{C}(I + \tilde{A})^{-1}\begin{bmatrix}I\\0\end{bmatrix} = -U\tilde{B}^\top(I + \tilde{A})^{-\top}\begin{bmatrix}I\\0\end{bmatrix}. \tag{6.54}$$

Observe that for $k = 0$, with $s_0 := \infty$, Theorem 6.5 reduces to Theorem 6.4 (Problem 6.13). The characterization of $\hat{G}(z)$ was first given by Glover (1984) for continuous-time systems. We also remark that there is no simple closed-form formula for an optimal solution $\hat{G}(z)$ in this general MIMO setting. Moreover, the general statement of the result for MIMO systems is weaker than that for SISO systems. Note also that the family $\mathcal{G}_k$ defined in (6.47) is not necessarily equivalent to the set $G^{[k]}$ defined in (5.1) for SISO systems in general. Finally, since $\hat{C}(zI - \hat{A})^{-1}\hat{B}$ in (6.51) is usually not strictly proper, the singular part $\hat{G}(z)$ in (6.51) is in general not equal to $\hat{C}(zI - \hat{A})^{-1}\hat{B}$ itself.

A proof of this theorem will be given in the next section.

6.3.3 Derivation of Results

We now establish the results stated in Sect. 6.3.2. Our proof is constructive in the sense that an explicit formulation of an optimal solution will be derived and part of it is even in closed-form. However, as Theorem 6.4 is contained in Theorem 6.5, it will not be discussed.

Since (A, B, C, D) is a stable minimal realization, the system is both completely controllable and observable with $H(z) \in H^\infty_{p\times p,s}$. It follows from the analysis in Sect. 6.1 that the system has a balanced realization $(\tilde{A}, \tilde{B}, \tilde{C}, \tilde{D})$ for which the s-numbers of the corresponding Hankel matrix Γ_H are $\{s_i\}_{i=1}^\infty$ with $s_1 \geq s_2 \geq \cdots \geq s_n > s_{n+1} = \cdots = 0$, where n is the McMillan degree of the transfer matrix $H(z)$. Recall from (6.48) that

$$s_1 \geq \cdots \geq s_k > s_{k+1} = \cdots = s_{k+r} > s_{k+r+1} \geq \cdots \geq s_n > 0$$

for some integer $r \geq 0$. We first establish the following lemma:

Lemma 6.4. Let $H(z)$ be a $p\times p$ stable rational transfer matrix of McMillan degree n, associated with a minimal realization (A, B, C, D), for an MIMO linear system. Then

$$\|H - G\|_\Gamma \geq s_{k+1} \qquad \text{for all} \quad G \in \mathcal{G}_k\,, \tag{6.55}$$

where $\mathcal{G}_k$ is defined in (6.47).

To prove the lemma, let $G(z) \in \mathcal{G}_k$ be arbitrarily given with a minimal realization $(\underline{A}, \underline{B}, \underline{C}, \underline{D})$. Define an augmented system by

$$A_0 = \begin{bmatrix} A & 0 \\ 0 & \underline{A} \end{bmatrix}, \quad B_0 = \begin{bmatrix} B \\ \underline{B} \end{bmatrix}, \quad C_0 = [C \quad -\underline{C}], \quad D_0 = D - \underline{D}. \tag{6.56}$$

It follows that the transfer matrix of this augmented system is given by

$$H_0(z) = D_0 + C_0(zI - A_0)^{-1}B_0 = H(z) - G(z)\,, \tag{6.57}$$

where

$$H(z) = D + C(zI - A)^{-1}B$$

and

$$G(z) = \underline{D} + \underline{C}(zI - \underline{A})^{-1}\underline{B}\,.$$

Let P and Q be the corresponding controllability and observability Gramians, respectively, for the augmented system (6.56). Then P and Q are positive definite symmetric matrices, satisfying the following Lyapunov equations:

$$P - A_0 P A_0^\top = B_0 B_0^\top$$

and

$$Q - A_0^\top Q A_0 = C_0^\top C_0 .$$

Now, we decompose P and Q into blocks as

$$P = \begin{bmatrix} P_{11} & P_{12} \\ P_{12}^\top & P_{22} \end{bmatrix} \qquad \text{and} \qquad Q = \begin{bmatrix} Q_{11} & Q_{12} \\ Q_{12}^\top & Q_{22} \end{bmatrix}$$

according to the dimensions of the block structure of the augmented system (A_0, B_0, C_0, D_0). Note that P is symmetric and non-negative definite. Let $\tilde{P} := P^{1/2}$ be the positive square root of P with $P = \tilde{P}\tilde{P}^\top$ where

$$\tilde{P} = \begin{bmatrix} \tilde{P}_{11} & \tilde{P}_{12} \\ 0 & \tilde{P}_{22} \end{bmatrix} .$$

Then, it can be easily verified that

$$\tilde{P}_{11}\tilde{P}_{11}^\top = P_{11} - \tilde{P}_{12}\tilde{P}_{12}^\top .$$

Consequently, it follows from Theorem 6.2 that

$$\begin{aligned} \|H - G\|_\Gamma^2 &= \lambda_{\max}(PQ) \\ &= \lambda_{\max}(\tilde{P}^\top Q \tilde{P}) \\ &\geq \lambda_{\max}\left([\tilde{P}_{11}^\top \;\; 0] \begin{bmatrix} Q_{11} & Q_{12} \\ Q_{12}^\top & Q_{22} \end{bmatrix} \begin{bmatrix} \tilde{P}_{11} \\ 0 \end{bmatrix} \right) \\ &= \lambda_{\max}(\tilde{P}_{11}^\top Q_{11} \tilde{P}_{11}) \\ &= \lambda_{\max}(Q_{11} \tilde{P}_{11} \tilde{P}_{11}^\top) \\ &= \lambda_{\max}(Q_{11}(P_{11} - \tilde{P}_{12}\tilde{P}_{12}^\top)) \\ &= \lambda_{\max}(Q_{11}^{1/2} P_{11} (Q_{11}^{1/2})^\top - Q_{11}^{1/2} \tilde{P}_{12} \tilde{P}_{12}^\top (Q_{11}^{1/2})^\top) \\ &\geq s_{k+1}^2(H(z)) , \end{aligned} \tag{6.58}$$

where the last inequality requires some algebraic manipulation. We leave the details to the reader (Problem 6.14). This completes the proof of the lemma.

We now give a proof of the theorem. As mentioned above, since (A, B, C, D) is a stable minimal realization, the system is stable and both completely controllable and observable with $H(z) \in H^\infty_{p\times p,s}$. From the analysis given in Sect. 6.1, we know that the system has a balanced realization $(\tilde{A}, \tilde{B}, \tilde{C}, \tilde{D})$. Let

$$s_1 \geq \cdots \geq s_k > s_{k+1} = \cdots = s_{k+r} > s_{k+r+1} \geq \cdots \geq s_n$$

be the s-numbers of the Hankel matrix associated with the transfer matrix of the system, and set

$$\Sigma = \begin{bmatrix} s_{k+1} I_{r\times r} & 0 \\ 0 & \Sigma_0 \end{bmatrix} , \tag{6.59}$$

where $\Sigma_0 = \text{diag}\{s_1, \cdots, s_k, s_{k+r+1}, \cdots, s_n\}$. Then, by the result of balanced realization we have

$$\begin{cases} \Sigma - \tilde{A}\Sigma\tilde{A}^\top = \tilde{B}\tilde{B}^\top , \\ \Sigma - \tilde{A}^\top\Sigma\tilde{A} = \tilde{C}^\top\tilde{C} . \end{cases} \tag{6.60}$$

Let $\tilde{U}$ be any unitary matrix such that $\tilde{U}\tilde{A} = \tilde{A}^\top$, and decompose $\tilde{A}, \tilde{B}$, and $\tilde{C}$ as

$$\tilde{A} = \begin{bmatrix} A_{11} & A_{12} \\ A_{21} & A_{22} \end{bmatrix}, \quad \tilde{B} = \begin{bmatrix} B_1 \\ B_2 \end{bmatrix}, \quad \text{and} \quad \tilde{C} = [C_1 \quad C_2],$$

where A_{11}, B_1, and C_1 are respectively $r \times r, r \times p$, and $p \times r$ matrices. Then, it can be verified from (6.60) that there is a unitary matrix U such that $U^\top U = I$ and

$$\tilde{C}(I + \tilde{A})^{-1} \begin{bmatrix} I \\ 0 \end{bmatrix} = -U\tilde{B}^\top(I + \tilde{A})^{-\top} \begin{bmatrix} I \\ 0 \end{bmatrix}, \tag{6.61}$$

(Problem 6.15). With these notation, define a new linear system $(\hat{A}, \hat{B}, \hat{C}, \hat{D})$ by (6.52) in terms of $(\tilde{A}, \tilde{B}, \tilde{C}, \tilde{D})$ and let $(\check{A}, \check{B}, \check{C}, \check{D})$ be the augmented system defined by

$$\check{A} = \begin{bmatrix} \tilde{A} & 0 \\ 0 & \hat{A} \end{bmatrix}, \quad \check{B} = \begin{bmatrix} \tilde{B} \\ \hat{B} \end{bmatrix}, \quad \check{C} = [\tilde{C} \quad -\hat{C}], \quad \check{D} = \tilde{D} - \hat{D} . \tag{6.62}$$

Then, as has been seen in (6.57), it follows that

$$\check{H}(z) = H(z) - \hat{H}(z) \tag{6.63}$$

where

$$\hat{H}(z) = \hat{D} + \hat{C}(zI - \hat{A})^{-1}\hat{B} . \tag{6.64}$$

We first establish (6.49). Observe that Lemmas 6.3 and 6.4 together imply that

$$\|H - G - F\|_{\mathcal{L}^\infty_{p\times p}} \ge \|H - G\|_\Gamma \ge s_{k+1}$$

for all $G(z) \in \mathcal{G}_k$ and $F(z) \in H^\infty_{p\times p,a}$. We will show that the matrix-valued rational function $\check{H}(z) = H(z) - \hat{H}(z)$ given in (6.63) is all-pass with

$$\check{H}(z)\check{H}(\bar{z}^{-1})^* = [H(z) - \hat{H}(z)][H(\bar{z}^{-1}) - \hat{H}(\bar{z}^{-1})]^* = s_{k+1}^2 I ,$$

which implies that

$$\|H - \hat{H}\|_{\mathcal{L}^\infty_{q\times p}} = s_{k+1} . \tag{6.65}$$

We will also verify that in the decomposition $\hat{H}(z) = \hat{G}(z) + \hat{F}(z)$ with $\hat{G}(z) \in H^\infty_{p\times p,s}$ and $\hat{F}(z) \in H^\infty_{p\times p,a}$, the singular part $\hat{G}(z)$ of $\hat{H}(z)$ belongs to $\mathcal{G}_k$. Once these two facts are established, then we have

$$\|H - \hat{G} - \hat{F}\|_{\mathcal{L}^{\infty}_{p\times p}} = s_{k+1} ,$$

so that

$$s_{k+1} = \|H - \hat{G} - \hat{F}\|_{\mathcal{L}^{\infty}_{p\times p}} \geq \|H - \hat{G}\|_{\Gamma} \geq s_{k+1} .$$

This yields (6.49) with the optimal solution $\hat{G}(z) \in \mathcal{G}_k$ and completes the proof of the theorem.

To show that $\check{H}(z)$ is all-pass, we apply Theorem 6.3 by verifying the conditions (a-d) stated therein. This verification is, however, fairly long. Some technical skills in the following matrix calculations can be found in Gu, Tsai, O'Young, and Postlethwaite [1989].

Using the notation given in (6.59), we first define

$$\check{P} = \begin{bmatrix} s_{k+1}I & 0 & 0 \\ 0 & \Sigma_0 & I \\ 0 & I & \Sigma_0(\Sigma_0^2 - s_{k+1}^2 I)^{-1} \end{bmatrix} \tag{6.66}$$

and

$$\check{Q} = \begin{bmatrix} s_{k+1}I & 0 & 0 \\ 0 & \Sigma_0 & -(\Sigma_0^2 - s_{k+1}^2 I) \\ 0 & -(\Sigma_0^2 - s_{k+1}^2 I) & \Sigma_0(\Sigma_0^2 - s_{k+1}^2 I) \end{bmatrix} . \tag{6.67}$$

Then by noting that $\Sigma_0(\Sigma_0^2 - s_{k+1}^2 I) = (\Sigma_0^2 - s_{k+1}^2 I)\Sigma_0$, we immediately have

$$\check{P}\check{Q} = s_{k+1}^2 I$$

which yields condition (c) in Theorem 6.3.

We now verify condition (b), namely,

$$\check{Q} - \check{A}^{\top}\check{Q}\check{A} = \check{C}^{\top}\check{C} ,$$

or equivalently,

$$\begin{cases} \Sigma - \tilde{A}^{\top}\Sigma\tilde{A} = \tilde{C}^{\top}\tilde{C} , \\ \begin{bmatrix} 0 \\ \Sigma_0^2 - s_{k+1}^2 I \end{bmatrix} - \hat{A}^{\top}\begin{bmatrix} 0 \\ \Sigma_0^2 - s_{k+1}^2 I \end{bmatrix}\hat{A} = \tilde{C}^{\top}\hat{C} , \\ \Sigma_0(\Sigma_0^2 - s_{k+1}^2 I) - \hat{A}^{\top}\Sigma_0(\Sigma_0^2 - s_{k+1}^2 I)\hat{A} = \hat{C}^{\top}\hat{C} . \end{cases} \tag{6.68}$$

Here, the first identity certainly holds since $(\tilde{A}, \tilde{B}, \tilde{C}, \tilde{D})$ is a balanced realization, see (6.60). To show that the other two equalities also hold, we first observe that the first identity in (6.60) implies

$$\begin{aligned}
&(I+\tilde{A})^{-1}\tilde{B}\tilde{B}^\top(I+\tilde{A})^{-\top} \\
&=(I+\tilde{A})^{-1}(\Sigma-\tilde{A}\Sigma\tilde{A}^\top)(I+\tilde{A})^{-\top} \\
&=\frac{1}{2}(I+\tilde{A})^{-1}[(I-\tilde{A})\Sigma(I+\tilde{A})^\top+(I+\tilde{A})\Sigma(I-\tilde{A})^\top](I+\tilde{A})^{-\top} \\
&=\frac{1}{2}[(I+\tilde{A})^{-1}(I-\tilde{A})\Sigma+\Sigma(I-\tilde{A})^\top(I+\tilde{A})^{-\top}] \\
&:=\frac{1}{2}[G\Sigma+\Sigma G^\top]
\end{aligned} \tag{6.69}$$

and similarly the second identity in (6.60) implies

$$\begin{aligned}
&(I+\tilde{A})^{-\top}\tilde{C}^\top\tilde{C}(I+\tilde{A})^{-1} \\
&=(I+\tilde{A})^{-\top}(\Sigma-\tilde{A}^\top\Sigma\tilde{A})(I+\tilde{A})^{-1} \\
&:=\frac{1}{2}[G^\top\Sigma+\Sigma G],
\end{aligned} \tag{6.70}$$

where

$$G=(I+\tilde{A})^{-1}(I-\tilde{A}). \tag{6.71}$$

From (6.69) and (6.70), on one hand we can verify that a unitary matrix U satisfying (6.54) exists (Problem 6.15), and on the other hand, we can easily deduce the following four identities:

$$\begin{aligned}
&[I\ 0](I+\tilde{A})^{-1}\tilde{B}\tilde{B}^\top(I+\tilde{A})^{-\top}\begin{bmatrix}0\\I\end{bmatrix} \\
&=\frac{1}{2}\left[[I\ 0]G\begin{bmatrix}0\\I\end{bmatrix}\Sigma_0+s_{k+1}[I\ 0]G^\top\begin{bmatrix}0\\I\end{bmatrix}\right],
\end{aligned} \tag{6.72}$$

$$\begin{aligned}
&[I\ 0](I+\tilde{A})^{-\top}\tilde{C}^\top\tilde{C}(I+\tilde{A})^{-1}\begin{bmatrix}0\\I\end{bmatrix} \\
&=\frac{1}{2}\left[[I\ 0]G^\top\begin{bmatrix}0\\I\end{bmatrix}\Sigma_0+s_{k+1}[I\ 0]G\begin{bmatrix}0\\I\end{bmatrix}\right],
\end{aligned} \tag{6.73}$$

$$\begin{aligned}
&[0\ I](I+\tilde{A})^{-1}\tilde{B}\tilde{B}^\top(I+\tilde{A})^{-\top}\begin{bmatrix}0\\I\end{bmatrix} \\
&=\frac{1}{2}\left[[0\ I]G\begin{bmatrix}0\\I\end{bmatrix}\Sigma_0+\Sigma_0[0\ I]G^\top\begin{bmatrix}0\\I\end{bmatrix}\right],
\end{aligned} \tag{6.74}$$

and

$$\begin{aligned}
&[0\ I](I+\tilde{A})^{-\top}\tilde{C}^\top\tilde{C}(I+\tilde{A})^{-1}\begin{bmatrix}0\\I\end{bmatrix} \\
&=\frac{1}{2}\left[[0\ I]G^\top\begin{bmatrix}0\\I\end{bmatrix}\Sigma_0+\Sigma_0[0\ I]G\begin{bmatrix}0\\I\end{bmatrix}\right].
\end{aligned} \tag{6.75}$$

Now, it follows from (6.52), (6.54), (6.73), and (6.72) successively that

$$
\begin{aligned}
&[I\ 0](I+\tilde{A})^{-\top}\tilde{C}^\top\hat{C} \\
&= [I\ 0](I+\tilde{A})^{-\top}\tilde{C}^\top\Big[\tilde{C}(I+\tilde{A})^{-1}\begin{bmatrix}0\\I\end{bmatrix}\Sigma_0 \\
&\qquad + s_{k+1}U\tilde{B}^\top(I+\tilde{A})^{-\top}\begin{bmatrix}0\\I\end{bmatrix}\Big](I+\hat{A}) \\
&= \Big[[I\ 0](I+\tilde{A})^{-\top}\tilde{C}^\top\tilde{C}(I+\tilde{A})^{-1}\begin{bmatrix}0\\I\end{bmatrix}\Sigma_0 \\
&\qquad - s_{k+1}[I\ 0](I+\tilde{A})^{-1}\tilde{B}\tilde{B}^\top(I+\tilde{A})^{-\top}\begin{bmatrix}0\\I\end{bmatrix}\Big](I+\hat{A}) \\
&= \frac{1}{2}\Big[[I\ 0]G^\top\begin{bmatrix}0\\I\end{bmatrix}\Sigma_0^2 + s_{k+1}[I\ 0]G\begin{bmatrix}0\\I\end{bmatrix}\Sigma_0 \\
&\qquad - s_{k+1}[I\ 0]G\begin{bmatrix}0\\I\end{bmatrix}\Sigma_0 - s_{k+1}^2[I\ 0]G^\top\begin{bmatrix}0\\I\end{bmatrix}\Big](I+\hat{A}) \\
&= \frac{1}{2}[I\ 0]G^\top\begin{bmatrix}0\\ \Sigma_0^2 - s_{k+1}^2 I\end{bmatrix} + \frac{1}{2}[I\ 0]G^\top\begin{bmatrix}0\\ \Sigma_0^2 - s_{k+1}^2 I\end{bmatrix}\hat{A}\,.
\end{aligned}
\tag{6.76}
$$

Since

$$
G = (I+\tilde{A})^{-1}(I-\tilde{A}) = 2(I+\tilde{A})^{-1} - I
$$

and

$$
[I\ 0]\begin{bmatrix}0\\ \Sigma_0^2 - s_{k+1}^2\end{bmatrix} = 0\,,
$$

(6.76) can be rewritten as

$$
\begin{aligned}
&[I\ 0](I+\tilde{A})^{-\top}\tilde{C}^\top\hat{C} \\
&= [I\ 0](I+\tilde{A})^{-\top}\begin{bmatrix}0\\ \Sigma_0^2 - s_{k+1}^2 I\end{bmatrix} + [I\ 0](I+\tilde{A})^{-\top}\begin{bmatrix}0\\ \Sigma_0^2 - s_{k+1}^2 I\end{bmatrix}\hat{A}\,.
\end{aligned}
$$

Observe, furthermore, that

$$
(I+\tilde{A})^{-\top} = I - (I+\tilde{A})^{-\top}\tilde{A}^\top\,. \tag{6.77}
$$

Applying this identity to the second term of the above result, we obtain

$$
\begin{aligned}
&[I\ 0](I+\tilde{A})^{-\top}\tilde{C}^\top\hat{C} \\
&= [I\ 0](I+\tilde{A})^{-\top}\begin{bmatrix}0\\ \Sigma_0^2 - s_{k+1}^2 I\end{bmatrix} - [I\ 0](I+\tilde{A})^{-\top}\tilde{A}^\top\begin{bmatrix}0\\ \Sigma_0^2 - s_{k+1}^2 I\end{bmatrix}\hat{A} \\
&= [I\ 0](I+\tilde{A})^{-\top}\Big[\begin{bmatrix}0\\ \Sigma_0^2 - s_{k+1}^2 I\end{bmatrix} - \tilde{A}^\top\begin{bmatrix}0\\ \Sigma_0^2 - s_{k+1}^2 I\end{bmatrix}\hat{A}\Big]\,.
\end{aligned}
\tag{6.78}
$$

Similarly, using the formula for $\hat{C}$ in (6.52), (6.75) with $G = 2(I+\tilde{A})^{-1} - I$, the formula for the matrix M [see (6.53)], and especially the definition for the matrix $\hat{A}$ in (6.52), we also have

$$
\begin{aligned}
&[0\ I](I+\tilde{A})^{-\top}\hat{C}^\top\hat{C} \\
&= [0\ I](I+\tilde{A})^{-\top}\tilde{C}^\top\Big[\tilde{C}(I+\tilde{A})^{-1}\begin{bmatrix}0\\I\end{bmatrix}\Sigma_0 \\
&\quad + s_{k+1}U\tilde{B}^\top(I+\tilde{A})^{-\top}\begin{bmatrix}0\\I\end{bmatrix}\Big](I+\hat{A}) \\
&= \Big[[0\ I](I+\tilde{A})^{-\top}\tilde{C}^\top\tilde{C}(I+\tilde{A})^{-1}\begin{bmatrix}0\\I\end{bmatrix}\Sigma_0 \\
&\quad + s_{k+1}[0\ I](I+\tilde{A})^{-\top}\tilde{C}^\top U\tilde{B}^\top(I+\tilde{A})^{-\top}\begin{bmatrix}0\\I\end{bmatrix}\Big](I+\hat{A}) \\
&= \Big[\frac{1}{2}\Big[[0\ I]G^\top\begin{bmatrix}0\\I\end{bmatrix}\Sigma_0^2 + \Sigma_0[0\ I]G\begin{bmatrix}0\\I\end{bmatrix}\Sigma_0\Big] \\
&\quad + s_{k+1}[0\ I](I+\tilde{A})^{-\top}\tilde{C}^\top U\tilde{B}^\top(I+\tilde{A})^{-\top}\begin{bmatrix}0\\I\end{bmatrix}\Big](I+\hat{A}) \\
&= \Big[[0\ I](I+\tilde{A})^{-\top}\begin{bmatrix}0\\I\end{bmatrix}\Sigma_0^2 - \frac{1}{2}\Sigma_0^2 + \Sigma_0[0\ I](I+\tilde{A})^{-1}\begin{bmatrix}0\\I\end{bmatrix}\Sigma_0 - \frac{1}{2}\Sigma_0^2 \\
&\quad + s_{k+1}[0\ I](I+\tilde{A})^{-\top}\tilde{C}^\top U\tilde{B}^\top(I+A)^{-\top}\begin{bmatrix}0\\I\end{bmatrix}\Big](I+\hat{A}) \\
&= \Big[[0\ I](I+\tilde{A})^{-\top}\begin{bmatrix}0\\ \Sigma_0^2 - s_{k+1}^2 I\end{bmatrix} + M - \Sigma_0^2\Big](I+\hat{A}) \\
&= [0\ I](I+\tilde{A})^{-\top}\begin{bmatrix}0\\ \Sigma_0^2 - s_{k+1}^2 I\end{bmatrix} + M - \Sigma_0^2 \\
&\quad + \Big[[0\ I](I+\tilde{A})^{-\top}\begin{bmatrix}0\\ \Sigma_0^2 - s_{k+1}^2 I\end{bmatrix} + M - \Sigma_0^2\Big]\hat{A} \\
&= [0\ I](I+\tilde{A})^{-\top}\begin{bmatrix}0\\ \Sigma_0^2 - s_{k+1}^2 I\end{bmatrix} + [0\ I](I+\tilde{A})^{-\top}\begin{bmatrix}0\\ \Sigma_0^2 - s_{k+1}^2 I\end{bmatrix}\hat{A} \\
&\quad - (M - s_{k+1}^2 I + \Sigma_0^2 - M)\hat{A} \\
&= [0\ I](I+\tilde{A})^{-\top}\begin{bmatrix}0\\ \Sigma_0^2 - s_{k+1}^2 I\end{bmatrix} \\
&\quad + [0\ I]\Big[((I+\tilde{A})^{-\top} - I)\begin{bmatrix}0\\ \Sigma_0^2 - s_{k+1}^2 I\end{bmatrix}\hat{A}\Big] \\
&= [0\ I](I+\tilde{A})^{-\top}\Big[\begin{bmatrix}0\\ \Sigma_0^2 - s_{k+1}^2 I\end{bmatrix} - \tilde{A}^\top\begin{bmatrix}0\\ \Sigma_0^2 - s_{k+1}^2 I\end{bmatrix}\hat{A}\Big]. \qquad (6.79)
\end{aligned}
$$

Hence, combining (6.78) and (6.79), we have actually verified the second identity in (6.68).

To verify the last identity in (6.68), we first apply the following three simple identities

$$\hat{A}(I+\hat{A})^{-1} = I-(I+\hat{A})^{-1}\,,$$
$$[\Sigma_0^2 - s_{k+1}^2 I]\Sigma_0 = \Sigma_0[\Sigma_0^2 - s_{k+1}^2 I]\,,$$

and

$$(I+\hat{A}) = [M - s_{k+1}^2 I]^{-1}\big[\Sigma_0^2 - s_{k+1}^2 I\big]\,, \tag{6.80}$$

see(6.52), to obtain

$$\begin{aligned}
&(I+\hat{A})^{-\top}\big[\Sigma_0\big[\Sigma_0^2 - s_{k+1}^2 I\big] - \hat{A}^\top \Sigma_0\big[\Sigma_0^2 - s_{k+1}^2 I\big]\hat{A}\big](I+\hat{A})^{-1} \\
&= (I+\hat{A})^{-\top}\Sigma_0\big[\Sigma_0^2 - s_{k+1}^2 I\big](I+\hat{A})^{-1} \\
&\quad + [(I+\hat{A})^{-\top} - I]\Sigma_0\big[\Sigma_0^2 - s_{k+1}^2 I\big][I-(I+\hat{A})^{-1}] \\
&= (I+\hat{A})^{-\top}\big[\Sigma_0^2 - s_{k+1}^2 I\big]\Sigma_0 + \Sigma_0\big[\Sigma_0^2 - s_{k+1}^2 I\big](I+\hat{A})^{-1} \\
&\quad - \Sigma_0\big[\Sigma_0^2 - s_{k+1}^2 I\big] \\
&= [M^\top - s_{k+1}^2 I]\Sigma_0 + \Sigma_0[M - s_{k+1}^2 I] - \Sigma_0[\Sigma_0^2 - s_{k+1}^2 I] \\
&= M^\top\Sigma_0 + \Sigma_0 M - \big[\Sigma_0^2 + s_{k+1}^2 I\big]\Sigma_0\,.
\end{aligned} \tag{6.81}$$

We can then similarly verify that (Problem 6.16)

$$\begin{aligned}
&(I+\hat{A})^{-\top}\hat{C}^\top\hat{C}(I+\hat{A})^{-1} \\
&= M^\top\Sigma_0 + \Sigma_0 M - \big[\Sigma_0^2 + s_{k+1}^2 I\big]\Sigma_0\,.
\end{aligned} \tag{6.82}$$

Combining (6.81) and (6.82), we obtain the last identity in (6.68). Hence, condition (b) in Theorem 6.3 has been verified.

It is easily seen that condition (a) in Theorem 6.3 may be verified in the same manner as condition (b), whose verification was given above, where formulas (6.74) and (6.75) may be used. We leave this to the reader (Problem 6.17).

Now, what is left in this part of the proof is to verify condition (d) stated in Theorem 6.3, namely,

$$\check{D}^\top\check{D} + \check{B}^\top\check{Q}\check{B} = s_{k+1}^2 I\,.$$

To do so, we first establish two auxiliary equalities

$$(\tilde{D}-\hat{D})^\top\tilde{C} + \tilde{B}^\top\Sigma\tilde{A} - \hat{B}^\top\big[0 \quad \Sigma_0^2 - s_{k+1}^2 I\big]\tilde{A} = 0 \tag{6.83}$$

and

$$(\tilde{D}-\hat{D})^\top\hat{C} + \tilde{B}^\top\begin{bmatrix}0\\ \Sigma_0^2 - s_{k+1}^2 I\end{bmatrix}\hat{A} - \hat{B}^\top\Sigma_0\big[\Sigma_0^2 - s_{k+1}^2 I\big]\hat{A} = 0\,. \tag{6.84}$$

This can be accomplished as follows. From the definition of $\hat{D}$ in (6.52), the first two equalities in (6.68), and the identity (6.77), we have

$$
\begin{aligned}
&(\tilde D-\hat D)^\top\tilde C+\tilde B^\top\Sigma\tilde A-\hat B^\top[0\quad \Sigma_0^2-s_{k+1}^2I]\tilde A\\
&=[s_{k+1}U^\top+\tilde B^\top(I+\tilde A)^{-\top}\tilde C^\top-\hat B^\top(I+\hat A)^{-\top}\hat C^\top]\tilde C\\
&\quad+\tilde B^\top\Sigma\tilde A-\hat B^\top[0\quad \Sigma_0^2-s_{k+1}^2I]\tilde A\\
&=s_{k+1}U^\top\tilde C+\tilde B^\top[(I+\tilde A)^{-\top}\Sigma+[I-(I+\tilde A)^{-\top}\tilde A^\top]\Sigma\tilde A]\\
&\quad-\hat B^\top[(I+\hat A)^{-\top}[0\quad \Sigma_0^2-s_{k+1}^2I]\\
&\quad+[I-(I+\hat A)^{-\top}\hat A^\top][0\quad \Sigma_0^2-s_{k+1}^2I]\tilde A]\\
&=s_{k+1}U^\top\tilde C+\tilde B^\top(I+\tilde A)^{-\top}\Sigma(I+\tilde A)\\
&\quad-\hat B^\top(I+\hat A)^{-\top}[0\quad \Sigma_0^2-s_{k+1}^2I](I+\tilde A),
\end{aligned}
$$

where the identity

$$
(I+\hat A)^{-\top}=I-(I+\hat A)^{-\top}\hat A^\top,
$$

which is similar to (6.77) and can be directly verified, has also been used. Then, by the definition of $\hat B$ in (6.52) and by using (6.83), we also have

$$
\begin{aligned}
\hat B^\top(I+\hat A)^{-\top}=&\left[U^\top\tilde C(I+\tilde A)^{-1}\begin{bmatrix}0\\ s_{k+1}I\end{bmatrix}\right.\\
&\left.+\tilde B^\top(I+\tilde A)^{-\top}\begin{bmatrix}0\\ \Sigma_0\end{bmatrix}\right][\Sigma_0^2-s_{k+1}^2I]^{-1},
\end{aligned}
$$

so that the above result can be further written as

$$
\begin{aligned}
&(\tilde D-\hat D)^\top\tilde C+\tilde B^\top\Sigma\tilde A-\hat B^\top[0\quad \Sigma_0^2-s_{k+1}^2I]\tilde A\\
&=s_{k+1}U^\top\tilde C+\tilde B^\top(I+\tilde A)^{-\top}\Sigma(I+\tilde A)-\left[s_{k+1}U^\top\tilde C(I+\tilde A)^{-1}\begin{bmatrix}0\\ I\end{bmatrix}\right.\\
&\quad\left.+\tilde B^\top(I+\tilde A)^{-\top}\begin{bmatrix}0\\ \Sigma_0\end{bmatrix}\right][0\ \ I](I+\tilde A)\\
&=s_{k+1}U^\top\tilde C(I+\tilde A)^{-1}\begin{bmatrix}I&0\\0&0\end{bmatrix}(I+\tilde A)\\
&\quad+s_{k+1}\tilde B^\top(I+\tilde A)^{-\top}\begin{bmatrix}I&0\\0&0\end{bmatrix}(I+\tilde A)\\
&=\left[s_{k+1}U^\top\tilde C(I+\tilde A)^{-1}\begin{bmatrix}I\\0\end{bmatrix}+s_{k+1}\tilde B^\top(I+\tilde A)^{-\top}\begin{bmatrix}I\\0\end{bmatrix}\quad 0\right](I+\tilde A)\\
&=0
\end{aligned}
$$

where the last equality follows from (6.54). This gives (6.83). Equality (6.84) can be similarly verified as follows:

$$
\begin{aligned}
&(\tilde D - \hat D)^\top \hat C + \tilde B^\top \begin{bmatrix} 0 \\ \Sigma_0^2 - s_{k+1}^2 I \end{bmatrix} \hat A - \hat B \Sigma_0 [\Sigma_0^2 - s_{k+1}^2 I] \hat A \\
&= [s_{k+1} U^\top + \tilde B^\top (I+\tilde A)^{-\top} \tilde C^\top - \hat B^\top (I + \hat A)^{-\top} \hat C^\top] \hat C \\
&\quad + \tilde B^\top \begin{bmatrix} 0 \\ \Sigma_0^2 - s_{k+1}^2 I \end{bmatrix} \hat A - \hat B \Sigma_0 [\Sigma_0^2 - s_{k+1}^2 I] \hat A \\
&= s_{k+1} U^\top \hat C + \tilde B^\top \left[(I+\tilde A)^{-\top} \begin{bmatrix} 0 \\ \Sigma_0^2 - s_{k+1}^2 I \end{bmatrix} - [(I+\tilde A)^{-\top} \tilde A^\top - I] \right. \\
&\quad \times \left. \begin{bmatrix} 0 \\ \Sigma_0^2 - s_{k+1}^2 I \end{bmatrix} \hat A \right] - \hat B^\top [(I+\hat A)^{-\top} \Sigma_0 [\Sigma_0^2 - s_{k+1}^2 I] \\
&\quad - [(I+\hat A)^{-\top} \hat A^\top - I] \Sigma_0 [\Sigma_0^2 - s_{k+1}^2 I] \hat A] \\
&= s_{k+1} U^\top \hat C + \tilde B^\top (I+\tilde A)^{-\top} \begin{bmatrix} 0 \\ \Sigma_0^2 - s_{k+1}^2 I \end{bmatrix} (I + \hat A) \\
&\quad - \hat B^\top (I+\hat A)^{-\top} \Sigma_0 [\Sigma_0^2 - s_{k+1}^2 I] (I + \hat A) \\
&= \tilde B^\top (I+\tilde A)^{-\top} \left[\begin{bmatrix} 0 \\ \Sigma_0^2 - s_{k+1}^2 I \end{bmatrix} - \begin{bmatrix} 0 \\ \Sigma_0^2 - s_{k+1}^2 I \end{bmatrix} \right] (I + \hat A) \\
&\quad + \sigma_1 U^\top \tilde C (I+\tilde A)^{-1} \begin{bmatrix} 0 \\ \Sigma_0 \end{bmatrix} - \sigma_1^2 U^\top \tilde C (I+\tilde A)^{-1} \begin{bmatrix} 0 \\ \Sigma_0 \end{bmatrix} \\
&= 0 .
\end{aligned}
$$

Now, using (6.83) and (6.84), we conclude that

$$
\begin{aligned}
&\check D^\top \check D + \check B^\top \check Q \check B \\
&= \check D^\top [s_{k+1} U + \tilde C (I+\tilde A)^{-1} \tilde B - \hat C (I+\hat A)^{-1} \hat B] \\
&\quad + [\tilde B^\top \hat B^\top] \begin{bmatrix} s_1 I & 0 & 0 \\ 0 & \Sigma_0 & -(\Sigma_0^2 - s_{k+1}^2 I) \\ 0 & -(\Sigma_0^2 - s_{k+1}^2 I) & \Sigma_0 (\Sigma_0^2 - s_{k+1}^2 I) \end{bmatrix} \begin{bmatrix} \tilde B \\ \hat B \end{bmatrix} \\
&= s_{k+1} \check D^\top U - \left[\tilde B^\top \Sigma - \hat B^\top [0 \quad \Sigma_0^2 - s_{k+1}^2 I] \right] \tilde A (I+\tilde A)^{-1} \tilde B \\
&\quad + \left[\tilde B^\top \begin{bmatrix} 0 \\ \Sigma_0^2 - s_{k+1}^2 I \end{bmatrix} - \hat B^\top \Sigma_0 [\Sigma_0^2 - s_{k+1}^2 I] \right] \hat A (I+\hat A)^{-1} \hat B \\
&\quad + \tilde B^\top \Sigma \tilde B - \hat B^\top [0 \quad \Sigma_0^2 - s_{k+1}^2 I] \tilde B \\
&\quad - \tilde B^\top \begin{bmatrix} 0 \\ \Sigma_0^2 - s_{k+1}^2 I \end{bmatrix} \hat B + \hat B^\top \Sigma_0 [\Sigma_0^2 - s_{k+1}^2 I] \hat B \\
&= s_{k+1} \check D^\top U - \tilde B^\top \Sigma [\tilde A (I+\tilde A)^{-1} - I] \tilde B + \hat B^\top [0 \quad \Sigma_0^2 - s_{k+1}^2 I] \\
&\quad \times [\tilde A (I+\tilde A)^{-1} - I] \tilde B + \tilde B^\top \begin{bmatrix} 0 \\ \Sigma_0^2 - s_{k+1}^2 I \end{bmatrix} [\hat A (I+\hat A)^{-1} - I] \hat B \\
&\quad - \hat B^\top \Sigma_0 [\Sigma_0^2 - s_{k+1}^2 I] [\hat A (I+\hat A)^{-1} - I] \hat B
\end{aligned}
$$

$$= s_{k+1}\check{D}^\top U + \tilde{B}^\top \Sigma (I+\tilde{A})^{-1}\tilde{B} - \hat{B}^\top \begin{bmatrix} 0 & \Sigma_0^2 - s_{k+1}^2 I \end{bmatrix} (I+\tilde{A})^{-1}\tilde{B}$$
$$- \tilde{B}^\top \begin{bmatrix} 0 \\ \Sigma_0^2 - s_{k+1}^2 I \end{bmatrix} (I+\hat{A})^{-1}\hat{B} + \hat{B}^\top \Sigma_0 \left[\Sigma_0^2 - s_{k+1}^2 I\right](I+\hat{A})^{-1}\hat{B}$$
$$= s_{k+1}^2 U^\top U + \tilde{B}^\top \Bigg[s_{k+1}(I+\tilde{A})^{-\top}\tilde{C}^\top U + \Sigma (I+\tilde{A})^{-1}\tilde{B}$$
$$- \begin{bmatrix} 0 \\ \Sigma_0^2 - s_{k+1}^2 I \end{bmatrix} (I+\hat{A})^{-1}\hat{B} \Bigg] - \hat{B}^\top \big[s_{k+1}(I+\hat{A})^{-\top}\hat{C}^\top U$$
$$+ \begin{bmatrix} 0 & \Sigma_0^2 - s_{k+1}^2 I \end{bmatrix} (I+\tilde{A})^{-1}\tilde{B} - \Sigma_0 \left[\Sigma_0^2 - s_{k+1}^2 I\right](I+\hat{A})^{-1}\hat{B}\big] .$$

Then, by (6.80) and the definition of $\hat{B}$ given in (6.52), we have

$$(I+\hat{A})^{-1}\hat{B}$$
$$= \left[\Sigma_0^2 - s_{k+1}^2 I\right]^{-1} \left[\Sigma_0 [0 \ I](I+\tilde{A})^{-1}\tilde{B} + s_{k+1}[0 \ I](I+\tilde{A})^{-\top}\tilde{C}^\top U\right],$$

so that the above result can be further written as

$$\check{D}^\top \check{D} + \check{B}^\top \check{Q} \check{B}$$
$$= s_{k+1}^2 I + \tilde{B}^\top \Bigg[s_{k+1}(I+\tilde{A})^{-\top}\tilde{C}^\top U + \Sigma (I+\tilde{A})^{-1}\tilde{B}$$
$$- \begin{bmatrix} 0 \\ I \end{bmatrix} [0 \quad \Sigma_0](I+\tilde{A})^{-1}\tilde{B} - s_{k+1} \begin{bmatrix} 0 \\ I \end{bmatrix} [0 \ I](I+\tilde{A})^{-\top}\tilde{C}^\top U \Bigg]$$
$$- \hat{B}^\top \big[s_{k+1}(I+\hat{A})^{-\top}(I+\hat{A})^\top \big[[0 \ \Sigma_0](I+\tilde{A})^{-\top}\tilde{C}^\top$$
$$+ s_{k+1}[0 \ I](I+\tilde{A})^{-1}\tilde{B}U^\top\big]U + \begin{bmatrix} 0 & \Sigma_0^2 - s_{k+1}^2 I \end{bmatrix}(I+\tilde{A})^{-1}\tilde{B}$$
$$- \Sigma_0 \left[\Sigma_0 [0 \ I](I+\tilde{A})^{-1}\tilde{B} + s_{k+1}[0 \ I](I+\tilde{A})^{-\top}\tilde{C}^\top U\right]\big]$$
$$= s_{k+1}^2 I + \tilde{B}^\top \Bigg[s_{k+1}(I+\tilde{A})^{-\top}\tilde{C}^\top U + \begin{bmatrix} s_1 I & 0 \\ 0 & 0 \end{bmatrix} (I+\tilde{A})^{-1}\tilde{B}$$
$$\begin{bmatrix} 0 & 0 \\ 0 & s_1 I \end{bmatrix} (I+\tilde{A})^{-\top}\tilde{C}^\top U \Bigg] - \hat{B}^\top \big[[0 \quad s_{k+1}\Sigma_0](I+\tilde{A})^{-\top}\tilde{C}^\top$$
$$+ [0 \ s_{k+1}^2 I](I+\tilde{A})^{-1}\tilde{B} + \begin{bmatrix} 0 & \Sigma_0^2 - s_{k+1}^2 I \end{bmatrix}(I+\tilde{A})^{-1}\tilde{B}$$
$$- \begin{bmatrix} 0 & \Sigma_0^2 \end{bmatrix}(I+\tilde{A})^{-1}\tilde{B} - \begin{bmatrix} 0 & s_{k+1}\Sigma_0 \end{bmatrix}(I+\tilde{A})^{-\top}\tilde{C}U\big]$$
$$= s_{k+1}^2 I + \tilde{B}^\top \begin{bmatrix} s_{k+1} I & 0 \\ 0 & 0 \end{bmatrix} \left[(I+\tilde{A})^{-\top}\tilde{C}^\top U + (I+\tilde{A})^{-1}\tilde{B}\right]$$
$$= s_{k+1}^2 I ,$$

where the last equality follows from (6.54). This is condition (d) stated in Theorem 6.3.

Note, moreover, that the constant square matrix $\check{D}$ is nonsingular (Problem 6.18). Hence, all the conditions stated in Theorem 6.3 have been verified. In other words, we have actually shown that the transfer matrix $\check{H}(z)$ defined in (6.63) is all-pass.

To complete the proof of the theorem, we have to show further that the matrix $\hat{G}(z)$ defined by the singular part of (6.51) belongs to the set $\mathcal{G}_k$. In doing so, we first show that $\hat{G}(z)$ has no poles on the unit circle. For this purpose, it is sufficient to show that no eigenvalue of $\hat{A}$ lies on the unit circle.

Suppose that the matrix $\hat{A}$ has an eigenvalue $\lambda = e^{j\theta}$ on the unit circle, and let $\mathbf{x}$ be a corresponding (nonzero) eigenvector. From the last identity in (6.68), we first note that

$$
\begin{aligned}
&\bar{\mathbf{x}}^\top \hat{C}^\top \hat{C}\mathbf{x} \\
&= \bar{\mathbf{x}}^\top \Sigma_0(\Sigma_0^2 - s_{k+1}^2 I)\mathbf{x} - \bar{\lambda}\lambda\bar{\mathbf{x}}^\top \Sigma_0(\Sigma_0^2 - s_{k+1}^2 I)\mathbf{x} \\
&= 0,
\end{aligned}
$$

which implies that $\hat{C}\mathbf{x} = 0$. Then, from this and the second identity in (6.68), we have

$$
\begin{aligned}
&\begin{bmatrix} 0 \\ \Sigma_0^2 - s_{k+1}^2 I \end{bmatrix}\mathbf{x} - \lambda\tilde{A}^\top \begin{bmatrix} 0 \\ \Sigma_0^2 - s_{k+1}^2 I \end{bmatrix}\mathbf{x} \\
&= \left[\begin{bmatrix} 0 \\ \Sigma_0^2 - s_{k+1}^2 I \end{bmatrix} - \tilde{A}^\top \begin{bmatrix} 0 \\ \Sigma_0^2 - s_{k+1}^2 I \end{bmatrix}\hat{A} - \tilde{C}^\top \hat{C}\right]\mathbf{x} \\
&= 0,
\end{aligned}
$$

which implies that

$$
\tilde{A}^\top \begin{bmatrix} 0 \\ [\Sigma_0^2 - s_{k+1}^2 I]\mathbf{x} \end{bmatrix} = \bar{\lambda}\begin{bmatrix} 0 \\ [\Sigma_0^2 - s_{k+1}^2 I]\mathbf{x} \end{bmatrix}.
$$

Since $\Sigma_0^2 - s_{k+1}^2 I$ is either positive or negative definite, we have $[\Sigma_0^2 - s_{k+1}^2 I]\mathbf{x} \neq 0$, so that $\bar{\lambda}$ is an eigenvalue of the matrix $\tilde{A}^\top$. However, since $\tilde{A}$ is stable, it has no eigenvalue on the unit circle, and hence we have a contradiction. Therefore, the singular part of (6.51) is in $H^\infty_{p\times p,s}$ as claimed.

Then, based on this result, it can be further verified that $\hat{G}(z) \in \mathcal{G}_k$ (Problem 6.20). This completes the proof of Theorem 6.5.

Problems

Problem 6.1. Suppose that the matrix $I + A$ is invertible. Show that the conformal mapping $s = (z-1)/(z+1)$ takes the transfer matrix $H(z) = D + C(zI - A)^{-1}B$ of a linear system (A, B, C, D) to $\tilde{H}(s) = \tilde{D} + \tilde{C}(sI - \tilde{A})^{-1}\tilde{B}$, where

$$\begin{cases} \tilde{A} = (I+A)^{-1}(A-I), \\ \tilde{B} = \sqrt{2}(I+A)^{-1}B, \\ \tilde{C} = \sqrt{2}C(I+A)^{-1}, \\ \tilde{D} = D - C(I+A)^{-1}B\,. \end{cases}$$

Problem 6.2. Let (A, B, C, D) be a continuous-time MIMO linear system with the controllability and observability Gramians defined by

$$P = \int_0^\infty e^{At}BB^\top e^{A^\top t}dt$$

and

$$Q = \int_0^\infty e^{A^\top t}C^\top C e^{At}dt\,,$$

respectively. Show that P and Q satisfy the following Lyapunov equations:

$$\begin{cases} AP + PA^\top = -BB^\top\,, \\ A^\top Q + QA = -C^\top C\,. \end{cases}$$

Problem 6.3. Show that for both discrete-time and continuous-time settings, the controllability and observability Gramians of a stable and completely controllable and observable linear system are both symmetric and positive definite.

Problem 6.4. Consider the matrix Lyapunov equation

$$PX + XQ = R\,, \qquad (*)$$

where P, Q, R are $m \times n, n \times n, m \times n$ constant matrices, respectively, and X is an $n \times n$ unknown constant matrix to be determined. Let

$$M = \begin{bmatrix} Q & 0 \\ R & P \end{bmatrix}$$

and let $\{\mathbf{w}_k\}_{k=1}^{m+n}$ be a set of linearly independent eigenvectors of M. For each $k = 1, 2, \cdots, m+1$, decompose $\mathbf{w}_k$ as $\mathbf{w}_k = [\mathbf{u}_k^\top\ \mathbf{v}_k^\top]^\top$, where $\mathbf{u}_k$ and $\mathbf{v}_k$ are respectively $n \times 1$ and $m \times 1$ vectors with

$$\left\{ \begin{bmatrix} \mathbf{u}_1 \\ \mathbf{v}_1 \end{bmatrix}, \cdots, \begin{bmatrix} \mathbf{u}_n \\ \mathbf{v}_n \end{bmatrix} \right\}$$

being a set of eigenvectors of M associated with the eigenvalues of Q. Verify the following statements:

(a) Every solution of the Lyapunov equation $(*)$ has the form

$$X = VU^{-1},$$

where

$$U = [\mathbf{u}_1 \ \cdots \ \mathbf{u}_n] \qquad \text{and} \qquad V = [\mathbf{v}_1 \ \cdots \ \mathbf{v}_n].$$

Conversely, if U and V are two constant matrices having the above property, then the matrix $X = VU^{-1}$ is a solution of the Lyapunov equation $(*)$.

(b) The Lyapunov equation $(*)$ has a solution if and only if the matrices

$$\begin{bmatrix} Q & 0 \\ R & -P \end{bmatrix} \qquad \text{and} \qquad \begin{bmatrix} Q & 0 \\ 0 & -P \end{bmatrix}$$

are similar in the sense that there exists a nonsingular matrix T such that

$$T^{-1} \begin{bmatrix} Q & 0 \\ R & -P \end{bmatrix} T = \begin{bmatrix} Q & 0 \\ 0 & -P \end{bmatrix}.$$

Problem 6.5. Verify that a nonsingular linear mapping T: $\mathbf{x} \to \mathbf{y}$ will transform the linear system

$$\begin{cases} \dot{\mathbf{x}} = A\mathbf{x} + B\mathbf{u}, \\ \mathbf{v} = C\mathbf{x} + D\mathbf{u}, \end{cases}$$

to

$$\begin{cases} \dot{\mathbf{y}} = \tilde{A}\mathbf{y} + \tilde{B}\mathbf{u}, \\ \mathbf{v} = \tilde{C}\mathbf{y} + \tilde{D}\mathbf{u}, \end{cases}$$

with $\tilde{A} = TAT^{-1}, \tilde{B} = TB, \tilde{C} = CT^{-1}$, and $\tilde{D} = D$. Moreover, it transforms the controllability and observability Gramians P and Q to $\tilde{P} = TPT^{\top}$ and $\tilde{Q} = T^{-\top}QT^{-1}$, respectively. Verify that this is also true for discrete-time linear systems.

Problem 6.6. Let A, B, C, D be real matrices such that (A, B, C, D) gives a minimal realization of a stable MIMO linear system, and let $P, Q, H(z)$ be the controllability and observability Gramians, and the transfer matrix of the system, respectively. Denote by Γ_H the Hankel matrix associated with $H(z)$. Show that

(a) $\Gamma_H^2\mathbf{x} = \lambda\mathbf{x}$ $(\lambda \neq 0, \mathbf{x} \neq 0)$ implies that

$$PQ(\widetilde{M}_{AB}\mathbf{x}) = \lambda(\widetilde{M}_{AB}\mathbf{x}), \quad \text{and}$$

(b) $QP\mathbf{y} = \lambda\mathbf{y}$ $(\mathbf{y} \neq 0)$ implies that

$$\Gamma_H^2(\widetilde{M}_{AB}^{\top}\mathbf{y}) = \lambda(\widetilde{M}_{AB}^{\top}\mathbf{y}),$$

where $\widetilde{M}_{AB} = [B \; AB \; \cdots \; A^n B \; \cdots \,]$. Based on the above results, show furthermore that

(c) the nonzero eigenvalues of Γ_H^2 are the same as those of QP, and

(d) if the linear system has a balanced realization with $P = Q = \Sigma$, then there exist two unitary matrices U and V such that

$$\Gamma_H = U \Sigma V^\top .$$

(See Silverman and Bettayeb [1980, Lemma 2.1.])

Problem 6.7. Under what conditions on the constants a and b does the discrete-time linear system with

$$A = \begin{bmatrix} 0 & b \\ a & 0 \end{bmatrix}, \quad B = \begin{bmatrix} 1 \\ 0 \end{bmatrix}, \quad C = [1 \; 0] \quad \text{and} \quad D = 0$$

admit a balanced realization?

Problem 6.8. (a) Show that a nonsingular transformation T exists such that (6.16) is satisfied. (b) Show that this nonsingular transformation T is unique if the linear system is a minimal realization.

Problem 6.9. Verify the second equality in (6.17) by imitating the proof for the first equation. This completes the proof of Lemma 6.1.

Problem 6.10. Let $M(z) = N(z)/d(z)$ be an $m \times n$ rational matrix-valued function in coprime form, where $d(z)$ is the least common multiple of the denominators of the entries of $M(z)$ and $N(z)$ is a polynomial matrix. Show that there exist two polynomial matrices $U_1(z)$ and $U_2(z)$ such that

$$M(z) = U_1(z) \begin{bmatrix} n_1(z)/d_1(z) & \cdots & 0 & 0 \\ \vdots & \ddots & \vdots & \vdots \\ 0 & \cdots & n_{p'}(z)/d_{p'}(z) & 0 \\ 0 & \cdots & 0 & 0 \end{bmatrix} U_2(z),$$

where $p' \leq \min(m, n)$ and all the entries on the diagonal are in coprime form and $n_i(z)$ divides $n_j(z)$ for each $i \leq j$. Show, moreover, that $d_i(z)$ divides $d_j(z)$ for each $i \geq j$. (See Kailath [1980, p.443])

Problem 6.11. Show that the McMillan degree of a strictly proper rational block matrix-valued function is equal to the rank (or, the number of nonzero s-numbers) of its corresponding block-Hankel matrix.

Problem 6.12. Find the Smith-McMillan form of the matrix

$$M(z) = \begin{bmatrix} z/(z-1) & 0 & 0 \\ 0 & 1/(z-1) & 0 \\ 0 & 0 & (z-1)^2 \end{bmatrix},$$

and determine its McMillan degree. This example shows that the Smith-McMillan form of a diagonal matrix (even if it is in coprime form) is not necessarily itself, and moreover, that its McMillan degree is necessarily equal to the sum of the degrees of the denominators in its original diagonal form.

Problem 6.13. Verify that for $k = 0$ with $s_0 := \infty$ the set $\mathcal{G}_0$ defined by (6.47) consists of only analytic functions and Theorem 6.5 can be restated in terms of the analytic part of the matrix-valued rational function $F(z)$ shown in (6.51), so that the result reduces to Theorem 6.4.

Problem 6.14. Prove that if A is an $n \times m$ complex constant matrix with a singular value decomposition $A = UDV$, where U and V are respectively $n \times n$ and $m \times m$ unitary matrices and

$$D = \mathrm{diag}\{s_1, s_2, \cdots, s_r, 0, \cdots, 0\}$$

with $s_1 \geq s_2 \geq \cdots \geq s_r > 0$, then, for any $0 \leq k \leq r$ it follows that

$$\inf_{\mathrm{rank}(\hat{A}) \leq k} \|A - \hat{A}\|_s = \|\mathrm{diag}(s_{k+1}, s_{k+2}, \cdots, s_r)\|_s \,,$$

where $\|\cdot\|_s$ is the spectral norm of the matrix. Moreover, the infimum is attained at

$$\hat{A} = U \begin{bmatrix} s_1 & & \\ & \ddots & \\ & & s_k \end{bmatrix} V \,.$$

(See Mirsky [1960, Theorem 2] or Glover [1984, Lemma 7.1])

Problem 6.15. Let X and Y be two constant matrices such that $XX^* = YY^*$, where * denotes the complex conjugate as usual. Show that

(a) there exists a unitary matrix U such that $Y = XU$; and

(b) there exists a unitary matrix U such that the identity (6.54) is satisfied. (See Glover [1984, Lemma 3.5])

Problem 6.16. By imitating the proof of the identity (6.81), establish (6.82).

Problem 6.17. In the proof of Theorem 6.5, we need to show that the matrix-valued rational function $\check{H}(z)$ defined in (6.63) is all-pass. In doing so, we have already verified that the second condition stated in Theorem 6.2 is satisfied. Verify, moreover, that the first condition stated in Theorem 6.2 is also satisfied.

Problem 6.18. Verify that the square constant matrix $\check{D}$ defined in (6.62) is nonsingular.

Problem 6.19. Let $E(M)$ be the number of eigenvalues of a square constant matrix M that are located inside the unit circle and let $\hat{A}$ and Σ_0 be defined as in (6.52) and (6.42), respectively. Show that if the matrix $\Sigma_0^2[\Sigma_0^2 - s_{k+1}^2 I]$ has no eigenvalues on the unit circle $|z| = 1$ then $E(\hat{A}) = E(-\Sigma_0[\Sigma_0^2 - s_{k+1}^2 I]) = k$. (See Glover [1984, Theorem 6.3 (3b)])

Problem 6.20. Verify that the singular part $\hat{G}(z)$ of the matrix-valued rational function $\hat{H}(z)$ defined in (6.51) belongs to the set $\mathcal{G}_k$ shown in (6.47). This completes the proof of Theorem 6.5. (Apply Problem 6.19 and see also Glover [1984, Theorem 7.2 (2)])

References

Chapter 1

1.1 W. Rudin (1966): *Real and Complex Analysis* (McGraw-Hill, New York)

1.2 N.K. Bose (1985): *Digital Filters: Theory and Applications* (North-Holland, New York)

Chapter 2

2.1 C.K. Chui, G. Chen (1989): *Linear Systems and Optimal Control* (Springer-Verlag, New York)

2.2 T. Kailath (1980): *Linear Systems* (Pentice-Hall, Englewood Cliffs, NJ)

2.3 F.R. Gantmacher (1966): *The Theory of Matrices*, Vols.I,II (Chelsea, New York)

2.4 V.M. Adamjan, D.Z. Arov, M.G. Krein (1971): Analytic properties of Schmidt pairs for a Hankel operator and he generalized Schur-Takagi problem. Math. USSR Sbornik, **15**, 31

2.5 C.K. Chui, X. Li, J.D. Ward (1989): Rate of uniform convergence of rational functions corresponding to best approximants of truncated Hankel operators. Math. Contr. Sign. Syst., to appear

2.6 C.K. Chui, X. Li, J.D. Ward (1991): System reduction via truncated Hankel matrices. Math. Contr. Sign. Syst. **4**, 161

2.7 K. Glover, R.F. Curtain, J.R. Partington (1988): Realisation and approximation of linear infinite dimensional systems with error bounds. SIAM J. Contr. Optim. **26**, 863

2.8 E. Hayashi, L.N. Trefethen, M.H. Gutknecht (1990): The CF Table. Constr. Approx. Th. **6**, 195

2.9 A. Bultheel (1987): *Laurent Series and Their Padé Approximation* (Birkhäuser, Boston)

2.10 C.K. Chui, A.K. Chan (1982): Application of approximation theory methods to recursive digital filter design. IEEE Trans ASSP. **30**, 18

2.11 W.M.Wonham (1979): *Linear Multivariable Control: A Geometric Approach.* 2nd Ed (Springer-Verlag, New York)

2.12 M. Vidyasagar (1985): *Control System Synthesis: A Factorization Approach* (MIT Press, MA)

2.13 G. Zames (1981): Feedback and optimal sensitivity: model reference transformations, multiplicative seminorms, and approximate inverses. IEEE Trans. Auto. Contr. **16**, 301

Chapter 3

3.1 P.J. Duren (1970): *Theory of H^p Spaces* (Academic Press, New York)

3.2 K. Hoffman (1962): *Banach Space of Analytic Functions* (Prentice-Hall, Englewood Cliffs, NJ)

3.3 W. Rudin (1966): *Real and Complex Analysis* (McGraw-Hill, New York)

3.4 A.Zygmund (1968): *Trigonometric Series* (Cambridge Univ. Press, New York)

3.5 C.K. Chui, A.K. Chan (1982): Application of approximation theory methods to recursive digital filter design. IEEE Trans. ASSP. **30**, 18

3.6 G. Szegö (1967): *Dynamical Systems Stability Theory and Applications* (Springer-Verlag, New York)

3.7 J.L. Walsh (1960): *Approximation by Bounded Analytic Functions.* Mémorial des Sciences Math. Vol.144. Paris

3.8 E.A. Robinson (1967): *Statistical Communication and Detection* (Hafner, New York)

3.9 J.L. Shanks (1967): Recursion for digital processing. Geophysics. **32**, 33

3.10 C.K. Chui (1980): Approximation by double least-squares inverses. J. Math. Anal. Appl. **75**, 149

3.11 A.J. Macintyre, W.W. Rogosinski (1950): Extremal problems in the theory of analytic functions. Acta Math. **82**, 275

3.12 D. Sarason (1967): Generalized interpolation in H^∞. Trans. AMS. **127**, 179

3.13 C.K. Chui, X. Li, L. Zhong (1989): "On computation of minimum norm tangent interpolation", in *Approximation Theory VI: Vol.I*, C.K. Chui et al. Eds. (Academic Press, New York) p. 137-140

3.14 M. Rosenblum, J. Rovnyak (1985): *Hardy Classes and Operator Theory* (Oxford Univ. Press, New York)

Chapter 4

4.1 V.M. Adamjan, D.Z. Arov, M.G. Krein (1971): Analytic properties of Schmidt pairs for a Hankel operator and the generalized Schur-Takagi problem. Math. USSR Sbornik. **15**, 31

4.2 V.M. Adamjan, D.Z. Arov, M.G. Krein (1978): Infinite Hankel block matrices and related extension problems. Amer. Math. Soc. Transl. **111**, 133

4.3 Z. Nehari (1967): On bounded bilinear forms. Ann. Math. **65**, 153

4.4 I. Gohberg, M.G. Krein (1969): *Introduction to the Theory of Linear Non-selfadjoint Operators* (AMS Colloq. Pub., Providence, RI)

4.5 J.R. Partington (1988): *An Introduction to Hankel Operators* (Cambridge Univ. Press, England)

4.6 W. Rudin (1973): *Functional Analysis* (McGraw-Hill, New York)

4.7 J.H. Wilkinson (1965): *The Algebraic Eigenvalue Problem* (Oxford Univ. Press, New York)

4.8 S.Y. Kung (1980): "Optimal Hankel-norm model reductions: scalar systems", in Proc. of Joint Auto. Contr. Conf., paper FA8D

4.9 A.C. Desoer, M. Vidyasagar (1975): *Feedback Systems: Input-Output Properties* (Academic Press, New York)

4.10 M. Vidyasagar (1985): *Control System Synthesis: A Factorization Approach* (MIT Press, MA)

Chapter 5

5.1 V.M. Adamjan, D.Z. Arov, M.G. Krein (1971): Analytic properties of Schmidt pairs for a Hankel operator and the generalized Schur-Takagi problem. Math. USSR Sbornik. **15**, 31

5.2 J.B. Conway (1985): *Functional Analysis* (Springer-Verlag, New York)

Chapter 6

6.1 V.M. Adamjan, D.Z. Arov, M.G. Krein (1978): Infinite Hankel block matrices and related extension problems. Amer. Math. Soc. Transl. **111**, 133

6.2 S.Y. Kung, D.W. Lin (1981): Optimal Hankel norm model reductions: Multivariable systems. IEEE Trans. Auto. Contr. **26**, 832

6.3 K. Glover (1984): All optimal Hankel-norm approximations of linear multivariable systems and their L^{∞}-error bounds. Int. J. Control, **39**, 1115

6.4 A.J. Laub (1980): "Computation of 'balancing' transformations", in Proc. of Joint Auto. Contr. Conf. San Francisco, CA, paper FA8-E

6.5 B. Moore (1981): Principal component analysis in linear systems: controllability, observability, and model reduction. IEEE Trans. Auto. Contr. **26**, 17

6.6 L. Pernebo, L.M. Silverman (1979): "Balanced systems and model reduction", in Proc. of 18th IEEE Conf. Decis. Contr. Fort Lauderdale, FL. p. 865-867

6.7 B. McMillan (1952): Introduction to formal realization theory. Bell. Syst. Tech. J. **31**, 217 and 541

6.8 R.E. Kalman (1965): Irreducible realizations and the degree of a rational matrix. SIAM J. Appl. Math. **13**, 520

6.9 R.E. Kalman (1966): "On structural properties of linear constant multivariable systems", in Proc. of the 3rd IFAC Congress. London, paper 6A

6.10 D.W. Gu, M.C. Tsai, S.D. O'Young, I. Postlethwaite (1989): State-space formulae for discrete-time H^{∞} optimization. Int. J. Contr. **49**, 1683

6.11 L.M. Silverman, M. Bettayeb (1980): "Optimal approximation of linear systems", in Proc. of Joint Auto. Contr. Conf. San Francisco, CA, paper FA8-A

6.12 T. Kailath (1980): *Linear Systems* (Prentice-Hall, Englewood Cliffs, NJ)

6.13 L. Mirsky (1960): Symmetric gauge functions and unitarily invariant norms. Q.J. Math. **2**, 50

Further Reading

[1] V.M. Adamjan, D.Z. Arov, M.G. Krein (1968): Infinite Hankel matrices and generalized problems of Carathéodory-Fejér and F. Riesz. Func. Anal. Appl. 2, 1-18

[2] V.M. Adamjan, D.Z. Arov, M.G. Krein (1968): Infinite Hankel matrices and generalized Carathéodory-Fejér and I. Shur problems. Func. Anal. Appl. 3, 269-281

[3] V.M. Adamjan, D.Z. Arov, M.G. Krein (1971): Analytic properties of Schmidt pairs for a Hankel operator and the generalized Schur-Takagi problem. Math. USSR Sbornik. 15, 31-73

[4] V.M. Adamjan, D.Z. Arov, M.G. Krein (1978): Infinite Hankel block matrices and related extension problems, Amer. Math. Soc. Transl. 111, 133-156

[5] A.C. Allison, N.J. Young (1983): Numerical algorithms for the Nevanlinna-Pick problem. Numer. Math. 42, 125-145

[6] B.D.O. Anderson (1986): Weighted Hankel-norm approximation: Calculation of bounds. Syst. Contr. Lett. 7, 247-255

[7] B.D.O. Anderson, S. Vongpanitlerd (1973): *Network Analysis and Synthesis: A Modern System Theory Approach* (Prentice-Hall, Englewood Cliffs, NJ

[8] A.C. Antoulas, B.D.O. Anderson (1986): On the scalar rational interpolation problem. IMA J. Math. Contr. Inform. 3, 61-88

[9] A.C. Antoulas, B.D.O. Anderson (1989): On the problem of stable rational interpolation. L. Alg. Appl. 124, 301-329

[10] A.C. Antoulas, J.A. Ball, J. Kang, J.C. Willems (1990): On the solution of the minimal rational interpolation problem. L. Alg. Appl. 137/138, 511-573

[11] J.A. Ball (1983): Interpolation problems of Pick-Nevanlinna and Loewner types for meromorphic matrix functions. Integ. Eq. Oper. Th. 6, 804-840

[12] J.A. Ball (1988): "Nevanlinna-Pick interpolation: Generalizations and applications", in *Surveys of Some Recent Results in Operator Theory*: Vol.1. J.B. Conway, B.B. Morrel eds (Longman, London) p. 51-94

[13] J.A. Ball, N. Cohen (1987): Sensitivity minimization in an H^∞ norm: Parameterization of all suboptimal solutions. Int. J. Contr. 46, 785-816

[14] J.A. Ball, I. Gohberg, L. Rodman (1988): "Realization and interpolation of rational matrix functions", in *Topics in Interpolation Theory of Rational Matrix-Valued Functions*. I. Gohberg ed (Birkhäuser, Boston) p. 1-72

[15] J.A. Ball, I. Gohberg, L. Rodman (1991): *Interpolation of Rational Matrix Functions* (Birhäuser, Boston)

[16] J.A. Ball, J.W. Helton (1982): Lie group over the field of rational functions, signed spectral factorization, signed interpolation, and amplifier design. J. Oper. Th. 8, 19-64

[17] J.A. Ball, J.W. Helton (1983); A beurling-Lax theorem for the Lie group $U(m, n)$ which contain most classical interpolation. J. Oper. Th. 9, 107-142

[18] J.A. Ball, J.W. Helton (1986): Interpolation problems of Pick-Nevanlinna and Loewner types for meromorphic matrix functions: Parametrization of the set of all solutions. Integ. Eq. Oper. Th. 9, 155-203

[19] J.A. Ball, D.W. Luse (1987): "Sensitivity minimization as a Nevanlinna-Pick interpolation problem", in *Modelling, Robustness and Sensitivity Reduction in Control Systems* (Springer-Verlag, New York) p. 451-462

[20] J.A. Ball, A.C.M. Ran (1987): Optimal Hankel norm model reduction and Wiener-Hopf factorization I: The canonical case. SIAM J. Contr. Optim. 25, 362-382

[21] J.A. Ball, A.C.M. Ran (1987): Optimal Hankel norm model reduction and Wiener-Hopf factorization II: The non-canonical case. J. Oper. Th. 10, 416-436

[22] J.A. Ball, A.C.M. Ran (1987): Local inverse spectral problems for rational matrix functions, Integ. Eq. Oper. Th. 10, 349-415

[23] D.S. Bernstein, W.M. Haddad (1989): LQG control with an H^∞ performance bound: A Riccati equation approach. IEEE Trans. Auto. Contr. 34, 293-305

[24] P. Boekhoudt (1990): Homotopy solution of polynomial equations in H_∞ optimal control. Int. J. Contr. 51, 721-740

[25] S. Boyd, V. Balakrishnan, P. Kabamba (1989): On computing the H_∞ norm of a transfer matrix. Math. Contr. Sign. Syst. 2, 207-219

[26] B.C. Chang, S.S. Banda (1989): Optimal H^∞ norm computation for multivariable systems with multiple zeros. IEEE Trans. Auto. Contr. 34, 553-557

[27] B.C. Chang, J.B. Pearson (1984): Optimal disturbance rejection in linear multivariable systems. IEEE Trans. Auto. Contr. 29, 880-887

[28] G. Chen, C.K. Chui, Y. Yu (1991): "On Hankel-norm approximation of large-scale Markov chains", in Proc. of Amer. Contr. Conf. Boston. 2, p. 1664-1665

[29] G. Chen, C.K. Chui, Y. Yu (1991): Optimal Hankel-norm approximation approach to model reduction of large-scale Markov chains. Int. J. Syst. Sci. To appear.

[30] T. Chen, B.A. Francis (1990): On the L_2-induced norm of a sampled-data system. Syst. Contr. Letts. 15, 211-219

[31] T. Chen, B.A. Francis (1991): Input-output stability of sampled-data systems. IEEE Trans. Auto. Contr. 36, 50-58

[32] T. Chen, B.A. Francis (1991): H_2-optimal sampled-data control. IEEE Trans. Auto. Contr. 36, 387-397

[33] T. Chen, B.A. Francis (1991): Linear time-varying H_2-optimal control of sampled-data systems. Automatica. To appear

[34] C.C. Chu, J.C. Doyle, E.B. Lee (1986): The general distance problem in H^∞ optimal control theory. Int. J. Contr. 44, 565-596

[35] C.K. Chui, A.K. Chan (1979): A two-sided rational approximation method for recursive digital filtering. IEEE Trans. ASSP. 27, 141-145

[36] C.K. Chui, A.K. Chan (1982): Application of approximation theory methods to recursive digital filter design. IEEE Trans. ASSP. 30, 18-24

[37] C.K. Chui, G. Chen (1987): *Kalman Filtering with Real-Time Applications.* 2nd Ed (1991) (Springer-Verlag, New York)

[38] C.K. Chui, G. Chen (1989): *Linear Systems and Optimal Control* (Springer-Verlag, New York)

[39] C.K. Chui, X. Li, J.D. Ward (1989): Rate of uniform convergence of rational functions corresponding to best approximants of truncated Hankel operators. Math. Contr. Sign. Syst. To appear

[40] C.K. Chui, X. Li, J.D. Ward (1990): On the convergence rate of s-numbers of compact Hankel operators. Cir. Syst. Sign. Proc. To appear

[41] C.K. Chui, X. Li, J.D. Ward (1991): System reduction via truncated Hankel matrices. Math. Contr. Sign. Syst. 4, 161-175

[42] C.K. Chui, X. Li, L. Zhong (1989): "On computation of minimum norm tangent interpolation", in *Approximation Theory VI*: Vol.I. C.K. Chui et al. eds (Academic Press, New York) p. 137-140

[43] R.F. Curtain ed (1987): *Modeling, Robustness and Sensitivity Reduction in Control Systems* (Springer-Verlag, New York)

[44] R.F. Curtain, K. Glover (1986): Robust stabilization of infinite dimensional systems by finite dimensional controllers. Syst. Contr. Lett. 7, 41-47

[45] M.A. Dahleh, J.B. Pearson (1987): ℓ^1-optimal feedback controllers for MIMO discrete-time systems. IEEE Trans. Auto. Contr. 32, 314-322

[46] M.A. Dahleh, J.B. Pearson (1988): Optimal rejection of bounded disturbances, robust stability and mixed sensitivity minimization. IEEE Trans. Auto. Contr. 33, 722-731

[47] P. Delsarte, Y. Genin, Y. Kamp (1979): The Nevanlinna-Pick problem for matrix-valued functions. SIAM J. Appl. Math. 36, 47-61

[48] P. Delsarte, Y. Genin, Y. Kamp (1981): On the role of the Nevanlinna-Pick problem in circuit and system theory. Circ. Th. Appl. 9, 177-187

[49] A.C. Desoer, M. Vidyasagar (1975): *Feedback Systems: Input-Output Properties* (Academic Press, New York)

[50] P. Dorato, Y. Li (1986): A modification of the classical Nevanlinna-Pick interpolation algorithm with applications to robust stabilization. IEEE Trans. Auto. Contr. 31, 645-648

[51] P. Dorato, H.B. Park, Y. Li (1989): An algorithm for interpolation with units in H^∞, with applications to stabilizations. Automatica. 25, 427-430

[52] J.C. Doyle, K. Glover, P.P. Khargonekar, B.A. Francis (1989): State-space solutions to standard H_2 and H_∞ control problems. IEEE Trans. Auto. Contr. 34, 831-847

[53] F. Fagnani, D. Flamm, S.K. Mitter (1987): "Some min-max optimization problems in infinite-dimensional control systems", in *Modelling, Robustness and Sensitivity Reduction in Control Systems* (Springer-Verlag, New York) p. 399-414

[54] A. Feintuch, P.P. Khargonekar, A. Tannenbaum (1986): On the sensitivity minimization problem for linear time-varying periodic systems. SIAM J. Contr. Optim. 24, 1076-1085

[55] S.D. Fisher (1983): *Function Theory on Planar Domains: A Second Course in Complex Analysis* (Wiley, New York)

[56] D.S. Flamm,d S.K. Mitter (1987): H^∞ sensitivity minimization for delay systems. Syst. Contr. Lett. 9, 17-24

[57] C. Foias (1991): *The Commutant Lifting Approach to Interpolation Problem* (Birkhäuser, Boston)

[58] C. Foias, A. Tannenbaum (1987): On the Nehari problem for a certain class of L^∞-functions appearing in control theory. J. Func. Anal. 74, 146-159

[59] C. Foias, A. Tannenbaum (1988): Optimal sensitivity theory for multivariate distributed plants. Int. J. Contr. 47, 985-992

[60] C. Foias, A. Tannenbaum (1988): On the four-block problem: I and II. Integ. Eq. Oper. Th. 32, 93-112 and 726-767

[61] C. Foias, A. Tannenbaum, G. Zames (1986): Weighted sensitivity minimization for delay systems. IEEE Trans. Auto. Contr. 31, 763-766

[62] C. Foias, A. Tannenbaum, G. Zames (1986): On decoupling the H^∞-optimal sensitivity problem for products of plants. Syst. Contr. Lett. 7, 239-245

[63] C. Foias, A. Tannenbaum, G. Zames (1987): On the H^∞-optimal sensitivity problem for systems with delays. SIAM J. Contr. Optim. 25, 686-706

[64] C. Foias, A. Tannenbaum, G. Zames (1987): Sensitivity minimization for arbitrary SISO distributed plants. Syst. Contr. Lett. 8, 189-195

[65] C. Foias, A. Tannenbaum, G. Zames (1988): Some explicit formulae for the singular values of certain Hankel operators with factorizable symbol. SIAM J. Math. Anal. 19, 1081-1091

[66] Y.K. Foo, I. Postlethwaite (1984): An H^∞-minimax approach to the design of robust control systems. Syst. Contr. Lett. 5, 81-88

[67] Y.K. Foo, I. Postlethwaite (1986): An H^∞-minimax approach to the design of robust control systems, II: All solutions, all-pass form solutions and the "best" solution. Syst. Contr. Lett. 7, 261-268

[68] B.A. Francis (1987): *A Course in H^∞-Control Theory* (Springer-Verlag, New York)

[69] B.A. Francis (1987): "A guide to H^∞-control theory", in *Modelling, Robustness and Sensitivity Reduction in Control Systems* (Springer-Verlag, New York) p. 1-30

[70] B.A. Francis, J.C. Doyle (1987): Linear control theory with an H_∞ optimal criterion. SIAM J. Contr. Optim. 25, 815-844

[71] B.A. Francis, J.W. Helton, G. Zames (1984): "H^∞ -optimal feedback controllers for linear multivariable systems", in *Mathematical Theory of Networks and Systems* P.A. Fuhrmann ed (Springer-Verlag, New York) p. 347-362

[72] B.A. Francis, J.W. Helton, G. Zames (1984): H^∞-optimal feedback controllers for linear multivariable systems. IEEE Trans. Auto. Contr. 29, 888-900

[73] B.A. Francis, G. Zames (1984): On H^∞ -optimal sensitivity theory for SISO feedback systems. IEEE Trans. Auto. Contr. 29, 9-16

[74] J.S. Freudenberg, D.P. Looze (1986): An analysis of H^∞-optimization design methods. IEEE Trans. Auto. Contr. 31, 194-200

[75] P.A. Fuhrmann (1991): A polynomial approach to Hankel norm and balanced approximations. L. Alg. Appl. 146, 133-220

[76] C. Ganesh, J.B. Pearson (1989): H^2-optimization with stable controllers. Automatica. 25, 629-634

[77] J.B. Garnett (1981): *Bounded Analytic Functions* (Academic Press, New York)

[78] Y.V. Genin, S.Y. Kung (1981): A two-variable approach to the model reduction problem with Hankel norm criterion. IEEE Trans. Circ. Syst. 28, 912-924

[79] K. Glover (1984): All optimal Hankel-norm approximations of linear multivariable systems and their L^∞-error bounds. Int. J. Contr. 39, 1115-1193

[80] K. Glover (1986): Robust stabilization of linear multivariable systems: relations to approximation. Int. J. Contr. 43, 741-766

[81] K. Glover (1989): "A tutorial on Hankel-norm approximation", in *From Data to Modle* J.C. Willems ed (Springer-Verlag, New York) p. 26-48

[82] K. Glover, R.F. Curtain, J.R. Partington (1988): Realisation and approximation of linear infinite dimensional systems with error bounds. SIAM J. Contr. Optim. 26, 863-898

[83] K. Glover, J.C. Doyle (1988): State-space formulae for all stabilizing controllers that satisfy an H_∞-norm bound and relations to risk sensitivity. Syst. Contr. Lett. 11, 167-172

[84] K. Glover, J.C. Doyle (1989): "A state space approach to H_∞ optimal control", in *Three Decades of Mathematical System Theory* H. Nijmeijer, J.M. Schumacher eds (Springer-Verlag, New York) p. 179-218

[85] K. Glover, J. Lam, J.R. Partington (1990): Rational approximation of a class of infinite-dimensional systems I: Singular values of Hankel operators. Math. Contr. Sign. Syst. 3, 325-344

[86] K. Glover, J. Lam , J.R. Partington (1991): Rational approximation of a class of infinite-dimensional systems II: Optimal convergence rates of L^∞ approximants. Math. Contr. Sign. Syst. 4, 233-246

[87] K. Glover, J. Lam, J.R. Partington (1991): Rational approximation of a class of infinite-dimensional systems: The L_2 case. J. of Approx. Th. To appear

[88] K. Glover, J.R. Partington (1987): "Bounds on the achievable accuracy in model reduction", in *Modelling, Robustness and Sensitivity Reduction in Control Systems* (Springer-Verlag, New York) p. 95-118

[89] I. Gohberg, ed (1986): *I. Schur Methods in Operator Theory and Signal Processing* (Birkhäuser, Boston)

[90] I. Gohberg, ed (1988): *Topics in Interpolation Theory of Rational Matrix-Valued Functions* (Birkhäuser, Boston)

[91] I. Gohberg, M.A. Kaashoek, L. Lerer (1987): On minimality in the partial realization problem. Syst. Contr. Letts. 9, 97-104

[92] I. Gohberg, M.A. Kaashoek, A.C.M. Ran (1988): "Interpolation problems for rational matrix functions with incomplete data and Wiener-Hopf factorization", in *Operator Theory: Advances and Applications* Vol.35, I. Gohberg ed (Birkhäuser, Boston)

[93] I. Gohberg, P. Lancaster, L. Rodman (1986): *Invariant Subspaces of Matrices with Applications* (Wiley, New York)

[94] I. Gohberg, S. Rubinstein (1987): Cascade decompositions of rational matrix functions and their stability. Int. J. Contr. 46, 603-629

[95] L.B. Golinskii (1983): On one generalization of the matrix Nevanlinna-Pick problem. Izv. Akad. Nauk. Arm. SSR Math. 18, 187-205

[96] A. Gombani, M. Pavon (1985): On the Hankel-norm approximation of linear stochastic systems. Syst. Contr. Lett. 5, 283-288

[97] A. Gombani, M. Pavon (1990): A general Hankel-norm approximation scheme for linear recursive filtering. Automatica. 26, 103-112

[98] M.J. Grimble (1986): Optimal H_∞ robustness and the relationship to LQG design problems. Int. J. Contr. 43, 351-372

[99] M.J. Grimble(1987): H_∞ robust controller for self-tuning control applications, I: Controller design. II: Self-tuning and robustness. Int. J. Contr. 46, 1429-1444 and 1819-1840

[100] M.J. Grimble (1988): Optimal H_∞ multivariable robust controllers and the relationship to LQG design problems. Int. J. Contr. 1, 33-58

[101] M.J. Grimble (1989): Minimization of a combined H_∞ and LQG cost-function for a two-degree-of-freedom control design. Automatica. 25, 635-638

[102] M.J. Grimble (1990): Design procedure for combined H_∞ and LQG cost problem. Int. J. Syst. Sci. 21, 93-128

[103] M.J. Grimble (1991): H_∞ fixed-lag smoothing filter for scalar systems. IEEE Trans. Sign. Proc. 39, 1955-1963

[104] D.W. Gu, M.C. Tsai, I. Postlethwaite (1989): An algorithm for super-optimal H^∞ design: The two-block case. Automatica. 25, 215-222

[105] D.W. Gu, M.C. Tsai, S.D. O'Young, I. Postlethwaite (1989): State-space formulae for discrete-time H^∞ optimization. Int. J. Contr. 49, 1683-1724

[106] L. Guo, L. Xia, Y. Liu (1988): Recursive algorithm for the computation of the H^∞-norm of polynomials. IEEE Trans. Auto. Contr. 33, 1154-1157

[107] M.H. Gutknecht (1986): "Hankel norm approximation of power spectra", in *Computational and Combinatorial Methods in Systems Theory* C.I. Byrnes, A. Lindquist eds (North-Holland, Amsterdam) p. 315-326

[108] M.H. Gutknecht, J.O. Smith, L.N. Trefethen (1983): The Carathéodory-Fejér method for recursive digital filter design. IEEE Trans. ASSP. 31, 1417-1426

[109] W.M. Haddad, D.S. Bernstein (1989): Combined L_2/H_∞ model reduction, Int. J. Contr. 49, 1523-1536

[110] W.M. Haddad, D.S. Bernstein (1990): Generalized Riccati equations for the full- and reduced-order mixed-norm H_2/H_∞ standard problem. Syst. Contr. Lett. 14, 185-198

[111] W.M. Haddad, D.S. Bernstein (1990): "Optimal reduced-order subspace-observer design with a frequency-domain error bound", in *Control and Dynamic Systems: Advances in Theory and Applications* C.T. Leondes ed Vol.32 (Academic Press, New York) p. 23-38

[112] S.D. Hara, H. Katori (1986): On constrained H^∞ optimization problem for SISO systems. IEEE Trans. Auto. Contr. 31, 856-858

[113] A.J. Helmicki, C.A. Jacobson, C.N. Nett (1991): "Control-oriented modeling and identification of distributed parameter systems", in *New Trends and Applications of Distributed Parameter Control Systems* G. Chen ed (Marcel Dekker, New York)

[114] A.J. Helmicki, C.A. Jacobson, C.N. Nett (1991): Fundamentals of control-oriented system identification and their application for identification in H_∞. IEEE Trans. Auto. Contr. To appear

[115] J.W. Helton (1980): The distance of a function to H^∞ in the Poincare metric: Electrical power transfer. J. Func. Anal. 38, 273-314

[116] J.W. Helton (1985): Worst case analysis in the frequency domain: The H^∞ approach to control. IEEE Trans. Auto. Contr. 30, 1154-1170

[117] J.W. Helton (1986): *Optimization in Operator Theory, Analytic Function Theory, and Electrical Engineering* (SIAM Pub, Philadelphia)

[118] J.W. Helton, A. Sideris (1989): Frequency response algorithm for H_∞ optimization with time domain constraints. IEEE Trans. Auto. Contr. 34, 427-434

[119] G. Heining, K. Rost (1984): *Algebraic Methods for Toeplitz-Like Matrices and Operators* (Birkhäuser, Boston)

[120] C.L. Huang, B.S. Chen (1988): Adaptive control of optimal model matching in H^∞-norm space. IEE Proc. D. Contr. Th. Appl. 4, 295-301

[121] Y.S. Hung (1988): H^∞ interpolation of rational matrices. Int. J. Contr. 48, 1659-1713

[122] Y.S. Hung (1989): H^∞ optimal control, I: Model matching; II: Solution for controllers. Int. J. Contr. 49, 1291-1360

[123] Y.S. Hung, K. Glover (1986): Optimal Hankel-norm approximation of stable systems with first-order stable weighting functions. Syst. Contr. Lett. 7, 165-172

[124] Y.S. Hung, M.A. Muzlifah (1990): Hankel-norm model reduction with fixed modes. IEEE Trans. Auto. Contr. 35, 373-377

[125] H.S. Hvostov (1990): Simplified H_∞ controller synthesis via classical feedback system structure. IEEE Trans. Auto. Contr. 35, 485-488

[126] I.S. Iohvidov (1982): *Hankel and Toeplitz Matrices and Forms* (Birkhäuser, Boston)

[127] E.A. Jonckheere, J.C. Juang (1987): "Hankel and Toeplitz operators in linear quadratic and H^∞-designs", in *Modelling, Robustness and Sensitivity Reduction in Control Systems* (Springer-Verlag, New York) p. 323-356

[128] E.A. Jonckheere, R. Li (1987): L^∞ error bound for the phase matching approximation. Int. J. Contr. 4, 1343-1354

[129] E.A. Jonckheere, M.S. Verma (1986): A spectral characterization of H^∞-optimal feedback performance and its efficient computation. Syst. Contr. Lett. 8, 13-22

[130] J.C. Juang, E.A. Jonckheere (1988): Vector-valued versus scalar-valued figures of merit in H^∞-feedback system design. J. Math. Anal. Appl. 133, 331-354

[131] M. Khammash, J.B. Pearson (1990): Robust disturbance rejection in ℓ^1 optimal control problems. Syst. Contr. Lett. 14, 93-102

[132] P.P. Khargonekar, T.T. Georgiou, A.M. Pascoal (1987): On the robust stabilizability of linear time-invariant plants with unconstructed uncertainty. IEEE Trans. Auto. Contr. 32, 201-207

[133] P.P. Khargonekar, H. Özbay, A. Tannenbaum (1989): A remark on the four block problem: Stable plants and rational weights. Int. J. Contr. 50, 1013-1023

[134] P.P. Khargonekar, I.R. Petersen, M.A. Rotea (1988): H_∞ optimal control with state-feedback. IEEE Trans. Auto. Contr. 33, 786-788

[135] P.P. Khargonekar, I.R. Petersen, K. Zhou (1990): Robust stabilization of uncertain linear systems: Quadratic stabilizability and H^∞ control theory. IEEE Trans. Auto. Contr. 35, 356-361

[136] P.P. Khargonekar, A. Tannenbaum (1985): Noneuclidean metrics and the robust stabilization of systems with parameter uncertainty. IEEE Trans. Auto. Contr. 30, 1088-1096

[137] H. Kimura (1984): Robust stabilizability for a class of transfer functions. IEEE Trans. Auto. Contr. 29, 788-793

[138] H. Kimura (1986): On interpolation-minimization problem in H^∞. Control: Th. Adv. Tech. 2, 1-25

[139] H. Kimura (1987): Directional interpolation approach to H^∞-optimization and robust stabilization. IEEE Trans. Auto. Contr. 32, 1085-1093

[140] V.C. Klema, A.J. Laub (1980): The singular value decomposition: Its computation and some applications. IEEE Trans. Auto. Contr. 25, 164-176

[141] C.K. Koc, G. Chen (1990): A fast algorithm for scalar Nevanlinna-Pick interpolation. Numer. Math. To appear.

[142] C.K. Koc, G. Chen (1991): Parallel algorithms for Nevanlinna-Pick interpolation: the scalar case. Int. J. Comp. Math. 40, 99-115

[143] J.M. Krause, P.P. Khargonekar, G. Stein (1990): Robust parameter adjustment with nonparametric weighted-ball-in-H^∞ uncertainty. IEEE Trans. Auto. Contr. 35, 225-229

[144] H. Kwakernaak (1985): Minimax frequency domain performance and robustness optimization of linear feedbk systems. IEEE Trans. Auto. Contr. 30, 994-1004

[145] H. Kwakernaak (1986): A polynomial approach to minimax frequency domain optimization of multivariable feedback systems. Int. J. Contr. 31, 117-156

[146] H. Kwakernaak (1987): "A polynomial approach to H^∞-optimization of control systems", in *Modelling, Robustness and Sensitivity Reduction in Control Systems* (Springer-Verlg, New York) p. 83-94

[147] S.Y. Kung (1980): "Optimal Hankel-norm model reductions: scalar systems", in Proc. of Joint Auto. Contr. Conf. paper FA8D

[148] S.Y. Kung, D.W. Lin (1981): Optimal Hankel norm model reductions: Multivariable systems. IEEE Trans. Auto. Contr. 26, 832-852

[149] S.Y. Kung, H.J. Whitehouse, T. Kailath, ed (1985): *VLSI and Modern Signal Processing* (Prentice-Hall, Englewood Cliffs, NJ)

[150] G.A. Latham, B.D.O. Anderson (1985): Frequency-weighted optimal Hankel-norm approximation of stable transfer functions. Syst. Contr. Lett. 5, 229-236

[151] K.E. Lenz, P.P. Khargonekar, J.C. Doyle (1988): When is a controller H^∞-optimal? Math. Contr. Sign. Syst. 1, 107-122

[152] X. Li (1991): Hankel approximation and its applications. Ph.D. thesis. Texas A & M Univ.

[153] D.J.N. Limebeer, B.D.O. Anderson (1988): An interpolation theory approach to H^∞ Controller design bounds. L. Alg. Appl. 93, 347-386

[154] D.J.N. Limebeer, Y.S. Hung (1987): An analysis of the pole-zero cancellation in H^∞-optimal control problems of the first kind. SIAM Contr. Optim. 25, 1457-1492

[155] D.W. Lin, S.Y. Kung (1982): Optimal Hankel-norm approximation of continuous-time systems. Circ. Syst. Sig. Proc. 1, 407-431

[156] Y. Liu, B.D.O. Anderson (1987): Model reduction with time delay. IEE Proc. D. 6, 349-367

[157] K.Z. Liu, T. Mita, R. Kawatani (1990): Parameterization of state feedback H^∞ controllers. Int. J. Contr. 51, 535-552

[158] D.W. Luse, J.A. Ball (1989): Frequency-scale decomposition of H^∞-disk problems. SIAM J. Contr. Optim. 27, 814-835

[159] T.A. Lypchuk, M.C. Smith, A. Tannenbaum (1988): Weighted sensitivity minimization: general plants in H^∞ and rational weights. L. Alg. Appl. 109, 71-90

[160] D.C. McFarlane, K. Glover (1990): *Robust Controller Design Using Normalized Coprime Factor Plant Description* (Springer-Verlag, New York)

[161] J. Meinguet (1983): "A simplified presentation of the Adamjan-Arov-Krein approximation theory", in *Computational Aspects of Complex Analysis* H. Werner et al eds (Reidel, Boston) p. 217-248

[162] J. Meinguet (1986): " On the Glover concretization of the Adamjan-Arov-Krein approximation theory", in *Modelling, Identification and Robust Control* C.I. Byrnes, A. Lindquist eds (Elsevier, Amsterdam) p. 325-334

[163] J. Meinguet (1988): "Once again: the Adamjan-Arov-Krein approximation theory", in *Nonlinear Numerical Methods and Rational Approximation* A. Cuyt ed (Reidel, Boston) p. 77-91

[164] D.G. Meyer (1988): Two properties of ℓ_1-optimal controllers. IEEE Trans. Auto. Contr. 9, 876-878

[165] B. Moore (1981): Principal component analysis in linear systems: controllability, observability, and model reduction. IEEE Trans. Auto. Contr. 26, 17-32

[166] J.B. Moore, T.T. Tay (1989): Loop recovery via H^∞/H^2 sensitivity recovery. Int. J. Contr. 49, 1249-1272

[167] D. Mustafa (1989): Relations between maximum-entropy/H_∞ control and combined H_∞/LQG control. Syst. Contr. Lett. 12, 193-204

[168] N.K. Nikol'skii (1986): *Treatise on the Shift Operators* (Springer-Verlag, New York)

[169] A.V. Oppenheim, R.W. Schafer (1975): *Digital Signal Processing* (Prentice-Hall, Englewood Cliffs, NJ)

[170] A.M. Ostrowski, H. Schneider (1962): Some theorems on the inertia of general matrices. J. Math. Anal. Appl. 4, 72-84

[171] H. Özbay, A. Tannenbaum (1990): A skew Toeplitz approach to H^∞ optimal control of multivariable distributed systems. SIAM J. Contr. Optim. 28, 653-670

[172] J.R. Partington (1988): *An Introduction to Hankel Operators* (Cambridge Univ. Press, England)

[173] J.R. Partington (1991): "The L_2 approximation of infinite-dimensional systems", in Proc. IMA Conf., Oxford University Press. To appear

[174] J.R. Partington (1991): Approximation of delay systems by Fourier-Laguerre series. Automatica. 27, 569-572

[175] J.R. Partington (1991): Approximation of unstable infinite-dimensional systems using coprime factors. Syst. Contr. Lett. 16, 89-96

[176] J.R. Partington (1991): Robust identification in H^∞. J. Math. Anal. Appl. To appear

[177] J.R. Partington and K. Glover (1990): "Robust stabilization of delay systems", in *Robust Control of Linear Systems and Nonlinear Control* M.A. Kaashoek et al eds (Birkhäuser, Boston) p. 609-616

[178] J.R. Partington and K. Glover (1990): Robust stabilization of delay systems by approximation of coprime factors. Syst. Contr. Lett. 14, 325-331

[179] J.R. Partington, K. Glover, H.J. Zwart, R.F. Curtain (1988): L^∞ approximation and nuclearity of delay systems. Syst. Contr. Lett. 10, 59-65

[180] J.B. Pearson, M.A. Dahleh (1987): "Control system design to minimize maximum errors", in *Modelling, Robustness and Sensitivity Reduction in Control Systems* (Springer-Verlag, New York) p. 415-424

[181] L. Pernebo, L.M. Silverman (1982): Model reduction via balanced state representation. IEEE Trans. Auto. Contr. 27, 382-387

[182] I.R. Petersen (1987): Disturbance attenuation and H^∞ optimization: A design method based on the algebraic Riccati equation. IEEE Trans. Auto. Contr. 32, 427-429

[183] A.C.M. Ran (1987): "Hankel norm approximation for infinite dimensional systems and Wiener-Hopf factorization", in *Modelling, Robustness and Sensitivity Reduction in Control Systems* (Springer-Verlag, New York) p. 57-70

[184] M. Rosenblum, J. Rovnyak (1985): *Hardy Classes and Operator Theory* (Oxford Univ. Press, New York)

[185] M. Saeki, M.J. Grimble, E. Korngoor, M.A. Johnson (1987): "H^∞-optimal control, LQG polynomial systems techniques and numerical solution procedures", in *Modelling, Robustness and Sensitivity Reduction in Control Systems* (Springer-Verlag, New York) p. 357-380

[186] M.G. Safonov (1980): *Stability Robustness of Multivariable Feedback Systems* (MIT Press, MA)

[187] M.G. Safonov (1986): Optimal diagonal scaling for infinity-norm optimization. Syst. Contr. Lett. 7, 257-260

[188] M.G. Safonov (1987): "Imaginary-axis zeros in multivariable H^∞-optimal control", in *Modelling, Robustness and Sensitivity Reduction in Control Systems* (Springer-Verlag, New York) p. 71-82

[189] M.G. Safonov, R.Y. Chiang, D.J.N. Limebeer (1990): Optimal Hankel model reduction for nonminimal systems. IEEE Trans. Auto. Contr. 35, 496-502

[190] M.G. Safonov, V.X. Le (1988): An alternative solution to the H^∞-optimal control problem. Syst. Contr. Lett. 10, 155-158

[191] M.G. Safonov, M.S. Verma (1985): L^∞ optimization and Hankel approximation. IEEE Trans. Auto. Contr. 30, 279-280

[192] M.G. Safonov, E.A. Jonckheere, M.S. Verma, D.J.N. Limebeer (1987): Synthesis of positive real multivariable feedback systems. Int. J. Contr. 45, 817-842

[193] M. Sampi, T. Mita, M. Nakamichi (1990): An algebraic approach to H_∞ output feedback control problems. Syst. Contr. Lett. 14, 13-24

[194] D. Sarason (1985): "Operator-theoretic aspects of the Nevanlinna-Pick interpolation problem", in *Operators and Function Theory* S.C. Power ed (Reidel, Dordrecht, Holland) p. 279-314

[195] A. Sideris(1990): H_∞ optimal control as a weighted Wiener-Hopf problem. IEEE Trans. Auto. Contr. 35, 361-366

[196] M.C. Smith (1990): Well-posedness of H^∞ optimal control problem. SIAM J. Contr. Optim. 28, 342-358

[197] G. Szegö (1967): *Dynamical Systems Stability Theory and Applications* (Springer-Verlag, New York)

[198] B. Sz-Nagy, C. Foias (1970): *Harmonic Analysis of Operators on Hilbert Space* (North-Holland, Amsterdam)

[199] G. Tadmor (1987): An interpolation problem associated with H^∞-optimal design in systems with distributed input lags. Syst. Contr. Lett. 8, 313-319

[200] G. Tadmor (1989): H^∞ interpolation in systems with commensurate input lags. SIAM J. Contr. Optim. 27, 511-526

[201] A. Tannenbaum (1982): Modified Nevanlinna-Pick interpolation of linear plants with uncertainty in the gain factor. Int. J. Contr. 36, 331-336

[202] L.N. Trefethen (1981): Rational Chebyshev approximation on the unit disk. Numer. Math. 37, 297-320

[203] S.R. Treil (1986): Vector variant of the Adamyam-Arov-Krein theorem. Func. Anal. Appl. 20, 74-76

[204] M.C. Tsai, D.W. Gu, I. Postlethwaite (1988): A state-space approach to super-optimal H^∞ control problems. IEEE Trans. Auto. Contr. 9, 833-843

[205] M.C. Tsai, D.W. Gu, I. Postlethwaite, B.D.O. Anderson (1990): Inner functions and pseudo-singular-value decomposition in super-optimal H^∞ control. Int. J. Contr. 51, 1119-1132

[206] K. Uchida, F. Matsumura (1990): Gain perturbation tolerance in H^∞ state feedback control. Int. J. Contr. 51, 315-328

[207] J. Vandewalle, L. Vandenberghe, M. Moonen (1989): "The impact of the singular value decomposition in system theory, signal processing, and circuit theory", in *Three Decades of Mathematical System Theory* H. Nijmeijer, J.M. Schumacher eds (Springer-Verlag, New York) p. 453-479

[208] M.S. Verma (1989): Synthesis of H^∞-optimal linear feedback systems, Math. Syst. Th. 21, 165-186

[209] M.S. Verma, E. Jonckheere (1984): L^∞ compensation with mixed sensitivity as a broadband matching problem. Syst. Contr. Lett. 4, 125-129

[210] M.S. Verma, J.C. Romig (1989): Reduced-order controllers in H^∞-optimal synthesis methods of the first kind. Math. Syst. Th. 22, 109-148

[211] M. Vidyasagar (1985): *Control System Synthesis: A Factorization Approach* (MIT Press, MA)

[212] M. Vidyasagar (1986): Optimal rejection of persistent bounded disturbances. IEEE Trans. Auto. Contr. 31, 527-534

[213] M. Vidyasagar, H. Kimura (1986): Robust controllers for uncertain linear multivariable systems. Automatica. 22, 85-94

[214] J.L. Walsh (1935): *Interpolation and Approximation by Rational Functions in the Complex Domain* (AMS Colloq. Pub., New York)
[215] J.H. Xu, M. Mansour (1988): H^∞−optimal robust regulation of MIMO systems. Int. J. Contr. 48, 1327-1341
[216] F.B. Yeh, T.S. Hwang (1988): Optimal sensitivity bound estimation and controller design. Int. J. Contr. 47, 979-984
[217] F.B. Yeh, T.S. Hwang (1989): A modification of the Nevanlinna algorithm on the solution of the $H^\infty_{m\times}$−optimization problem. IEEE Trans. Auto. Contr. 34, 627-629
[218] F.B. Yeh, C.D. Yang (1987): Singular value decomposition of an infinite block-Hankel matrix and its applications. Numer. Func. Anal. Optim. 9, 881-916
[219] F.B. Yeh, C.D. Yang (1987): A new algorithm for the minimal balanced realization of a transfer function matrix. Syst. Contr. Lett. 9, 25-31
[220] F.B. Yeh, C.D. Yang, M.C. Tsai (1988): Unified approach to H^∞-optimization, Hankel approximation and balanced realization problems. Int. J. Contr. 47, 967-978
[221] N.J. Young (1983): The singular value decomposition of an infinite Hankel matrix. L. Alg. Appl. 50, 639-656
[222] N.J. Young (1985): "Interpolation by analytic matrix functions", in *Operators and Function Theory* S.C. Power ed (Reidel, Boston) p. 351-383
[223] N.J. Young (1986): The Nevanlinna-Pick problem for matrix-valued functions. J. Oper. Th. 15, 239-265
[224] N.J. Young (1987): "Super-optimal Hankel norm approximations", in *Modelling, Robustness and Sensitivity Reduction in Control Systems* (Springer-Verlag, New York) p. 47-56
[225] N.J. Young (1990): A polynomial method for the singular value decomposition of block Hankel operators. Syst. Contr. Lett. 14, 103-112
[226] N.J. Young (1990): *An Introduction to Hilbert Space* (Cambridge Univ. Press, New York)
[227] S.D. O'Young, B.A. Francis (1985): Sensitivity tradeoffs for multivariable plants. IEEE Trans. Auto. Contr. 30, 625-632
[228] S.D. O'Young, I. Postlethwaite, D.W. Gu (1989): A treatment of $j\omega$-axis model-matching transformation zeros in the optimal H^2 and H^∞ control designs. IEEE Trans. Auto. Contr. 34, 551-553
[229] G. Zames (1981): Feedback and optimal sensitivity: model reference transformations, multiplicative seminorms, and approximate inverses. IEEE Trans. Auto. Contr. 16, 301-320
[230] G. Zames, B.A. Francis (1983): Feedback, minimax sensitivity, and optimal robustness. IEEE Trans. Auto. Contr. 28, 585-600
[231] K. Zhou, P.P. Khargonekar (1987): On the weighted sensitivity minimization problem for delay systems. Syst. Contr. Lett. 8, 307-312
[232] K. Zhou, P.P. Khargonekar (1988): Robust stabilization of linear systems with norm bounded time varying uncertainty. Syst. Contr. Lett. 10, 17-20
[233] K. Zhou, P.P. Khargonekar (1988): An algebraic Riccati equation approach to H^∞ optimization. Syst. Contr. Lett. 11, 85-91
[234] H.J. Zwart, R.F. Curtain, J.R. Partington, K. Glover (1988): Partial fraction expansions for delay systems. Syst. Contr. Lett. 10, 235-244
[235] K. Zhu (1990): *Operator Theory in Function Spaces* (Dekker, New York)

List of Symbols

Subject Index